# BIOLOGY TEACHERS' GUIDE II

## PART THREE INHERITANCE AND DEVELOPMENT

## PART FOUR ECOLOGY AND EVOLUTION

**Revised Nuffield Advanced Science**
Published for the Nuffield–Chelsea Curriculum Trust
by Longman Group Limited

General editor
Revised Nuffield
Advanced Biology
**Grace Monger**

Editors of Part Three,
'Inheritance and
development'
**John A. Barker**
**Grace Monger**
**Ianto Stevens**

Editors of Part Four,
'Ecology and
evolution'
**Dr T. J. King**
**Grace Monger**

**Contributors to this book**
John A. Barker
John Bradley
S. L. Brooks
Dr Alun Brown
Dr A. F. Dyer
Peter L. Forey
Wilma George
Graham Goldsworthy
Professor Brian Goodwin
Professor A. Hallam
Carolyn Halliday
Dr Tim Halliday
Dr J. W. Hannay
Dr Jennifer Jones
Dr T. J. King
David Kirkland
Sophie McCormick
Grace Monger
Dr Alan Radford
Pete Richardson
Mary Sangster
Ianto Stevens
Dr J. A. Bird Stewart
Dr Peter Thorogood
Cheryll Tickle
Tim Turvey
Professor L. Wolpert, F.R.S.

*Adviser on safety matters*
John Wray

The General Editor would like to acknowledge with thanks the helpful advice of Professor John Harper on Chapters 26, 27, 28, and 29 and that of Dr Geoffrey Harper on Chapter 30.

**The Nuffield–Chelsea Curriculum Trust is grateful to the authors and editors of the first edition:**

**Organizers**, P. J. Kelly, W. H. Dowdeswell; **Editors**, John A. Barker, John H. Gray, P. J. Kelly, Margaret K. Sands, C. F. Stoneman; **Contributors**, John A. Barker, L. C. Comber, J. F. Eggleston, Dr P. Fleetwood-Walker, W. H. Freeman, Peter Fry, Dr R. Gliddon, John H. Gray, Stephen W. Hurry, P. J. Kelly, R. E. Lister, Dr R. Lowery, Diana E. Manuel, Brian Mowl, M. B. V. Roberts, Margaret K. Sands, C. F. Stoneman, K. O. Turner, Dr A. Upshall.

# CONTENTS

# FOREWORD

When the Nuffield Advanced Science series first appeared on the market in 1970, they were rapidly accepted as a notable contribution to the choices for the sixth form science curriculum. Devised by experienced teachers working in consultation with the universities and examination boards, and subjected to extensive trials in schools before publication, they introduced a new element of intellectual excitement into the work of A-level students. Though the period since publication has seen many debates on the sixth form curriculum, it is now clear that the Advanced Level framework of education will be with us for some years in its established form. That period saw various proposals for change in structure which were not accepted but the debate to which we contributed encouraged us to start looking at the scope and aims of our A-level courses and at the ways they were being used in schools. Much of value was learned during those investigations and has been extremely useful in the planning of the present revision. The time since first publication has also seen a remarkable expansion in the number of candidates taking A-level biology and it is encouraging to us to know that we helped in this development.

The revision of the biology series under the general editorship of Grace Monger has been conducted with the help of a committee under the chairmanship of Arthur Lucas, Professor of Curriculum Studies, CSME, Chelsea College, University of London. We are grateful to him and to the committee. We also owe a considerable debt to the Joint Matriculation Board which for many years has been responsible for the special Nuffield examinations in biology, and to the representatives of the Board who sat on the advisory committee and who have given help in many other ways.

The Nuffield–Chelsea Curriculum Trust is also grateful for the advice and recommendations received from its Advisory Committee, a body containing representatives from the teaching profession, the Association for Science Education, Her Majesty's Inspectorate, universities, and local authority advisers; the committee is under the chairmanship of Professor P. J. Black, academic adviser to the Trust.

Our appreciation also goes to the editors and authors of the first edition of Nuffield Advanced Biological Science, who worked under the joint direction of W. H. Dowdeswell and P. J. Kelly, the project organizers. Their team of editors and writers included John A. Barker, John H. Gray, Margaret K. Sands, and C. F. Stoneman. The present revision has only been possible because of their original work.

I particularly wish to record our gratitude to Grace Monger, the General Editor of the revision. This is the second occasion on which we have asked her to undertake the revision of one of our biology series, as she was responsible for the highly successful O-level Biology revision. We are therefore doubly grateful to Miss C. M. Holland, Headmistress of the Holt School, Wokingham and the Berkshire Education Authority for agreeing to her secondment. Grace Monger has had a particularly onerous task because the many topics that biology covers have been subject to an exceptional number of changes and new discoveries in recent years. She and her team of editors have been fortunate in being able to draw on the help, as writers and consultants, of experts in their fields in universities, teaching hospitals, and other institutions of learning. To Grace Monger and her editors, John A. Barker, T. J. King, M. B. V. Roberts, Ianto Stevens, Tim Turvey, and Colin Wood-Robinson, and to the many contributors, we offer our most sincere thanks.

I would also like to acknowledge the work of William Anderson, publications manager to the Trust, his colleagues, and our publishers, the Longman Group, for their assistance in the publication of these books. The editorial and publishing skills they contribute are essential to effective curriculum development.

K. W. Keohane
Chairman, Nuffield–Chelsea Curriculum Trust

# PREFACE TO THE FIRST EDITION

The materials produced by the Nuffield Advanced Biological Science Project do not represent a rigid syllabus. They have been devised after careful evaluation of the results of extensive school trials so they can be used in a variety of ways related to the different circumstances found in schools and the varied abilities, backgrounds, and aspirations of students.

The work has three major objectives:

To develop in students the intellectual and practical abilities which are fundamental to the understanding of biological science.

To introduce students to a body of biological knowledge relevant to modern requirements, through investigating living things and studying the work of scientists. In doing so, students will consider the processes of research and the implications of science for society.

To develop in students the facility for independent study, especially how to learn through critical evaluation rather than memorizing by rote.

These aims have been central not only to the design of the publications and other materials but also to the complementary examinations that have been prepared.

## Abilities

It is intended to develop abilities in the following kinds of work which are assumed to provide the basis for learning biological science.

**1** Acquiring a knowledge of living things and an understanding of the techniques used to study them.
**2** Making observations and asking relevant questions about them.
**3** Analysing biological data and synthesizing them into conclusions and principles.
**4** Handling quantitative information and assessing error and degree of significance.
**5** Critical judgment of hypothetical statements in the light of their origin and application.
**6** Making use of acquired knowledge for identifying and investigating problems with unfamiliar materials.
**7** Evaluating the implications of biological knowledge for human society.
**8** Communicating biological knowledge both coherently and with relevance.

## Subject matter

The subject matter covers four units, each of approximately 90 periods (of 40 minutes) of class work and parallel homework or preparation. The units can be taken in various sequences and there are opportunities for a flexible treatment within each unit. The outlines of the units are as follows:

*Maintenance of the organism*
Interaction and exchange between organisms and their environment
Gas exchange systems
Transport inside organisms
Transport media
Digestion and absorption
Enzymes and organisms
Photosynthesis
Metabolism and the environment

*The developing organism*
Sexual reproduction
Early development
Cell development and differentiation
The nature of genetic material
Gene action
Development and the internal environment
Development and the external environment

*Organisms and populations*
Variation in a community
Inheritance and the origin of variation
The cell nucleus and inheritance
Population genetics and selection
Population dynamics
Organisms and their physical environment
Organisms and their biotic environment
The community as an ecosystem
Evolution and the origin of species

*Control and co-ordination in organisms*
The organism and water
The cell and water
Control by the organism
Stimuli and their influence
Nerves and movement
Structure and function in the nervous system
Social behaviour

We have attempted to provide a comprehensive, balanced, and integrated coverage of the main fields of biological science, both pure and applied (including biological technology). Where it is relevant, we have considered aspects of the physical sciences and mathematics in a biological context. Also, under a number of topics, we consider the nature of biological investigation and the social implications of the subject.

The units cover all levels of biological organization, that is, molecular levels, cellular levels, organ and tissue, organism and population. However, the focus of each is on the whole organism. They also illustrate many of the major themes or concepts of biology, such as variety and adaptation; structure in relation to function; organisms in relation to their environment; the similarity of many processes of physiology and behaviour; the genetic and evolutionary continuity of life; matter and energy cycles; homeostasis; development; and the uniqueness of the individual.

We hope that students will also be encouraged to appreciate the aesthetic and humanitarian aspects of the subject, although this is clearly something for which the teacher must be primarily responsible.

## Approach

We suggest that investigations should play a major part in the work.

These can involve either practical work or exercises of an investigatory nature based on secondhand data, mainly in the form of excerpts from published literature, the results of experiments carried out by the project, or visual aids.

The practical work is described in the *Laboratory guides*, non-practical work in the *Study guide*.

The work in the *Laboratory guides* and *Study guide* is complementary and, in some cases, alternative. This makes it possible for students to do practical and non-practical work in varying proportions at the discretion of the teacher. At the same time, through the use of cross-references, questions, and bibliographies (which mention the specially prepared Topic Reviews), it is possible to make discussion, reading, demonstration, and written work an integral part of the investigations.

The work in the *Laboratory guides* and *Study guide* is of three types. *Preliminary work* aims to cover possible deficiencies in the students' background and provides reinforcement material for the less able student. *Main work* is within the capacity of the majority of A-level students. The amount of practical work has been carefully limited to fit into a reasonable allowance of time.

*Extension work* contains broader and more rigorous treatments of some of the topics included as main work. It is intended for the more able and advanced student and consists of a range of exercises from which a choice can be made.

The materials have been devised so that they can be used for work with a class or small group or by individual students. However, we recommend that students should be encouraged gradually to rely more and more on their own resources. This is particularly important for Extension work and for projects. The latter are small-scale, open-ended, individual investigations described in *Projects in biological science*.

In short, this is not a specialized course, nor is it harnessed only to the needs of future biologists. It is an attempt to provide a way of presenting biological science as an interesting and important subject, relevant to scientist and non-scientist alike. We also hope that it will help those who undertake it to cultivate the abilities and attitudes which are necessary if they are to understand and evaluate the contribution that biological science makes to our society.

# THE REVISED EDITION

The Nuffield Advanced Biological Science materials were first published in the 1970s. Ten years later it was obviously necessary that a decision about their future should be taken. A working party representing various interests was set up and it is as a result of its conclusions that this revision has come about.

It was quite clear that there was much in the original materials that was worth preserving although a revision must be forward-looking. At the same time it appeared that the original intention of the books may have been misunderstood and that the biological ideas were not being seen clearly enough. In addition practical work had become the dominant form of student activity and had tended to overburden the scheme. When the working party reported to the Nuffield–Chelsea Curriculum Trust it recommended that a revision of the materials should be undertaken. It stated that the original aims of the scheme should be kept and given even more emphasis; however, it did suggest that different types of books were needed.

All schools which entered candidates for the Nuffield Advanced level biology examination, which is administered by the Joint Matriculation Board, were written to and teachers were asked for their views. The teachers who sent their comments, many of them very detailed, made a greatly valued contribution to the form this revision has taken.

The major feature of this revision, then, is a change in the form and content of the books. The *Study guide* is in two volumes and contains a substantial amount of descriptive text in addition to a collection of biological data presented in Study items. There is no specified extension work although in some areas many teachers may think the detail goes beyond that required for Advanced level. This has been done deliberately where the subject is of such interest as to justify it and for the benefit of the more able students who may appreciate a greater depth of knowledge.

The subject matter still covers four areas arranged as follows:

*Volume I*
Maintenance of the organism
Control and co-ordination in organisms

*Volume II*
Inheritance and development
Ecology and evolution

The outline of the areas covered is as follows:

**Maintenance of the organism**
Gas exchange
Breathing and gas exchange in Man
The circulatory systems of animals and plants
Blood and the transport of oxygen
Cells and chemical reactions

Heterotrophic nutrition
Photosynthesis

**Control and co-ordination in organisms**
The plant and water
The cell and water
Control by the organism
Co-ordination and communication
The response to stimuli
Behaviour
The human brain and the mind

**Inheritance and development**
Cell development and differentiation
The cell nucleus and inheritance
Variation and its causes
The nature of genetic material
Gene action
Population genetics and selection
Principles and applications of biotechnology
Methods of reproduction
Patterns of development
Control and integration through the internal environment
Development and the external environment

**Ecology and evolution**
The organism and its environment
Organisms and their biotic environments
Population dynamics
Communities and ecosystems
Evolution

The *Study guide* therefore contains the core of the work of the revised scheme. The recommendations made by the working party, which gained support from the teachers too, have also been implemented in the revision. These are:

**1** That the process of biology, rather than the content, should be emphasized (the lack of content has been criticized in the past but the new form of the *Study guide* should remedy the deficiency of factual material).
**2** That the revised material must be appropriate to, and appeal to, those students who will not continue with an academic study after A-level, as well as to those who will do so.
**3** That the central theme should be the whole organism, although there are several subsidiary themes.
**4** That there would be a modest development of learning through historical aspects so that students will appreciate how present day knowledge and ideas have been reached.

The working party proposed few changes in content, but as a result of teachers' comments it was realized that certain areas needed more attention. In the revised books, more detail about cells, for instance, is included and there is a separate booklet, *Systematics and classification.* There is also a new chapter on biotechnology, and one on human behaviour and the mind. Other areas have been expanded and brought up to date. These include photosynthesis, water relations of cells, immunology, and control by hormones. The chapters on genetics and development have been extensively rewritten and the ecology chapters have a new look. Evolution is presented in a novel way. There is a second supporting booklet, *Mathematics for biologists.*

The practical investigations are now contained in seven separate *Practical guides* and are fully cross-referenced to the *Study guide.* The parallel *Teachers' guides* contain, as before, principles, assumptions, and answers to questions. In addition there is a second part to each chapter which is the guide to the *Practical guide.*

The Nuffield Advanced level biology examination will continue to be administered by the Joint Matriculation Board. It will, however, be based on a syllabus which will be prepared by (and available from) the Joint Matriculation Board.

Although we have used modern nomenclature based on the International System of Units for laboratory chemicals, we have not attempted to do so for biochemical pathways, because of the confusion it would cause.

Attention is drawn to the joint statement of the Association for Science Education, the Institute of Biology, and the Universities' Federation for Animal Welfare, entitled 'The use of animals and plants in school science'. This is reproduced in the Appendix to *Teachers' Guide I.*

We would also emphasize that students should not be required to carry out dissections against their wishes. There may be instances where students would prefer to watch the investigation being carried out on video.

Obviously a major revision such as this is has involved a lot of hard work by a great many people. I should also like to thank the editors mentioned in the Foreword and the authors whose names appear on the books. This revision has relied very heavily on the large number of authors who have written material for us and who have shown such an interest in the whole exercise. But it is the editors who have carried the main responsibility for the final version of the text. They have been a most professional team and it has been a pleasure to work with them. The contribution made by the staff of the Publications department of the Nuffield–Chelsea Curriculum Trust cannot be overestimated and the final result owes a great deal to their skill and dedication.

I should also like to thank personally Joyce Moate who has done all my typing and finally I should like to record my appreciation of the two people who have made it possible for me to continue with this task after the one year full-time secondment ended. Firstly, to Miss C. M. Holland, headmistress of the Holt School, Wokingham, who has supported me in many ways, and has provided me with facilities which have made the task so much easier. Secondly to Mary Sangster, who has shared a teaching timetable with me on a regular basis and on whom I have always been able to rely to give extra time. It would have been unthinkable to complete the task in the time available but for the calm and efficient way in which she has assisted me.

Many people therefore have cooperated to produce these revised materials. We hope the result will be seen as an interesting and exciting approach to studying biology at Advanced level whether it is the intention to enter for the Nuffield A-level examination or not.

Grace Monger
General editor

# PART THREE INHERITANCE AND DEVELOPMENT

In references to figures and tables, **'S'** denotes the **Study guide**.
**'P'** refers to the **Practical guide**.
Example: 'figure (S)2'.

□ denotes the end of a Study item.

# CHAPTER 15 CELL DEVELOPMENT AND DIFFERENTIATION

*A review of the chapter's aims and contents*

1 The continuity of life, resulting from mitosis, and cell diversity, resulting from differentiation, provide the major themes.
2 Cell diversity is a consequence of a multicellular organization.
3 The concept of genetic material acting as a store of information in the zygote is introduced. The term gene is used without elaboration as a unit of the genetic material.
4 Two different mechanisms for cell differentiation are discussed – the selective destruction of irrelevant genes on the one hand and the selective expression of certain genes from a full complement of genetic material on the other.
5 Evidence from nuclear transplantation, in vertebrates and in *Acetabularia*, indicates that the genetic material resides in the nucleus and that it remains intact during differentiation.
6 The grafting together of parts of two cells in *Acetabularia* yields evidence that the nucleus is in turn under cytoplasmic control.
7 The process of mitosis is described in detail and it is emphasized that mitosis provides the means for achieving genetic continuity.
8 The significance of various types of changes taking place during differentiation is discussed.
9 The example of haemoglobin is used to show that different genes become active at different stages of development and that genes are responsible for the structure of proteins. This theme will be further emphasized in Chapter 19.

# PART I The *Study guide*

## 15.1 Introduction

*Assumptions*

1 An understanding of the terms unicellular and multicellular.

*Principles*

1 Multicellular organization opens up the possibility of division of labour.
2 Even a limited specialization of function renders a cell incapable of an independent existence.
3 The zygote must contain genetic material which specifies the materials the future organism will synthesize, the range of cell types it will produce, and the organization of these cells into a functioning whole.
4 Differentiation may be the result either of destruction of irrelevant parts of the genetic material or of selective gene expression. The latter hypothesis is supported by much experimental evidence outlined in the

Study items below and in Chapter 19. Destruction of or alteration to the genetic material do take place, for example in the development of *Ascaris* sp. and in the differentiation of mammalian lymphocytes.

### STUDY ITEM
### 15.11 Plant tissue culture

*Assumptions*

1. An awareness of the need for sterile techniques.
2. Some knowledge of the range of nutrients produced and required by cells.
3. Understanding of the terms phloem, pith, and storage tissue.

*Principles*

1. Differentiated plant cells may divide if placed in the correct environment.
2. Under certain circumstances these cells may produce complete plants, showing that some at least of the original cells possessed an intact genome.

*Questions and answers*

a ***Why do carrot cells in culture require many organic compounds?***

Genes needed to synthesize these compounds are inactive in the cells.

b ***What does this show about the genetic material of the original differentiated cells?***

It must have remained unaltered by the differentiation process which produced the original cells founding the culture.

c ***Suggest a practical application for plant tissue culture.***

If a highly desirable plant has no straightforward means of vegetative propagation, it might be used to produce a large family of genetically identical plants (a clone).

d ***Why do you think that a period of adjustment is needed?***

Substances needed for cell division may have to be synthesized and the level of activity of genes may need to change. □

### STUDY ITEM
### 15.12 The influence of nucleus and cytoplasm

*Principles*

1. The nucleus contains information needed for the development of the cytoplasm.
2. It acts by means of material passing from the nucleus to the cytoplasm.
3. The cytoplasm may in turn control the activity of the nucleus.
4. The nucleus is the seat of inherited differences between species.

*Questions and answers*

a **From this experiment, what can be concluded about the location of the genetic material in the Acetabularia cell?**

It must reside in the rhizoid section since only this develops completely. (From this experiment alone, one might conclude that some genetic material was to be found in a developing stem tip.) The nucleus cannot be implicated with certainty since it has not been isolated from rhizoid cytoplasm.

b **How does the result of this investigation compare with that of the previous one?**

The stem acquires the ability to develop a 'hat' by being attached to the rhizoid and nucleus.

c **Suggest a hypothesis to explain this result.**

Some influence, probably chemical, passes from the rhizoid up the stem and results in the stem tip acquiring the ability to develop a 'hat'.

d **In what ways do the results of the investigation illustrated in figure [S]6 support or modify the hypothesis you put forward in answer to question a?**

The nucleus and not the rhizoid must control the development of the complete cell. Any part of the cell containing the nucleus can grow into a complete cell.

e **Compare the behaviour of an isolated rhizoid of a mature Acetabularia with an isolated, differentiated cell of a multicellular organism.**

Both seem to have lost the ability to develop and grow into a complete organism.

f **Can differentiated cytoplasm control the activity of the nucleus?**

Yes. (This can be related in discussion to the switching on and off of genes during differentiation. It is important to remember that *Acetabularia* is a unicellular organism and that its development may differ in important ways from the differentiation of cells in a metazoan. In *Acetabularia* the nucleus always gives rise in the end to gametes, while in a gut enterocyte differentiation is not reversed.)

g **Explain how this final investigation validates and extends the conclusions already reached.**

The development of an intermediate type of 'hat' as a result of the interspecific graft shows that substances or other influences from the nucleus (messengers, see *Study guide II*, page 112) accumulate at the growing tip. The regeneration of a 'hat' characteristic of the species donating the nucleus, after the intermediate 'hat' has been removed, shows that the inherent character of each species of cell is determined by its nucleus.

It is important to emphasize that *A. mediterranea* and *A. crenulata* are distinct species, not alternative states of differentiation. Unlike metazoan cells, *Acetabularia* cells can only differentiate in one way. □

## STUDY ITEM
## 15.13 Transplantation of nuclei

*Assumptions*

**1** Some elementary knowledge of sexual reproduction in animals is helpful.

**2** Some acquaintance with names for stages of embryonic development and with the term endoderm. It is not important for students to understand these terms fully in order to appreciate the significance of the data.

**3** An ability to plot several graphs on the same axes.

**4** An understanding of the term plasma membrane.

*Principles*

**1** The genetic material of a differentiated animal cell must be placed in undifferentiated cytoplasm if its state of differentiation is to be reversed.

**2** At least in some cases, the genetic material of a differentiated cell remains intact.

*Questions and answers*

**a** ***Indicate the nature of the control experiments.***

**1** The zygote's own nucleus could be removed and replaced, using a similar micro-pipette.
**2** Small portions of differentiated cytoplasm from the donor cells could be injected into zygotes.

**b** ***Using the information in table 1, plot six graphs, on the same axes, of the percentage survival of the operated eggs against the developmental stage they had reached. There should be one line for each stage of embryo donating nuclei.***

The graphs should show a decrease in the survival of transplants as the donor cells become progressively more differentiated.

**c** ***Put forward an alternative hypothesis to this idea.***

They become more easily damaged as differentiation proceeds.

**d** ***Why is this colour contrast an advantage?***

If the animal resulting from the experiment is the colour of the donor of the eggs, it might be supposed that the egg nucleus had not been destroyed. If it turns out to be the colour of the animal from which the differentiating nucleus was obtained, this is good evidence that this nucleus did indeed act to produce the complete organism. □

## 15.2 Cell division during growth and development

**Practical investigation. *Practical guide* 5, investigation 15A, 'Mitosis in a plant meristem'.**

*Assumption*

1 Some knowledge of eukaryote cell structure and an understanding of the terms eukaryote and prokaryote.

*Principles*

1 The genetic material replicates during interphase.
2 It is organized in a linear manner into chromosomes which become more or less visible as they spiralize and despiralize.
3 Mitosis is a process which ensures that each chromosome duplicates and that one duplicate is delivered to each of two daughter nuclei.
4 The study of cell division by fixation and staining must be supplemented by methods which allow living cells to be observed.
5 The phases into which mitosis is divided are artificial but useful for reference.

The process of mitosis is described in some detail, and the students' understanding of it is tested in Study item 15.51. Practical investigation 15A, 'Mitosis in a plant meristem', provides reinforcement for this section. A film or video recording to show the process in unfixed and unstained cells would be most helpful. The parts of the cell relevant to a description of cell division are introduced.

*Questions and answers*

**a** ***There are dyes which stain all types of nuclei, from whatever type of cell, in a similar way. What can be deduced from this?***

That nuclei from all types of cell have structures and biochemical components which are similar.

**b** ***What are the parts* [in figure (S)12] *to which the arrows point?***

Labels for figure (S)12:
Reading clockwise from the left: chromatin; cell membrane; nuclear membrane; cytoplasm.

**c** ***Suggest why there are different quantities of genetic material found in a population of interphase nuclei.***

Some nuclei have replicated their genetic material and so have twice as much as others in which it has not yet replicated. A minority is in the process of replicating the genetic material.

The students are told that some cells do not have centrioles.

**d** ***What does this suggest about the role of centrioles in mitosis?***

Centrioles are associated with the spindle in some way but do not cause its essential structure and behaviour.

e ***Do you know of any of these exceptions, that is, structures having many nuclei in a common cytoplasm?***

Some fungi, such as *Mucor*. (Striated and cardiac muscle fibres arise by fusion of cells having single nuclei.)

## STUDY ITEM
## 15.21 Evidence for mitosis

*Questions and answers*

a ***List the photographs in the order in which they should appear as a cell goes through the mitotic cycle starting with interphase.***

(S)19c, (S)19e, (S)19g, (S)19d, (S)19b, (S)19h, (S)19i, (S)19a, (S)19j, (S)19f.

b ***If a photograph has no corresponding drawing, make one and label it.***

The drawings produced should be interpretive, rather than attempts at exact copies. (It might be wise not to attempt drawings of figures (S)19e and f.) Students should label constrictions, centromeres, trailing chromosome arms, and evidence for duplication or for spiral structure. Each drawing could be titled by the named phase of mitosis into which it falls, *e.g.* (S)19b: metaphase; (S)19h: early anaphase.

c ***Explain the appearance of the pair of chromatids labelled 'r' in*** **figure [S] 19h.**

The pair of separating chromatids has folded over so that the centromere region was out of focus when the photograph was taken. (This shows the importance of good squash preparations – see the introduction to *Practical guide 5*, investigation 15A, 'Mitosis in a plant meristem'.)

d ***How many chromosomes are there in the cells of this plant?***

Six.

e ***What is the evidence that spindle fibres (microtubules) attach to chromosomes only at their centromeres?***

The arms trail behind the centromeres at anaphase.

f ***The spindle is a transparent structure and may not show up if a cell is stained to reveal chromosomes. Make two diagrams to show how the chromosomes and their centromeres would appear at metaphase if viewed:***
***1 from above one of the spindle poles***
***2 from the side of the spindle in the same plane as its equator.***

**1** See *figure 1*.
**2** See *figure 2*.

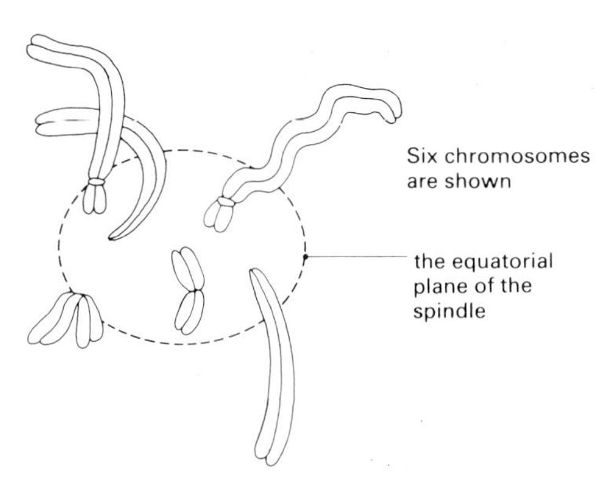

**Figure 1**
The six chromosomes viewed from above one of the spindle poles.

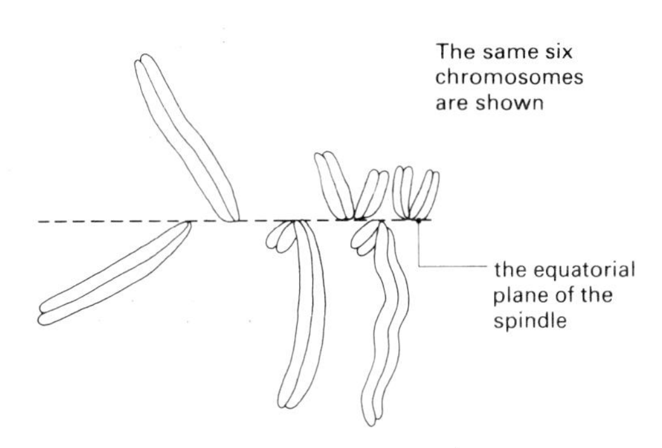

**Figure 2**
The same six chromosomes viewed from the side of the spindle in the same plane as its equator.

**g** ***What would be the result of failure of a centromere to divide at metaphase in terms of chromosome content of the two daughter cells?***

Both chromatids would pass to one of the daughter nuclei (non-disjunction). One would thus gain an extra chromosome while the other would be one short.

**h** ***Of all cell components, only the chromosomes have a mechanism to ensure regular and precisely equal division into the daughter cells. Why should this be so?***

Because they carry the genetic information, that is, they are responsible for genetic continuity. □

## 15.3 A review of cell diversity

**Practical investigations.** ***Practical guide*** **5, investigation 15B, 'Differentiation in stems and roots' and investigation 15C, 'Differentiation in cells'.**

*Assumption*

**1** A knowledge of the ultrastructure of generalized plant and animal cells.

*Principles*

**1** As a cell differentiates there are likely to be visible changes, and also more subtle ones, in the nucleus, cytoplasm, plasma membrane, and external matrix or wall.

**2** These changes may be a consequence of the process of differentiation but will also have a functional significance.

Space does not permit large numbers of cell types and tissue types to be described. Many examples will have been encountered in studies of the lung, the blood vascular system, the leaves and vascular systems of flowering plants, the liver, the kidney, and the gut. This section might be an opportunity to recall and revise some of these examples. Atlases of histology and books of micrographs will afford additional examples if time permits, as will the large range of transparencies marketed by Philip Harris (Biological) Ltd, Griffin & George Ltd, and other suppliers.

*Question and answer*

a **Why is it difficult to obtain samples of plant cell membranes?**

They are firmly stuck to the polysaccharide wall. (The techniques of growing isolated plant protoplasts in culture could be mentioned here.)

## 15.4 The 'one gene–one polypeptide' hypothesis

*Assumptions*

1 An elementary knowledge of enzymes.
2 An understanding of the nature of peptide bonds.
3 A knowledge of haemoglobin dissociation curves.

*Principles*

1 Differentiation involves biochemical changes. These can be related to changes in the levels of enzymes and other proteins.
2 Proteins from different species have characteristic primary structures which must be genetically determined.
3 This leads logically to the notion that for every polypeptide which a cell can make there is a specific gene.

These ideas are developed by the Study items on haemoglobin which follow and are further refined in Chapters 18 and 19.

### STUDY ITEM
### 15.41 Sequential activation of haemoglobin genes in humans

*Principles*

1 Mammals produce different types of red blood cell at different stages of their intra-uterine development.
2 Associated with each stage is a different type of haemoglobin.
3 Embryonic, foetal, and adult haemoglobins have different dissociation curves and thus enable the individual to obtain oxygen under different circumstances.
4 Each of these types of haemoglobin is produced by a different gene.
5 The genes are switched on and off in a sequence during development.

*Questions and answers*

a ***From* figure [S]22, *what percentage of cells of each type would be expected in an individual:***
*1* ***2 months after conception?***
*2* ***6 months after conception?***
*3* ***2 months after birth?***

See table 1.

| *Age of foetus or neonate* | *Embryonic cells/%* | *Foetal cells/%* | *Adult cells/%* |
|---|---|---|---|
| 2 months after conception | 60 | 40 | 0 |
| 6 months after conception | 0 | 90 | 10 |
| 2 months after birth | 0 | 40 | 60 |

**Table 1**

**b** ***Why are three different haemoglobins, each with a different oxygen-dissociation curve, an advantage?***

In intra-uterine life oxygen must be obtained from the maternal blood. In the first stages of development, the vascular system of embryo and uterus will be less well developed and haemoglobin able to hold oxygen at comparatively low $pO_2$ will be an advantage. As the foetus matures, haemoglobin with a shallower dissociation curve allows a higher $pO_2$ in the foetal tissues but the curve is still to the left of the adult haemoglobin of the mother and so allows efficient transfer of oxygen to the foetus.

**c** ***It has been suggested that each type of red blood cell produces its own type of haemoglobin. How could this hypothesis be tested?***

Pure samples of the three types of cell would need to be analysed. (Adults sometimes have quite high proportions of foetal haemoglobin in their adult red blood cells.)

**d** ***What is the implication of the appearance of three types of haemoglobin at different stages during development for the hypothesis that genetic material is selectively activated or inactivated as cells differentiate?***

The hypothesis is supported. The zygote must carry information to make all three types of haemoglobin, but only during the appropriate stage of development is each type produced in quantity. □

## STUDY ITEM
## 15.42 Haemoglobins and inherited variation

*Assumptions*

1 An understanding of the terms cathode and anode.
2 A knowledge of the term buffer solution.

*Principles*

1 The technique of gel electrophoresis is described and explained.
2 Proteins including haemoglobins can be separated and identified by this technique.
3 Inherited differences between individuals can be detected if blood samples are subject to electrophoresis.

*Questions and answers*

**a** ***Why is this [the ice pack] necessary?***

To cool the gel and prevent the haemoglobin from becoming denatured.

**b** ***Discuss these results*** **[figure (S)24]** ***briefly.***

The bands caused by non-haemoglobin proteins should be ignored. (Students may need to be reminded that they are looking at a black and white photograph and that the haemoglobin is coloured.) Young babies show a band of foetal haemoglobin. Foetal haemoglobin persists to varying extents in adults. Individuals carrying the allele for an unusual type of haemoglobin will usually be heterozygotes and will show the haemoglobin A band and a second band.

c ***If blood from a person is shown by electrophoresis to contain an unusual type of haemoglobin, what investigations might follow up this discovery – assuming the cooperation of the individual concerned?***

**1** Test members of his or her family to show that the condition was inherited.
**2** Investigate the morphology of the red blood cells.
**3** Investigate the dissociation curve and oxygen-carrying capacity.
**4** Analyse the haemoglobin to find which of the two types of peptide was different and in what way. □

## PART II The *Practical guide*

The investigations related to this chapter appear in *Practical guide 5*.

*Introduction*
The investigations chosen have been arranged in a sequence which parallels the sequence of topics and concepts found in *Study guide II*, Chapters 15 to 20. If the investigations are used in a different sequence it may be necessary to explain terminology more fully and/or to modify the questions asked.

It would be unlikely that anyone would wish to use every investigation. Some are clearly alternatives and if resources permit, investigations could be carried out by different groups of students. Much can be learned from attempts to communicate results and ideas between groups.

In several cases, for example where an organism has to be cultured or a tissue fixed for staining, it will be possible for teachers or technicians to carry out much of the time-consuming preparatory work. A balance must be struck between the need to save time and fit the work into the timetable and the need to involve students in the complete investigation.

To avoid repetition two alternative staining techniques, toluidine blue and ethano-orcein, are given at the beginning of the *Practical guide* as are two alternative means of achieving a squash preparation.

The ethanoic alcohol used to fix tissues before staining with orcein or Feulgen must be freshly prepared.
**Toluidine blue.** Dissolve 0.5 g of solid toluidine blue stain in 100 $cm^3$ distilled water.
**Ethano-orcein** (*take care*)
Ethanoic acid, glacial, 90 $cm^3$
Orcein, solid, 1.5 g
Water, distilled, 110 $cm^3$

Grind the orcein thoroughly with pestle and mortar. Mix the ethanoic acid and water and bring to the boil. Pour the boiling mixture over the orcein and stir very thoroughly (use a fume chamber). Leave the solution overnight, then filter and store in a tightly stoppered dark bottle. The orcein solution should be used as concentrated as possible. The stock solution as prepared above is usually satisfactory. If it does overstain dilute with 45 % ethanoic acid and start again.

Add 1.0 mol dm$^{-3}$ hydrochloric acid to the stain shortly before use (*take care*). Keep some stain free of hydrochloric acid for the final mount.

Toluidine blue and ethano-orcein are best freshly prepared but will keep for a time. Toluidine blue keeps much better than orcein.

Propionic orcein (made with propionic acid rather than ethanoic) has the advantage that it evaporates more slowly when warmed with the specimen. Lacmoid ethanoic is another satisfactory stain for chromosomes. If time is short teachers may wish to use Feulgen stain instead of orcein or toluidine blue in one of the early cytological investigations – details are on page 53 of the *Practical guide* and page 95 of this *Teachers' guide*.

**Ethanoic alcohol**
Ethanoic acid, glacial, 25 cm$^3$
Ethanol or industrial alcohol, 75 cm$^3$
Mix just before use, adding the acid to the alcohol.

## INVESTIGATION
## 15A Mitosis in a plant meristem

ITEMS NEEDED

Root tips, suitable (see *Practical guide* 5, page 13) 1/1

Ethanoic alcohol fixative (if required)
Hydrochloric acid, 1 mol dm$^{-3}$
Sodium hypochlorite solution, 1 %
Stains;
ethano-orcein
*or*
Feulgen (page 95)
*or*
toluidine blue
Water, distilled (if required)

Filter paper or blotting-paper
Forceps
Hotplate at 50–60 °C or Bunsen burner 1/group
*or*
Water bath at 60 °C 1/class
Measuring cylinder, 20 cm$^3$ or syringe, 20 cm$^3$ 1/1
Microscope, monocular 1/1
Microscope slides and coverslips
Mounted needles
Pipettes, dropping
Razor blade or scalpel 1/1
Scissors
Watch-glass or small beaker 1/1

(*Study guide* 15.2 'Cell division during growth and development'.)

**Ethanoic alcohol** (see above)
**Sodium hypochlorite solution** (*take care*)
Sodium hypochlorite solution, concentrated, 2 cm$^3$
Water, distilled, 198 cm$^3$

*Assumptions*

1 That the student knows how to use a microscope correctly with HP objective.
2 That the student can recognize a root cap and chromosomes.
3 That reference work is available showing clear photographs and/or drawings or representative stages of mitosis.

*Principles*

1 By general staining it becomes possible to observe and examine directly all mitotic stages.
2 A single representative photograph or drawing cannot show the complete range of manifestations of an event.
3 Mitosis is a dynamic and transient process.
4 During nuclear division chromosome behaviour is exactly and carefully regulated; a regulation which is essential to the survival of the daughter cells.

*Procedure: bulbs*

1 Balance each bulb over the opening of a specimen tube or glass jar.

The opening must be just too small to allow it to fall in.

2 Fill each jar with tap water until it just touches the base of the bulb.

3 Set the jars on a convenient shelf out of direct sunlight and examine them daily for signs of root growth.

4 As soon as root growth starts, lower the water level so that the base of the bulb is in air and only the tip of the longest root touches water. Use the roots as soon as they are growing fast.

*Procedure: seedlings*

1 Sterilize the peas or beans by soaking them in 1 % sodium hypochlorite solution for not more than three minutes and then rinsing twice in tap water.

2 Soak the seeds in a tray of tap water overnight.

3 Place the layer of damp sawdust, peat, vermiculite or gravel about 10 cm deep in a plastic tray; pack the material loosely.

4 Distribute the soaked seeds over the surface so that they do not touch.

5 Cover with damp medium and press down very lightly. (The depth of the covering material should be approximately the same as the seed diameter.)

6 Keep in a damp condition until green shoots have appeared. Each seedling can be carefully pulled up and will have several lateral roots growing from the elongated radicle.

If many small bulbs need to be grown a good alternative procedure is to cut holes of the right size in a thick piece of expanded polystyrene. Fit the bases of the bulbs (for example, cloves of garlic) into the holes and float the polystyrene in water. (This method is also good for growing plants in nutrient solutions.)

Instructions for growing the plants have not been provided in the *Practical guide* but this does not mean that students should not be made aware of the procedures used. They could be asked to grow the plants themselves if time permits.

Some students may have come across mitosis in previous courses or the teacher may decide to study the process using photographs, diagrams, or a film before embarking on this exercise. Students will need help in finding suitable cells in their squash preparations, and need to be encouraged to draw what they see as simply as possible. It is important to prevent this from becoming a dull exercise in confirmation (the temptation is to find photographs of ideal stages and to draw these). It is helpful to use a plant species with a different karyotype to the one shown by photographs.

The vital points of the technique are:

**1** the fixative must be freshly made up (when using orcein or Feulgen)
**2** not more than 1 mm of the root tip should be used
**3** the preparation must not be overheated.

Roots may be pretreated by soaking them, after excision and before fixation, for 4–6 hours in saturated paradichlorobenzene in the dark in order to inhibit spindle formation. This inhibitor is somewhat less toxic than the more commonly used colchicine. This technique could be used either as an 'unknown' treatment in order to test the students' ability to see and interpret mitotic stages correctly and/or as evidence of the role of the spindle in the mitotic mechanism.

*Questions and answers*

**a** ***What structure of the cell, essential to the process of mitosis, is not visible in your preparations? Suggest three reasons why this may be so.***

The spindle fibres are not visible. This may be because:
they are too small to be resolved at this magnification and with this instrument;
they are not stained by this staining technique;
they are destroyed by the treatment given to the roots.

**b** ***Suggest a reason why more cells appear to be in interphase than in any of the active stages of mitosis. What can you infer from your counts of the number of cells seen at each of the four stages of mitosis?***

The greater frequency of interphase suggests that mitosis occurs in relatively few cells in the root tip. This preponderance of interphase is worth stressing. In fact the distinction between interphase and early prophase is a hard one to judge and is to some extent arbitrary. The observation of the greater number of interphase nuclei is also supporting evidence for the hypothesis that the stages of the cell cycle where chromosomes are visible are of shorter duration than the interphase. Authorities suggest that the duration of prophase – telophase – is between one and three hours in onion, while interphase may last between 24 and 60 hours. Temperature and other environmental variables play a part.

The duration of the stages of mitosis has been recorded and the data (see table 2) could be used to compare the observed frequencies of the different stages as recorded by the students.

| *Stage* | *Observed occurrence in representative microscope fields* | | *Duration* | |
|---|---|---|---|---|
| | Number of cells | % of total cells | minutes | % of total time |
| Prophase | 216 | 85.0 | 71 | 85.0 |
| Metaphase | 17 | 6.7 | 6.5 | 7.7 |
| Anaphase | 8 | 3.3 | 2.4 | 2.9 |
| Telophase | 13 | 5.1 | 3.8 | 4.4 |

**Table 2**

**c** ***You have been asked to classify cells into different stages of mitosis. Do your slides provide evidence that these stages are real and that they represent a sequence of processes which each cell would undergo if it had not been killed?***

Chromosomes undergo very significant changes in appearance during prophase and metaphase, and early anaphase or late anaphase and early telophase are rather arbitrary divisions (especially if one only has a student microscope). This suggests that the stages are convenient reference points in a continuous cycle of change. If dead stained cells are examined the dynamic process undergone by the nucleus must remain a hypothesis, but it can be confirmed if living cells are observed by using phase contrast.

## INVESTIGATION
## 15B Differentiation in stems and roots

(*Study guide* 15.3 'A review of cell diversity'.)

### ITEMS NEEDED

Seeds, tobacco or other fine seeds
Agar, 2–3 % in Petri dish 1/group
Hydrochloric acid, 1.0 mol dm$^{-3}$
Sodium hypochlorite solution, 1 %
Toluidine blue stain
Water, distilled
Beaker for soaking seeds 2/group
Bunsen burner 1/group
Cloth, fine, small piece 1/group
Forceps
Graticule, eyepiece 1/1
Incubator at 20–25 °C 1/class
Microscope, monocular 1/1
Microscope slides and coverslips
Mounted needle or fine scalpel 1/1
Pipettes, dropping 2/group
Stage micrometer 1/class
Tape, adhesive

*Procedure A*
**Sodium hypochlorite solution.** See page 13.
**Toluidine blue.** See page 12.

*Assumptions*
1 A knowledge of the terms root cap and root hair.
2 Ability to use an eyepiece graticule to measure an object under a microscope.

*Principles*
1 The age of tissues increases with distance from a stem or root apex.
2 Cells are produced in a meristem, enlarge, and finally differentiate visibly.
3 The ratio of cell volume to nuclear volume increases markedly during growth and differentiation.

Any very small seed that germinates easily on agar will be suitable. Tobacco is ideal. Panicum millet may be purchased from pet shops (it produces beautiful fine coleoptiles ideal for tropism experiments). Cress is rather large and coarse but would serve.

If time permits students can be asked to grow the plants themselves.

*Procedure: seeds*
The whole exercise can be over in about forty minutes if the seeds are sown by a technician and presented already germinated to the class. (Take care during steps **1** and **2**.)

1 Sterilize some seeds of tobacco or panicum millet by soaking them in 1 per cent sodium hypochlorite solution (see page 13) for not more than 3 minutes and then washing twice in tap water.
2 Use a flamed mounted needle to place seeds on the surface of a sterile Petri dish containing plain agar. Close the dish and seal with two pieces of sticky tape.
3 Prop the closed dish up on its edge in a warm place out of direct sunlight, for instance in an incubator at 20–25 °C, until the seeds have germinated. Minute roots will grow vertically downwards over the agar surface.

It is important not to heat the root too long in acid or stain as it will be disrupted.

Seeds should not be kept too warm or a fungal growth will occur. The vertical position of the Petri dish prevents the roots growing into the agar.

The eyepiece graticule can be used as a transect line for selecting more or less random cells (step **9**).

If students are unfamiliar with the idea that enlargement of cells takes place behind the tip in a root then this could be demonstrated using a broad bean radicle marked with Indian ink. Alternatively, the principle of this demonstration could be mentioned.

*Questions and answers*

**a** ***Did you observe any direct evidence of meristematic activity in the region of very small densely packed cells behind the cap? If not refer to* figure [P]3.**

Mitosis may have been observed. (It is shown in figure (P)3d.) The objective of this exercise is not to look for mitotic stages. Higher magnification and a squash preparation would be needed to see these clearly.

**b** ***Comment on your table of cell lengths and nuclear diameters.***

Cell length increases greatly as one moves further from the root apex. There is a change in nuclear volume but this is not great. The ratio of cell volume to nuclear volume therefore increases with the age of cells.

**c** ***Have you gathered any evidence for cell wall or cytoplasm changes during the growth of cells?***

No, since a stain designed to colour nuclei has been employed. Cell walls of differentiated cells may be visibly thicker.

## ITEMS NEEDED

Pea plant, young healthy with 5–8 foliage leaves
Hydrochloric acid, concentrated
Phloroglucinol in hydrochloric acid
Water
Beaker, small 1/group
Glass marking pencil or pen 1/1
Graticule, eyepiece 1/1
Microscope, monocular 1/1
Microscope slides and coverslips
Mounted needle 1/1
Pipettes, dropping 2/group
Razor blade, new 1/group
Stage micrometer 1/class
Watch-glasses or Petri dishes 3/group

*Procedure B*

**Phloroglucinol**
Ethanol, 95 %, 100 $cm^3$
Hydrochloric acid, concentrated, 1.5 $cm^3$
Phloroglucinol, 5 g
Dissolve the phloroglucinol in the ethanol with stirring and slowly add the concentrated hydrochloric acid (*take care*), stirring all the time until the mixture begins to precipitate.

*Assumptions*

1 A knowledge of the terms xylem, vascular bundle, and internode.
2 Ability to use an eyepiece graticule to measure an object under a microscope.

*Principles*

1 Proceeding backwards from the shoot apex is equivalent to moving in time, to show progressive developmental changes in the stem.

Students will have little difficulty in cutting sections from the base of the stem if they have a high quality, new blade and use a slicing movement which does not crush the tissue. The youngest internode is more difficult as the epidermis and the developing bundles easily drag. The teacher should cut a good supply of sections from a well chosen plant before the lesson so that members of the class can be given a reasonable set of three sections, if necessary, in time to complete the staining and observation. Students usually enjoy cutting the sections and are much more involved if allowed to do so.

A similar exercise can be carried out on a rapidly elongating woody stem, for example *Fuchsia, Tilia* sp., or *Cotoneaster.* The amount of lignification can be related to rigidity. The exercise would need to be done in late spring or early summer or preserved material would have to be used. It is difficult to be sure that one is dealing entirely with primary growth in a woody stem but the sectioning may prove easier. Young pea stems have a very determinate growth with four major bundles and very little secondary growth. They can be germinated in a tray of compost at any time of the year (see page 14).

Care should be taken to ensure that the concentrated acid does not come into contact with the microscope.

*Questions and answers*

a ***Use your data to argue whether growth between the youngest and the oldest internodes which you observed has been largely by cell division or largely by cell enlargement.***

Students should argue from the ratio of cell number to cell diameter which they observe. One would expect the growth observed to be by cell enlargement.

b ***What evidence do you have that lignification of cells increases progressively with the age of the stem?***

The number of lignified cells is greater in the older sections. The number of lignified cells increases much more than does the stem diameter when young and old sections are compared.

## INVESTIGATION
## 15C Differentiation in cells

ITEMS NEEDED

Pear, hard fleshed 1/class

Hydrochloric acid, concentrated
Iodine in potassium iodide solution
Phloroglucinol in hydrochloric acid
Sulphuric acid, concentrated
Water

Absorbent paper
Electronmicrographs of 2 cell types 1/1
Microscope, monocular 1/1
Microscope slides and coverslips
Mounted needles 2/1
Razor blade 1/1
Watch-glass 1/1

(*Study guide* 15.3 'A review of cell diversity'.)

**Iodine in potassium iodide solution**
Iodine crystals, 3 g
Potassium iodide, 6 g
Water, distilled
Dissolve the potassium iodide in about 200 $cm^3$ of distilled water and then add the iodine. Make the solution up to 1 $dm^3$ with distilled water. Leave to stand overnight, filter, and store.
**Phloroglucinol in hydrochloric acid.** See page 17.
**Electronmicrographs.**
Suitable examples are micrographs numbers 1 and 4 from the Electron Micrograph Transparency Set by Philip Harris Biological, Oldmixon, Weston-super-Mare, Avon. These show liver cell, × 25 000 (number 1) and skeletal muscle, × 25 000 (number 4).

*Assumptions*

1. The ability to use a microscope and to cut hand sections of a 'solid' tissue.
2. Some understanding of the nature of the primary cell wall in plants.
3. Sufficient experience of electronmicrographs to make it possible to identify major cytoplasmic features in an animal cell.

*Principles*

1. Cell differentiation has occurred in plants if the primary wall of the cell has been added to by the deposition of lignin.
2. Staining of tissue can be a means of positive identification of the chemical nature of parts of that tissue.
3. In animal cells, walls are absent and differentiation of the cytoplasm can only be observed with the electron microscope. The changes shown in animal cell cytoplasm have parallels in plant cell cytoplasm.
4. Differentiation allows cells to become specialized, and specialization leads to improved efficiency of the whole organism.

*Procedure A: plant cells*

There are few problems with this investigation. The section cutting need not be done with great finesse, although observation is easier if the sections have thin edges. Students should be cautioned about the use of the concentrated acids, and it is wise to prevent these liquids from being added to the slide while it is on the stage of the microscope. Microscope stages should be kept horizontal throughout this investigation.

It is possible to add these stains in succession to the same section, but the blue stain on the cellulose disappears in the presence of concentrated hydrochloric acid.

*Questions and answers*

**a** ***It is easier to explain the occurrence of a homogeneous differentiated tissue than it is to explain how the scattered patches of sclerenchyma (sclereids) as seen here may have arisen. What are the problems that have to be solved in order to explain the distribution of the sclereids?***

Students should realize that a group of sclereids arises from an individual cell that has divided *in situ* to provide one of the scattered groups of 2–10 sclereids. It is to be hoped that some will see the problem of explaining how certain cells in the very young pear fruit could have been 'instructed' to differentiate as sclereids while others remain essentially parenchymatous. It is hard to find a simple answer to this, but at least students may be encouraged to suggest both hormonal and genetic influences on the cell, and interaction between these influences.

**b** ***The lines on the cell wall that you observed in step 11, above, are called 'pits'. Suggest a function that they might perform and explain why they are tubular.***

These are branched pits, serving to communicate between the remnants of a functional protoplasm and the exterior of the cell. They are present from early on in the cell's primary growth, and as a succession of layers of lignification is deposited, what was a depression becomes tubular.

*Procedure B: animal cells*
Electron micrographs in the form of photographs reproduced in books, OHP transparencies, 35 mm transparencies, or special card-mounted sets are readily available. Any of these would be suitable for this study. The idea should be to present one micrograph that shows the cytoplasm of a relatively unspecialized cell (for example, a liver cell) and one with a highly specialized cytoplasm (for example, striated muscle, thyroid gland, etc.). The students could be instructed to draw the features that they observe, although this is not strictly necessary. Their interpretation of the electronmicrographs will depend largely upon their previous experience with such material. If they have not met electronmicrographs before, simple observations described verbally will suffice.

With more experienced students it would be instructive to suggest that they read the article by Hillman and Sartory, 1980 (see the Bibliography), once they have completed this investigation, and then to initiate some discussion on the problems that are raised by these authors' views about the reliability of electronmicrographs.

*Questions and answers*

**c** ***Explain the difference between the type of differentiation observed in the plant cells such as those seen in part A and in these animal cells.***

In the animal cells differentiation is expressed as major cytoplasmic differences whereas in the plant cells one can see merely the differences in the thickness and chemical nature of the cell wall. This difference is, of course, largely a result of the difference in magnification, and the viewer's inability, in part A, to see any cytoplasmic structures. It will also probably be suggested that plants, because they have no excretory or locomotory functions, and no cell-mediated integration throughout their bodies, have fewer cytoplasmic specializations and are physiologically less complex than animals. This may sound a naive assessment, but it is in fact a fruitful line of discussion, since the relative lack of complexity in all respects is fundamentally linked to the plants' autotrophic habit (Students can be referred to *Study guide II* page 22.)

**d** ***What does differentiation achieve 1 for a cell; and 2 for an organism?***

**1** Differentiation is capable of turning a very ordinary cell of general functional capabilities into a most specialized unit, adapted to perform one function or a particular group of functions particularly efficiently. This is achieved at the expense of some more basic functions for which it relies upon other, differently specialized, cells.
**2** Cell specialization is a prerequisite of tissue formation; tissue formation implies and allows cooperation between cells, and thus the development of a larger organism, with all the advantages that may bring, because many cells can together perform a function that they could not achieve were they spatially widely separated.

# PART III Bibliography

ALBERTS, BRUCE *et al. The molecular biology of the cell.* Garland Publications, 1983.

ESAU, K. *Anatomy of seed plants.* 2nd edn. John Wiley, 1977.

HILLMAN, H. and SARTORY, P. 'A re-examination of the fine structure of the living cell and its implications for biological education', *School science review*, **62**, 1980.

HURRY, S. W. *The microstructure of cells.* John Murray, 1965.

LEVINE, L. *Papers in genetics*, The CV Mosby Company, 1971. (A collection of classic papers covering whole field of genetics.)

PRESCOTT, D. M. Carolina Biology Readers No. 96, *The reproduction of eukaryotic cells.* Carolina Biological Supply Company, distributed by Packard Publishing Ltd, 1978.

ZUCKERKANDL, E. 'The evolution of haemoglobin'. *Scientific American*, **212**(5), 1965. pp. 110–18.

# CHAPTER 16 THE CELL NUCLEUS AND INHERITANCE

*A review of the chapter's aims and contents*

1. The general contribution by two gametes to a zygote was mentioned in the last chapter and this concept is developed.
2. The need for a reduction from the diploid to the haploid state at some stage in any life cycle is stressed.
3. Evidence is accumulated at all stages of the chapter that the chromosomes carry the genetic material of the zygote and hence of the whole organism.
4. The appearance and structure of chromosomes as seen with a light microscope is explored in some detail, so that the concept of homologous chromosomes may be thoroughly understood, and the hypothesis of a constant linear order of genes will seem a logical development.
5. Polyploidy and aneuploidy are considered:
   to reinforce the concept of homologous chromosomes;
   to stress the importance of chromosome pairing for a balanced meiosis
   to provide further evidence that chromosomes are the site of genetic material;
   to point out that chromosomes ultimately influence the structure and behaviour of the whole organism;
   to lay a foundation for an understanding of flowering plant evolution.
6. A factual account of meiosis is provided.
7. The mechanism of determination of sex by chromosomal systems is explored, and the genetic function of chromosomes further stressed by introducing a case of sex linkage.
8. We return to the idea that the nucleus controls the development of the cytoplasm at the end of this chapter, when fungal genetics is introduced to verify the concept that genetic material segregates in a pattern directed by meiosis.
9. It is assumed throughout that the student has no knowledge of Mendelian genetics.
10. A thorough understanding of principles has been demanded, and to this end, technical terms have been reduced to a minimum; for example, the names of substages of prophase I are omitted.
11. The implications of our understanding of cytology for society have been briefly examined.

## PART I The *Study guide*

### 16.1 Do male and female contribute equally to inheritance?

*Assumptions*

1. A knowledge of mitosis, as outlined in the previous chapter.

2 Some knowledge of sexual reproduction.
3 Some knowledge of life cycles (see page 165).

*Principles*

1 We are culturally conditioned to believe that inheritance takes place according to certain rules, but the mechanism of inheritance, like other biological processes, can only be understood by carefully sifting the evidence. The similar results of reciprocal crosses is one piece of evidence.
2 Cytoplasmic inheritance exists and is likely to involve mainly the female parent.
3 Most genetic material is chromosomal.
4 Gametes each contribute a set of chromosomes to the zygote. This accounts for the equal contribution of the two sexes to inheritance. A special type of cell division is needed to reduce the chromosome number from the diploid to the haploid state.

*Question and answer*

a ***Do the results of these crosses support ideas of matrilineal or patrilineal inheritance?***

Neither; an equal contribution is suggested.

## 16.2 Evidence that chromosome changes are associated with inherited characteristics

The concept of a karyotype is introduced.

*Principles*

1 Chromosomes have individual morphology which can be recorded in detail. Once a chromosome has been described, homologues of it may be observed in cells of other individuals of the same species, or even in cells of related species.
2 Histochemical techniques, such as treatment with colchicine and the use of different types of stain, can help us to visualize details of structure.
3 Chromosome number is not entirely fixed. Aberrations of cell division can result in changes which may be perpetuated by subsequent cell division. Changes in chromosome number may result in changes in phenotype, particular changes in development being associated with particular chromosomes. This is evidence that they contain genetic material.

### STUDY ITEM
### 16.21 Chromosome banding

*Questions and answers*

a ***How many chromosomes are there?***

12.

b ***How many sets of chromosomes are there?***

2 (one would expect a root tip cell to be diploid).

c ***Suggest a hypothesis to explain this [variation in band sequence].***

Chromosomes have broken and the ends have rejoined in a different arrangement from that originally found. Chromosome replication has perpetuated the new and the old structure in different individuals or populations. The genetic consequences of inversion and translocation are mentioned on page 59.

d ***Suggest an explanation for this [chromosome similarity between human and chimpanzee].***

If human and chimpanzee have evolved from a common ancestor, one would expect changes in chromosome structure (and in genetic material) to have occurred. Some chromosomes might, however, have been replicated with no great alteration in structure in both species, and retain their similarity to those found in the common ancestor.

e ***What does this observation suggest? [Bar eye associated with chromosome duplication.]***

It suggests that the genetic material found in the band which had undergone duplication has an influence on eye development. This is evidence that particular bands carry particular types of genetic material. It is, however, not strong enough for us to claim that a chromosome band is identical with a single gene. Stronger evidence will appear in Chapter 5. Able students may point out the danger of confusing correlation with causation in this case. □

## 16.3 Polyploidy and aneuploidy

*Principles*

1 The mechanism of cell division can fail, and mitosis perpetuates the consequence of the failure if the resulting karyotype allows survival.
2 Environmental influences can produce chromosome changes.

The term mutation has been avoided as it could not be used without a discussion of point mutation, for which the students are unprepared.

### Down syndrome: an example of human aneuploidy

A sensitive approach must be adopted since a student may have a close relative with Down syndrome. The changes in phenotype must be a result of changes in interaction between genetic material brought about by an increased dosage of genes on chromosome 21.

Teachers may wish students to consider the rarer form of the syndrome caused by translocation of a large part of chromosome 21 to chromosome 15, and also the possibility of a mosaic individual produced by non-disjunction of chromosome 21 at an early embryonic mitosis. The Abal project materials (Abal-Genetics Section 4) may be useful here.

The possibility of early diagnosis of genetic abnormality by amniocentesis is considered here, and in Chapter 23. It is important that students are aware of the social and moral issues involved.

*Question and answer*

a **_How could the tendency of trisomics to undergo miscarriage be discovered?_**

The karyotypes of a large sample of spontaneously aborted foetuses would need to be determined and compared with those of a sample of live births.

## STUDY ITEM
## 16.31 Chromosome changes in a plant

*Questions and answers*

a **_What is the basic number of chromosomes in_ Taraxacum?**

Since $3n = 21$, $n = \frac{21}{3} = 7$.

b **_How many different varieties with a chromosome number of 20 could be produced by aneuploidy alone from an ancestral variety with a chromosome number of 21?_**

Seven varieties – each of the seven different chromosomes could be lost.

c **_Would it be correct to argue from this that dandelion chromosomes carry fewer genes than human chromosomes?_**

No. As the dandelion is triploid, loss or gain of a chromosome is offset by the increased number of whole sets of chromosomes. The effect of genetic balance is less drastic than in diploid *Homo sapiens*. □

## 16.4 Meiosis and its significance

*Assumption*

1 A knowledge of mitosis and chromosome structure.

*Principles*

1 Meiosis involves the production of four haploid nuclei from one diploid nucleus.
2 The chromosomes replicate once but the nucleus divides twice.
3 There is a similarity in the way that the process occurs in all eukaryotes.
4 Chiasmata hold bivalents together and are the physical expression of genetic exchange between homologous chromosomes of maternal and paternal origin.
5 Random orientation of bivalents allows independent assortment of the chromosomes. This principle is reinforced in Chapter 17. At this stage, the students should become aware that the haploid products of meiosis do not contain the same sets of chromosomes as those which came together in the zygote that gave rise to the meiotic mother cell.

Practical investigations 16A, 'Gamete production by a plant', and 16B, 'Gamete production in an animal', can be associated with this section. These practicals allow the students to investigate meiosis for themselves. These exercises might be regarded as alternatives and could either be used as a preliminary investigation before considering the factual details or they could be used as confirmation. Philip Harris Biological Ltd publish sets of prints of mitosis and meiosis. These are unmarked photographs and could be used in association with this section.

**Practical investigations. *Practical guide* 5, investigation 16A, 'Gamete production by a plant', and investigation 16B, 'Gamete production in an animal'.**

*Question and answer*

**a** ***What would be the genetic consequence if the first pair of chromosomes to orientate on the metaphase plate determined the orientation of the other pairs?***

Certain combinations of chromosomes would be produced in the haploid products of meiosis, resulting from the organizing activity of the first pair. If all the pairs behave independently, a random distribution of chromosomes will result. (Evidence that this is in fact what happens will come from the study of dihybrid ratios in Chapter 17.)

## STUDY ITEM

### 16.41 Meiosis summarized

*Questions and answers*

**a** ***Study these two figures and make a table such as the one which follows, which best matches the numbered drawings of* figure [S]37 *with the photographs of* figure [S]38. *Not every drawing has a corresponding photograph.***

| *Drawing number (figure [S]37)* | *Photograph number (figure [S]38)* |
|---|---|
| **1** | **i** |
| **2** | **a** |
| **3** | **f** |
| **4** | **c** |
| **5** | **g** |
| **6** | **l** |
| **7** | **b** |
| **8** | **d, h** |
| **9** | **j** |
| **10** | **e** |
| **11** | no corresponding photograph |
| **12** | **k** |
| **13** | no corresponding photograph |

Photograph **e** (second division metaphase) is taken looking down from the pole of the spindle and may not immediately be recognized by students. Photographs **d** and **h** correspond to stages between drawings **7** and **8** more closely than to **8** itself.

**b** ***What is the haploid chromosome number in drawing 3 of figure [S]37?***

Two

**c** ***Consider a chromosome with a centromere near the middle and a second, non-homologous chromosome with the centromere at one end. Copy and complete table [S]4 for these chromosomes.***

| *Chromosome type* | *Number of chromatid arms trailing at anaphase of* | | |
|---|---|---|---|
| | *mitosis* | *meiosis division I* | *meiosis division II* |
| Centromere near middle | 2 | 4 | 2 |
| Centromere at one end | 1 | 2 | 1 |

**d** ***Give two reasons why it is important that chromosomes come together in their homologous pairs at meiosis and why this is not necessary at mitosis.***

**1** Pairing ensures that each product of division of the nucleus gets one of the two homologues. This is not necessary in mitosis since each chromosome merely has to replicate.
**2** Pairing allows genetic exchange between homologues. This would not normally be desirable during mitosis, the function of which is to replicate the genome unchanged. (The whole subject of somatic pairing and the transposition of genes should be avoided at this stage.)

**e** ***For a diploid gamete to result, when would meiosis need to fail?***

Second anaphase. Failure of first anaphase and second anaphase would result in a tetraploid gamete. Failure of first anaphase with a successful second anaphase would give a pair of diploid gametes. □

## Meiosis in hybrids and polyploids

*Assumption*

**1** A knowledge of polyploidy and of meiosis.

*Principles*

**1** Exact pairing of homologues must occur if a regular meiosis is to result.
**2** The greater the heterozygosity of the cell which becomes tetraploid, the more likely it is that regular pairing and hence sexual fertility will result.
**3** Large sudden changes in chromosome number have been important in plant evolution and not in animal evolution because many plants have a vegetative reproduction process, while most animals do not.

The whole subject of fertility in polyploids is complex. The effect on meiosis is likely to vary with the basic chromosome number and with the size of the chromosomes. In *Triticum aestivum* a locus on one of the chromosomes derived from *Aegilops speltoides* suppresses pairing

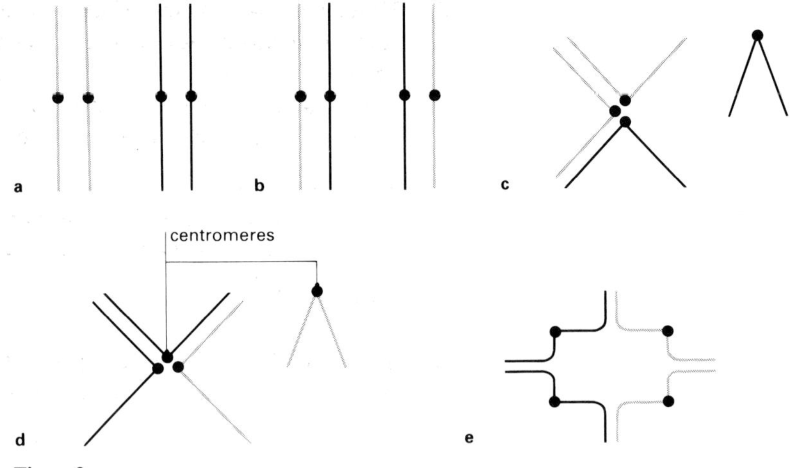


**Figure 3**
Some possible associations between homologues at 1st prophase in a tetraploid. (Chromosomes that are genetically identical have the same shading.)

between homologues in the three sets of chromosomes involved in the allohexaploid and this results in a very regular segregation. Some ferns, for example in the genus *Dryopteris*, are highly polyploid.

The classic examples of *Spartina townsendii* and *Primula kewensis* could be mentioned here. (See Darlington, C. D., 1963, listed in the Bibliography to this chapter.)

*Questions and answers*

**b** **Figure [S]39** ***shows possible associations of chromosomes in a triploid at 1st prophase of meiosis. Make similar drawings to show the possible associations of four homologous chromosomes in a tetraploid. Show the two identical maternally derived ones in one colour and the two paternally derived ones in another colour. (Do not include a 'secondary constriction'.)***

See *figure 3*.

**c** ***Why has polyploidy been unimportant in the evolution of animals?***

A sterile karyotype may survive by vegetative reproduction (as a clone) until further genetic change results once again in sexual fertility. Plants often show vegetative reproduction. Few metazoan animals have any significant power of vegetative reproduction. Vegetative reproduction pre-adapts a species for evolution by major chromosome mutation.

## 16.5 The inheritance of sex

*Principles*

1 Either sex can be heterogametic and this will result in a 1:1 ratio of males to females in the offspring as a result of meiosis.
2 Aneuploidy of the sex chromosomes occurs and may lead to abnormal development of sexuality in the adult – further evidence that there is a causal link between chromosomes and inheritance.

3 Abnormal development of sexual characteristics can occur for reasons other than chromosomal abnormality.

4 The sex chromosomes may determine characters which are not sexual. These characters are sex-linked and their inheritance is further verification of the chromosomal theory of inheritance.

*Questions and answers*

a ***Consider meiosis of a single human male cell. How many X-bearing and how many Y-bearing haploid nuclei would result?***

Two of each.

b ***If a man liberated 3 $\times$ $10^9$ sperm cells at sexual intercourse, how many of these would you expect to be Y-bearing?***

Half of them, that is, 1.5 $\times$ $10^9$ of each type.

c ***What ratio of male to female children will be expected in any population? Explain your answer and state what assumptions you have made.***

The argument used by the student is much more crucial than the answer. If one assumes that X- and Y-bearing sperm are equally likely to fertilize the X-bearing eggs then XY and XX zygotes are equally likely, that is, a 1:1 ratio of males to females. This ratio will only be seen at birth if both types of zygotes are equally viable.

d ***What types of haploid cell would meiosis produce in a male grasshopper and which of these types would produce male offspring after fertilization?***

The unpaired X chromosome will replicate once while the cell will divide twice resulting in two haploid cells with a single X chromosome and two with no sex chromosome. These later cells will produce X0, that is male zygotes, when they fuse with X-bearing eggs.

e ***How many Barr bodies would you expect in nuclei of cheek cells from a girl with Turner's syndrome.***

None. Each cell has a single X chromosome and this cannot spiralize to produce a Barr body since it seems that one active (unspiralized) X must be present.

f ***From which parent must any male mammal inherit its X chromosome?***

From the female parent. It is XY itself and must have inherited the Y from its male parent since females do not have Y chromosomes.

g ***Explain why tortoiseshell male cats are not found (with the exception of rare infertile XXY individuals).***

Tortoiseshell seems to be a mixture of colour. This depends on genetic material determining black and genetic material determining ginger being present in the same individual. These two types of genetic material must be alternative conditions of the X chromosome. Only a female can have both since she has two X chromosomes.

This last question and its answer are crucial to students' understanding and are worthy of a good deal of class discussion. They lead directly to the allele concept and it is important that those members of a class who have already understood the concept from previous courses, or from background reading, are made to explain matters very clearly to the rest. Notice that an example of sex linkage with no dominance has been used. The existence of the heterochromatic X (the Barr body) in female mammals and the case of random patches of red and black colour (tortoiseshell) will be returned to in Chapter 19.

## STUDY ITEM

### 16.51 The sex ratio in human beings

*Principles*

1 The assumption (made in section 16.5) that X- and Y-bearing sperm have an equal chance of achieving fertilization is not entirely valid.

2 Male and female zygotes do not have equal survival chances.

*Questions and answers*

**a** ***What factors might change the expected ratio of the sexes, at birth, and in a population?***

X- and Y-bearing sperm do not necessarily have an equal chance of achieving fertilization, neither do male and female zygotes have equal chances of survival.

**b** ***What are the advantages for humans (and other organisms) in having a 1:1 sex ratio? In what circumstances would it be a disadvantage?***

**1** It allows monogamy and the sharing of the care of offspring between two parents (in humans and in many birds.)
**2** In polygamous species such as sealions it seems to be a waste of resources to produce equal numbers of male and female offspring. If a single male fertilizes all available eggs, all of the next generation will be half-sibs. This leads to inbreeding unless it is balanced by a tendency of males to disperse as in, for example, lions. Teachers may wish to defer discussion of these matters to Chapter 20.

**c** ***There is positive correlation between the excess of males over females at birth and the affluence of the group of mothers sampled. Put forward a hypothesis to explain this.***

Affluent mothers are well nourished, rest, and have good medical care during pregnancy. They are less liable to miscarry or have a still-birth, and it seems that male embryos are more delicate than female ones.

**d** ***How might such changes in the proportion of males and females affect the size and composition of future generations?***

**1** Size of population could be affected if changes in sex ratio resulted in earlier (or later) marriage. The size of individual families can be influenced greatly by the parents' desire for children of a particular sex.
**2** Composition of future generations cannot be affected by the existing sex ratio in a population. It will be affected only by meiosis, fertilization, and survival.

**e** ***If the procedure of determining karyotype from cells of the amniotic fluid becomes routine, what might be the influence on the sex ratio and on the growth of populations?***

**1** If parents are able to choose the sex of their children (having an abortion if they produce a foetus of the wrong sex) then the sex ratio of at least some societies might be greatly disturbed.
**2** Ability to choose the sex of offspring might result in smaller families since parents might no longer have a number of children while trying to get one of a particular sex. □

Understanding of, and control over, fertilization are an alternative means whereby parents might influence the sex of a child (see Chapter 23).

## 16.6 Inheritance in fungi

**Practical investigation. *Practical guide* 5, investigation 16C, 'Inheritance of ability to synthesize starch in *Zea mays*' and investigation 16D, 'Spore colour in *Sordaria fimicola*'.**

The reasons why fungal genetics is so important are listed in this section. In the context of this chapter, their haploid state and the ability of the experimenter to observe the consequences of a single meiosis should be stressed.

*Principles*

1. The idea of a heterokaryon is introduced.
2. In *Coprinus* four basidiospores arise from a single meiosis and so two of them will have one allele and two will have the other.
3. Each gene has a locus on a chromosome.
4. Alternative forms of a gene, occupying the same locus, are alleles.
5. Genes and their alleles cannot, like chromosomes, be seen but their existence can be inferred from patterns of inheritance.

*Questions and answers*

**a** ***Explain how the inheritance of an ability to produce choline in*** **Coprinus** ***illustrates the allele concept.***

There should be a brief review of meiosis, perhaps something on the lines of figure (S)33, page 40 of the *Study guide*, and a statement that the pattern of two wild type and two choline-requiring spores on each basidium indicates that these alternative states are caused by separation of a pair of homologous chromosomes.

**b** ***The pollen grains of a flowering plant are the haploid products of meiosis. Explain why pollen is much less useful than basidiospores as a means of establishing the allele concept.***

Mature pollen grains do not remain in tetrads, so there is no conclusive proof that a single meiosis produces two cells of each contrasting type. It is unusual to find two contrasting types of pollen in the same species of plant. (The example of the starchy and waxy pollen of maize is an exception.)

Basidiospores could easily be demonstrated to a class by getting them to examine pieces of gill from a fungal fruiting body with a dissecting microscope. It is important not to use cultivated mushrooms as these have a pair of (binucleate) basidiospores instead of four which will be confusing!

Investigation 16C in *Practical guide 5*, 'Inheritance of ability to synthesize starch in *Zea mays*', provides a very good reinforcement for this section and could be used before the students meet question b above.

## STUDY ITEM

## 16.61 Spore formation in *Neurospora*

This section might be regarded as alternative to, or reinforcement for, practical investigation 16D, 'Spore colour in *Sordaria fimicola*' in *Practical guide 5*.

*Questions and answers*

**a** ***If the wild-type allele is represented by + and the lysine-requiring allele by L, what would be the genetic constitution of the diploid cell which produced each ascus?***

+L.

**b** ***What would be the possible genetic constitutions of the nuclei in stage B of* figure [S] 50*?***

B is the end of meiosis division I so each nucleus would have pairs of chromatids joined by centromeres. The two nuclei could be ++ and LL or +L and +L. Some discussion will be needed to make clear that these alternatives depend on whether a chiasma has occurred between the centromere and the lysine-requiring mutant.

**c** ***'It is clear from these rather limited data that this inability to synthesize lysine is transmitted as it should be if it were differentiated from the normal by a single gene.' Explain this statement.***

The wild-type fungus has genetic material which produces a product which allows lysine synthesis. The corresponding genetic material of the lysine-requiring strain does not function to produce lysine. Spores have one of these alternative units of genetic material because they are carried at corresponding positions on homologous chromosomes which separate during meiosis. The alternative states of the genetic

material are alleles. No intermediate phenotype is observed. (We are dealing with haploids so dominance can be discounted.)
Students may raise the problem of whether the lysine-requiring allele has a real existence or whether it is the absence of the wild-type allele. (Compare with the XO system of sex determination.) This question could lead to a discussion of deletion mutants, but the data given provide no evidence for or against the idea.

d ***Account for the difference in sequence between the spores in ascus number 3 and ascus number 4 in table [S]5.***

In number 3 the alleles segregated at 2nd anaphase while in number 4 they segregated at 1st anaphase. One expects segregation at 1st anaphase unless a chiasma has occurred so that alleles have become attached to different centromeres.

e ***Between which two stages shown in figure [S]50 must this difference in sequence have arisen when the two asci developed?***

□ Between A and B, that is, during division I of meiosis.

# PART II The *Practical guide*

The investigations related to this chapter appear in *Practical guide 5, Inheritance.*

## INVESTIGATION
## 16A Gamete production by a plant

### ITEMS NEEDED

Developing inflorescence or bulb, suitable

Ethanoic alcohol fixative (if required)
Hydrochloric acid, 1.0 mol dm$^{-3}$
Stains:
ethano-orcein
*or*
Feulgen
*or*
toluidine blue
Water, distilled

Beaker or watch-glass 2/1
Blotting-paper or filter paper
Forceps 2/1
Hot plate at 50–60 °C or Bunsen burner 1/group,
*or*
Water bath at 60 °C (as appropriate) 1/class *(continued)*

(*Study guide* 16.4 'Meiosis and its significance'.)

**Ethanoic alcohol.** See page 13.
**Ethano-orcein.** See page 12.
**Feulgen.** See page 95.
**Toluidine blue.** See page 12.

*Assumptions*
1 An elementary knowledge of flower structure.
2 Ability to use a microscope correctly.
3 Ability to use a stereomicroscope.

*Principle*
1 The manufacture of pollen by an angiosperm requires meiosis to occur, producing haploid nuclei.

There are numerous possible sources of meiotic cells in angiosperms, but in all cases the crucial point is to find very young anthers. If the anther contains pollen, it is usually possible to squeeze the loose grains

## ITEMS NEEDED *(continued)*

Measuring cylinder, 20 $cm^3$, or syringe, 20 $cm^3$ 1/1
Mounted needles 2/1
Pipettes, dropping 2/1
Microscope, monocular 1/1
Microscope, stereo 1/group
Microscope slides and coverslips
Razor blade or scalpel 1/1

out of the loculi during the staining process, so that with a squash of the anther wall its gametogenic cells will be produced. Some anthers do not seem to take up stain very readily, and it is wise to use a number of different sources of material.

Generally, the smaller the number and the larger the size of the chromosomes, the better is the preparation and the easier is the interpretation.

Possible sources of material are:
Corms or bulbs of *Crocus*, *Narcissus*, hyacinth, tulip, and *Lilium*.
Very young buds of *Pelargonium* sp., *Tradescantia* sp., and *Impatiens* sp.
Winter buds of *Aesculus* (horse chestnut) and *Fraxinus* (ash).
Bulbs should be used in late summer and early autumn, since they will be rather over-ripe later in the year. They should be avoided if they have started to sprout.

Here are some practical points.
The anthers used must not be at all yellow, but should be transparent.
It is better to have very little material on the slide than too much since, if the cells are to be sufficiently separated, the squashing process must be done correctly and thoroughly. Excessive material prevents successful squashing.
The lighting of the slide is crucial and the student should be shown how to use the condenser correctly, in order to be able to achieve the best contrast.
It will be found that in some areas of a slide the chromosomes are not clearly defined. Also in many cases the squashing is not performed carefully enough, the cells do not squash, and the chromosomes are viewed end on – this causes the nuclei to appear to be full of small circular structures.
Very often the act of drawing a nucleus reveals the stage it has reached, while merely observing it leaves it uninterpreted. The judicious use of the fine focus (and, where available, an oil immersion lens) will usually allow ready interpretation.
Sometimes the cells show a tendency to break up somewhat, in which case free chromosomes may be seen in the field of view.
No prior knowledge about meiosis is assumed in the instructions and questions. However, it is quite possible to use this work as a confirmatory practical exercise, in which case the extra question **d**, given in this *Guide* at the end of the investigation, could be added.

*Questions and answers*

**a** ***You will almost certainly see some maturing pollen grains on your slide. Compare these with the cells that are clearly in the process of division. What are the major differences between the two types of cell?***

Mature pollen grains have a denser cytoplasm than meiotic cells. The volume of cytoplasm in the maturing pollen grain relative to the volume of nucleus is greater than in a meiotic cell. The cell wall of the maturing pollen grain is somewhat thicker than that of meiotic cells.

**b** ***State any evidence which you saw in your preparation which led you***

***to the conclusion that the cells were undergoing an orderly process rather than a random division of contents. (Cell division in gamete-producing cells is called meiosis.)***

Patterns of chromosome alignment or distribution within the nuclei are repeated in many cells. The existence of such recognizable patterns suggests that an ordered process of chromosome distribution is taking place. This can be confirmed by observation of living cells, using phase contrast and other illumination techniques.

c ***Compare the actively dividing cells which you see in this preparation with those which you observed in a root tip preparation. Can you see anything which would lead you to believe that a different process was occurring in the two types of cell?***

Because of the limitations of resolution that are placed upon observation through the use of the light microscope, it is unlikely that inexperienced students will easily be able to notice any significant differences. It may be that the configurations of 1st prophase, showing chiasmata, can be identified as occurring in this preparation but never in the root tip. Equally, the occasional sighting of a tetrad of cells whose nuclei are clearly at 2nd telophase might lead the student to this conclusion. Also, in this preparation there are quite patently a large number of stages of the division where the chromosomes are clearly 'doubled', and this contrasts nicely with mitotic preparations where it is usually hard to see double strands at any stage. Configurations in the form of a cross at later stages of prophase and at metaphase indicate bivalents. (*Study guide II* page 42.)

*Additional question*

**d** ***State any evidence which you saw, or hoped to see, in your preparation which would lead you to conclude that the cells were undergoing meiosis rather than mitosis.***

This question should produce a rather more tightly argued answer than the one to question **c**. The student would be expected to argue from the known facts about the behaviour of chromosomes at meiosis and mitosis in order to interpret the observed stages in the preparation.

## INVESTIGATION
## 16B Gamete production in an animal

ITEMS NEEDED

Male locust nymph 3–5 instar, freshly killed

Ethanoic alcohol fixative
Hydrochloric acid, 1.0 mol $dm^{-3}$
Saline, insect
Stains:
ethano-orcein
*or*
Feulgen *(continued)*

(*Study guide* 16.4 'Meiosis and its significance'.)

**Ethanoic alcohol.** See page 13.
**Ethano-orcein.** See page 12.
**Feulgen.** See page 95.

**Ringer's solution, insect.** Mix the following ingredients together:
Calcium chloride, anhydrous, 0.22 g
Magnesium chloride, 0.20 g
Potassium chloride, 0.74 g

## ITEMS NEEDED *(continued)*

Beaker or watch-glass 2/1
Blotting paper or filter paper
Dissecting dish with cork or wax 1/1
Forceps 2/1
Hotplate at 50–60 °C or Bunsen burner 1/group
*or*
Water bath at 60 °C 1/class (as appropriate)
Measuring cylinder, 20 $cm^3$
Syringe, 20 $cm^3$ 1/1
Microscope, monocular 1/1
Microscope slides and coverslips
Mounted needles 2/1
Pins
Pipettes, dropping 2/1
Scissors

Sodium chloride, 8.10 g
Sodium dihydrogen phosphate, 0.78 g
Sodium hydrogen carbonate, 0.33 g
Water, distilled, 1 $dm^3$

*Assumptions*

1 Sufficient familiarity with cells to be able to recognize chromosomes.
2 Ability to use a microscope at H.P.

*Principles*

1 Gamete-producing organs contain high proportions of cells actively undergoing cell division.
2 The orientation of the chromosomes as revealed by this technique represents reliably the meiotic stages, not merely the effects of the treatment given to the cells.

The age of the locust is not too critical and good results may be obtained from 3rd or even 2nd instar animals as well as the more customary 4th and 5th instars. Even young adults, immediately post-ecdysis, are good, and they show maturing sperm as well as the sperm mother cells. If there is sufficient time or skill available to a class it is an interesting exercise for students to try to compare the follicles from locust nymphs of two different ages.

The important point of the technique is to produce a thin layer of cells without actually breaking the cells open. After the staining treatment used here, smearing with knife blades tends to disrupt the cells. Provided that only four or five follicles are squashed, they spread very well using the method written here and the work is very straightforward. Prepared slides of locust testis are available, but many lack the clarity of fresh, home-produced slides.

These preparations do not keep very well, although, if the coverslip is generously ringed with, for example, euparal mountant, and the preparation is not disturbed, it can be kept for a few days without too much loss of contrast or distortion in appearance.

*Questions and answers*

**a** ***What evidence did you see that indicated that these cells were undergoing an ordered process of division rather than division of a random nature?***

See answer to question **b** on page 35.

**b** ***Suggest reasons why it is of advantage to the animal for its testis to mature during its nymphal life rather than to begin its maturing process only after the animal is adult.***

The main advantage of precocious gametogenesis is that the male imago is fertile almost as soon as the final ecdysis is complete. It can therefore compete with other males to be the first (and possibly only) mate of a female which can store sperm and need mate only once.

**c** ***If it is possible, arrange the nuclei that you observed into some sort of order representing progressive change. What is the significance of***

***the order into which the nuclei fit?***

It should be possible for interphase nuclei, many first prophase (leptotene), and a few of each of the other stages to be located. The significance is that any sequence suggests ordered change. The division of the complete chromosome set is very precisely ordered to permit one member of each homologous pair of chromosomes to go to each first telophase nucleus and to allow one member of each pair of chromatids to go into each second telophase nucleus.

## INVESTIGATION
## 16C Inheritance of ability to synthesize starch in *Zea mays*

(*Study guide* 16.6 'Inheritance in fungi'.)

### ITEMS NEEDED

Maize, spikelet 1/1

Iodine in potassium iodide solution
Water, distilled

Forceps 2/1
Microscope, monocular 1/1
Microscope slides and coverslips
Mounted needle 1/1
Pipettes, dropping

**Iodine in potassium iodide solution**
Iodine crystals, 1.5 g
Potassium iodide, 3 g
Water, distilled, 1 $dm^3$
Dissolve the potassium iodide in 100 $cm^3$ of the water and add the iodine crystals. Make the solution up to 1 $dm^3$. Leave this mixture to stand for 24 hours. Filter and store, preferably in a dark cupboard.

*Assumptions*

1 Some knowledge of flowering plant reproduction is helpful.
2 The ability to use the $\chi^2$ test.
3 That students have not formally considered Mendelian genetics but are familiar with meiosis, that is, that the sequence of ideas presented in *Study guide II* has been followed to near the end of Chapter 16 but not into Chapter 17.

*Principles*

1 The two chromosomes carrying genetic material for starch production segregate at meiosis.

It is essential to use a dilute aqueous iodine solution to differentiate between the two types of pollen. It is also imperative that the illumination of the grains is adjusted to give clear distinction. If viewed solely by transmitted light, the two types of grain look very similar in colour, since their contents are very dense. The nearer the student can get to dark ground illumination, the easier it will be to differentiate between the two types of pollen.

The state of maturity of the pollen and the length of time it has been stored dry or preserved in ethanol, both seem to affect the amount of stain it will take up. If a sample stains very darkly with the concentration of iodine suggested, try serial dilution. A little preliminary work by the teacher is advisable.

Sufficient detail about the morphology of the plant is given to ensure that correct terminology may be used in description and also that the nature and relationships of the parts being examined is understood. (Suppliers sometimes send pollen preserved in ethanol rather than a portion of inflorescence.)

'Starch', the food reserve of most plants, is a mixture of two polymers of α-D-glucopyranose. One consists of unbranched chains of sugar units joined by α-1-4-glycosidic bonds; this is amylose and it forms a dark blue–purple–black complex with iodine in solution. The other consists of branched chains of the sugar units, with α-1-4-glycosidic bonds joining the sugars within the chains and α-1-6-glycosidic bonds found at the branching points; this is amylopectin and gives a red–brown colour when treated with iodine in solution. In the mixture called starch, the blue of the amylose/iodine complex masks the red of the amylopectin.

In this maize variety, a recessive allele (wx) causes the production of a food reserve that is 100 % amylopectin, whereas the dominant allele (Wx) leads to the synthesis of a starch that contains only about 25 % amylopectin. Hence the two types of haploid pollen grains (wx and Wx) stain respectively red–brown and blue–black, and are colloquially referred to as 'waxy' and 'starchy'.

Quite clearly, if Mendel's 'Law of segregation' holds true, then as a result of meiosis in the heterozygous plant, half of the pollen grains will carry the dominant allele and half will carry the recessive allele, and this will result in an unexpected phenotypic ratio of 1:1 in step 9.

Suppliers also market dry maize cobs from heterozygous (Wxwx) plants which have been selfed. If the fruits are firmly embedded in the cob they can be cut or filed with a carpenter's file to expose their endosperms. Alternatively they may be cut or filed after extraction from the cob. The exposed endosperm will stain either brown or black with aqueous iodine (the concentration is not as critical as with pollen) and should give a 3:1 starchy:waxy ratio. This is a very quick and easy demonstration and could be carried out as a reinforcement exercise with Study item 17.31 (*Study guide II* page 69) or with Practical investigation 17C, 'Patterns of inheritance'.

If not heavily stained, the 'seeds' may be stored exposed to air and the iodine will sublime away in the course of a year so that they can be used again.

A sample can be planted in late April on a south-facing site and the pollen of each plant sampled. One vigorous plant produces a 'tassel' with many spikes at its apex.

It may be appropriate to point out to students that an endosperm is triploid. This could lead to useful discussion on the dosage effect of alleles in relation to dominance. (Some starchy endosperms should be of genotype WxWxwx and others wxWxwx.)

*Questions and answers*

a ***If separation of two homologous chromosomes, one maternal and the other paternal, has occurred during meiosis you will expect the two types of pollen grain to occur in a particular ratio. Use the $\chi^2$ test to determine whether the observed numbers of the two pollen grain types are in accordance with your expected ratio. State in words the null hypothesis you are testing.***

The null hypothesis would be along the lines of 'There is no significant difference between the observed numbers of the two types of pollen grain and the numbers of each type that would be expected if separation has occurred and the two chromosomes have been distributed equally to the gametes, that is, a 1:1 ratio.' It is worth stressing the conscious formulation of the null hypothesis to reinforce the importance of the null statement to the use of the $\chi^2$ test; it should help to prevent that test from becoming a rather poorly thought-out, mechanical tool.

**b** ***If the $\chi^2$ test shows that the observed distribution does not agree with your predictions, assuming those predictions to be genetically 'correct', where would you expect that the discrepancy had first arisen – in the anther during meiosis, during your preparation of the slide, or during the counting of the pollen grains?***

Any discrepancy between observed and expected results is almost certainly not a result of either faulty genetic theory or faulty meiosis. The most usual causes of such discrepancy are either too small a sample size, or slide preparation techniques that have in some way introduced bias into the sampling procedure. Immature pollen grains stain brown and are a common cause of bias.

## INVESTIGATION
## 16D Spore colour in *Sordaria fimicola*

### ITEMS NEEDED

Cornmeal agar, sterile, in Petri dish 1/group
*Sordaria fimicola* cultures, pigmented and unpigmented
Water

Bunsen burner 1/group
Glass marking pen 1/group
Incubator at 25 °C 1/class
Microscope, monocular 1/1
Microscope slides and cover slips
Mounted needle 1/1
Tape, adhesive

(*Study guide* 16.6 'Inheritance in fungi'.)

**Cornmeal agar** (sufficient for about 20 plates)
Cornmeal agar (Oxoid), 5.4 g
Water, distilled, 320 $cm^3$
Yeast extract, 0.3 g
Disperse the ingredients in a conical flask, plug, and autoclave.

Or:
Maize meal, 12 g
Water, distilled, 400 $cm^3$
Boil the meal for 15 minutes in the water and then decant off the clear liquid. Add 2.0 g agar to each 100 $cm^3$ of clear liquid and autoclave.

*Assumptions*
1. A knowledge of meiosis.
2. An understanding of the term allele.
3. A knowledge of elementary sterile techniques.

*Principles*
1. The asci of *Ascomycetes* hold together the products of a single meiosis and so enable the meiotic basis of a 1:1 segregation of alleles to be established.
2. The pattern of distribution of black and white spores in hybrid asci supports the hypothesis that chromosomes orientate randomly with segregation of two genes on different, i.e. non-homologous, chromosomes.
3. Both first and second division segregation of alleles can be seen, showing that recombination between the spore colour locus and the centromere is occurring.

Plates with hybrid asci can be kept for months in a refrigerator if sealed with adhesive tape to prevent desiccation.
Cornmeal agar is a rather nutrient-deficient medium and encourages the sexual reproduction of the fungus.
It is important to acquire black and white spored strains of compatible mating types or very few hybrid perithecia will be obtained.
If time is short, thc cross could be carried out for the students and they could merely examine hybrid perithecia.
It is desirable to have available plates inoculated with each of the strains separately, so that the students can assure themselves that the segregation of spores results from a cross.

*Questions and answers*

a **_Make a series of simple diagrams to show the different arrangements of ascospores that are apparent within the asci. Number your diagrams 1, 2 etc._**

See *figure 4*.

b **_Which of your diagrams can be explained by the separation of b and $b^+$ at the first anaphase of meiosis?_**

Numbers 1 and 2 of *figure 4*.

c **_Which of your diagrams can be explained by the separation of b and $b^+$ at the second anaphase of meiosis?_**

Numbers 3, 4, 5, and 6 of *figure 4*.

d **_How do the class data provide evidence that, during meiosis, the orientation of chromosomes on the spindle is random with respect to the two poles of the spindle? (See_ Study guide II, figure 35.*)***

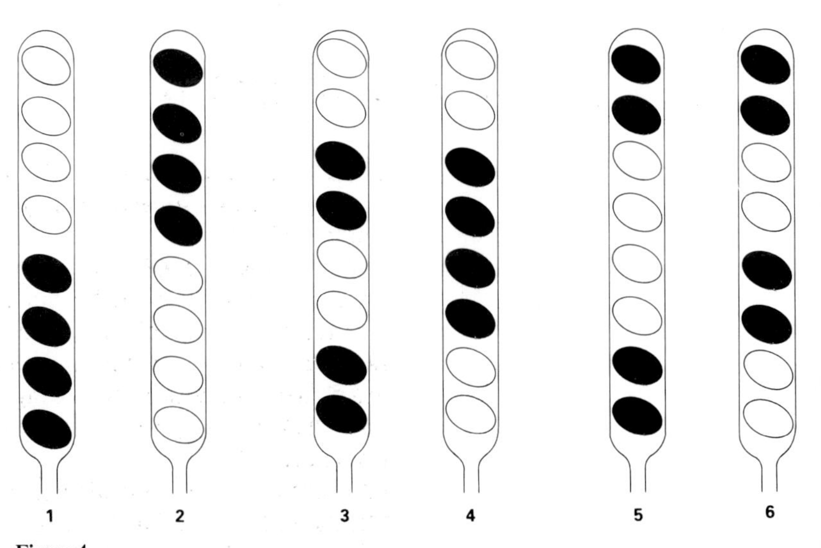


**Figure 4**
Arrangements of ascospores in hybrid asci of *Sordaria fimicola*. Drawings **1** and **2** are first division segregations. Drawings **3**, **4**, **5**, and **6** are second division segregations.

In the case of drawing number 1 of *figure 4* the centromere attached to two chromatids carrying b has orientated to the tip of the ascus and the centromere attached to two chromatids carrying $b^+$ has orientated to the base of the ascus. In the case of drawing number 2, the opposite orientation has occurred. The frequency of these two types of asci are the same, within the limits of chance. This can be checked by a $\chi^2$ test. Similar arguments apply to other pairs of ascus types such as number 3 and number 4.

e ***How can the separation of b and $b^+$ at the first anaphase of meiosis in some asci, and at the second anaphase of meiosis in other asci, be explained?***

By the occurrence of a chiasma between the spore colour gene (locus) and the centromere in the case of second division segregations but not in the case of first division segregations. (Able students may realize that double crossing over would restore the first division segregation pattern of spores.)

f ***In what percentage of asci did separation occur at the second anaphase of meiosis?***

$$\frac{\text{sum of 2nd division segregations}}{\text{total number of asci counted}} \times 100$$

Tetrad analysis of this sort, in which the cross-over frequency between loci and the centromere is calculated, can be used to map loci using the centromere as a reference point. As only two of the four chromatids are involved in a chiasma, chiasma frequency is half the frequency of second division segregation asci. If more than one marker at a time is employed, double crossing over can be detected and allowed for.

# PART III Bibliography

BOYCE, A. J. (ed.) Symposia of the Society for the Study of Human Biology Vol. 14, *Chromosome variations in human evolution.* Taylor & Francis, 1975.

DARLINGTON, C. D. *Chromosome botany and the origins of cultivated plants.* 2nd edn. Allen and Unwin, 1963.

FINCHAM, J. R. S. Carolina Biology Readers No. 2, *Genetic recombination in fungi.* 2nd edn. Carolina Biological Supply Company distributed by Packard Publishing, 1983.

GARBER, E. D. *Cytogenetics – an introduction.* McGraw-Hill Book Company, 1972.

HARRISON, D. *Patterns in biology.* Edward Arnold, 1975.

McDERMOTT, A. Outline Studies in Biology, *Cytogenetics of man and other animals.* Chapman & Hall, 1975.

# CHAPTER 17 VARIATION AND ITS CAUSES

*A review of the chapter's aims and contents*

The first two chapters in this volume introduce the student to genetic concepts by way of a fairly detailed study of cytology. This chapter is largely concerned with Mendelian inheritance and the analysis of the results of breeding experiments.

1. Reasons why variation in living organisms is interesting and important are listed.
2. A distinction is established between continuous and discontinuous variation.
3. The roles of genotype and environment in producing the phenotype are introduced and discussed.
4. The ideas and experimental approaches existing at and before the time of Mendel are briefly sketched as is his contribution.
5. Some important Mendelian terminology is introduced.
6. Monohybrid inheritance is explained and the methods of carrying out a rigorous genetic analysis are established.
7. Sex linkage has been introduced in Chapter 2 and is now further explained.
8. Dihybrid examples showing independent assortment and autosomal linkage are explained and related to meiosis and chromosome behaviour.
9. The inheritance of continuous variation is discussed.
10. The chapter concludes with an account of mutation and the effects of mutagens.

# PART I The *Study guide*

## 17.1 Introduction

**STUDY ITEM**

### 17.11 Continuous and discontinuous variation

**Practical investigation. *Practical guide* 5, investigation 17A, 'Describing variation'.**

*Assumptions*

1. Ability to prepare frequency tables and histograms.
2. Understanding of the terms mean, median, and mode.
3. Knowledge of standard deviation and standard error of difference.
4. Understanding of use of keys.

*Principles*

1. Characters vary continuously and discontinuously.
2. Statistical procedures allow the range of continuous variation to be described and the extent of variation between samples to be compared.

*Questions and answers*

a ***Prepare two lists of characters which show variation in some familiar organisms. In one list, place the continuously varying characters and in the second, those showing discontinuous variation (or use the results of Practical investigation 17A, 'Describing variation').***

The answer will depend on the example chosen. Some characters, for example, sensitivity to phenylthiocarbamide in humans, vary continuously and discontinuously. See also page 61.

b ***Prepare a table which groups the plants of each variety separately into 3 cm classes.***

See table 3.

**Sutton's Purple Podded**

| *Class intervals* (cm) | 37–39.9 | 40–42.9 | 43–45.9 | 46–48.9 | 49–51.9 | 52–54.9 | 55–57.9 | 58–60.9 | 61–63.9 | 64–66.9 |
|---|---|---|---|---|---|---|---|---|---|---|
| *Numbers in class* | 1 | 0 | 3 | 1 | 3 | 3 | 5 | 0 | 3 | 1 |

**Meteor**

| *Class intervals* (cm) | 10–12.9 | 13–15.9 | 16–18.9 | 19–21.9 |
|---|---|---|---|---|
| *Numbers in class* | 1 | 7 | 6 | 6 |

**Table 3**

(The results are from an actual class experiment. These varieties are strikingly different and also differ in flower colour and pod colour.)

c ***Use this class frequency table to show the size distribution of Meteor and Sutton's Purple Podded peas as a pair of histograms. Show these on the same set of axes, using different colours.***

As there is no overlap, the class intervals need not be continuous between the two distributions. Scales should be such as to use the paper effectively.

d ***Calculate and list the mean, the median, and the modal class values for each variety.***

See table 4.

| | *Sutton's purple podded* | *Meteor* |
|---|---|---|
| Mean | 52.5 | 16.8 |
| Median | 53 | 17 |
| Modal class (From table 3) | 55–57.9 | 13–15.9 |

**Table 4**

e ***In presenting data of this kind, explain why it is correct to use histograms and not continuous graphs?***

Because the exact shape of the distribution depends on the class intervals chosen and because minor variations in the form of the distribution may be due to chance. A continuous graph is by convention only used for a very large sample which approximates to an infinite population. The class intervals would need to be very small for a continuous graph.

f ***The data have been grouped into 3 cm classes. Give the advantages and disadvantages of using a smaller class interval. How can the class interval chosen for a histogram be related to sample size?***

**Advantage.** Subtle variations in the shape of the distribution, for example, minor modes, may be revealed by a small class interval.
**Disadvantage.** Chance variations in the shapes of distributions are likely to obscure the general trends if the interval is too small.
**Sample size.** The larger the sample the less important this disadvantage will be.

g ***What calculation would be needed to evaluate the hypothesis that the two groups of peas differed significantly in height?***

Standard error of difference.

h ***Is height a continuously or a discontinuously varying character in these varieties of peas?***

It varies discontinuously between varieties but continuously within each variety.

i ***Making allowance for sampling error, is it probable that height is normally distributed in peas?***

Yes as mode, median, and mean are all close together.

j ***Give two reasons why discontinuous rather than continuous variables are referred to in the construction of a key.***

They are more convenient as no measuring is involved. There is less likelihood of ambiguity as intermediate types are by definition unlikely with discontinuous variation. (An example of non-overlapping continuous variation could be used in a key if necessary.) □

## 17.2 The role of inheritance and environment

**Practical investigation. *Practical guide* 5, investigation 17B, 'The influence of environment on development'.**

*Principles*

1 All characters develop in an environment.
2 Phenotype = genotype + environment.
3 We are usually concerned with the range of variation shown rather than the absolute existence of a feature.
4 The external environment influences the development of some features of an organism more than others.

**5** Both continuous and discontinuous variation are subject to genetic and environmental influence.

*Questions and answers*

**a** ***Design an experiment to test the hypothesis that the ratio of wing length to abdominal length of locusts is determined by the density at which newly hatched (first instar) hoppers are kept.***

Adequate samples of hoppers, preferably siblings, would be separated on hatching into two groups. One group would be reared in individual containers, the other in a single container. Food would be *ad libertum* and, as far as possible, temperature, lighting, humidity, and other factors would be similar for both groups. When the locusts had undergone their final ecdysis, each locust would be measured in millimetres and the ratio of wing length to body length calculated. A table of the ratios grouped in suitable classes would be prepared for each group of locusts. The SED would be calculated (or the t-test used if the samples were small).

**b** ***Suggest some examples where variation usually depends on genetic differences and the environment seems to have little if any influence.***

Coat or plumage colour differences in animals or flower colour in plants and blood groups are all good examples. It should be made clear to students that we are discussing the external environment. If the internal environment can be altered during development the phenotype may be greatly affected. For example, if a patch of hair is shaved from the white coat of a Himalayan rabbit and the fur allowed to grow under unusually cold conditions, it will be black, as it usually is on the ears, paws, and nose which are always cooler than the rest of the skin.

**c** ***Two groups of mice are found to differ significantly in mean mass. How would you seek to investigate whether this difference is totally or partially the result of inherent (genetic) differences between the groups?***

Several females of each group would be mated to males from their own group at the same time. They would all be housed in the same room in similar cages and fed in the same way during the period of gestation. Litter sizes would need to be adjusted to a standard number per female shortly after birth. If the two groups differed in colour, pairs of females, one of each colour, could rear their litters in a communal nest or a system of cross-fostering could be adopted (mice readily accept this). The young mice would be weighed at a standard age and the masses handled as in the earlier example of pea heights.

## 17.3 Mendel and his contempories

**Practical investigation. *Practical guide* 5, investigation 17C, 'Patterns of inheritance'.**

*Assumptions*

1 That students have some idea of the meaning of the term species – they need not have discussed the difficulties of defining the term in any detail.

2 An understanding of the processes of mitosis, meiosis, and fertilization.

*Principles*

1 No theory for the origin of species can be complete and valid if it does not rest upon a valid theory of inheritance.

2 Darwin and his predecessors were handicapped by a lack of understanding of gametogenesis, fertilization, meiosis, and other processes involved in sexual reproduction.

3 Mendel was not working in an intellectual vacuum but was seeking to refine and test ideas developed by other plant hybridizers.

4 His success depended largely on his very careful experimental design and on his understanding and use of statistical ideas.

Many texts give a brief biography of Mendel and many introductions to genetics have begun with an account of his work on peas. If the historical approach is to be of much value he must be presented in the context of the ideas current when he was planning his work.

*Questions and answers*

**a** ***How could the idea of pangenesis have been tested?***

Since the theory suggested that the blood carried information from the organs to the gonads, blood transfusion should result in alteration of the ability of an animal to transmit a character such as coat colour. This test was tried by Francis Galton and the coat colour of the offspring of a rabbit was unaffected by blood from a variety of different colour.

**b** ***Did Gärtner believe in evolution?***

No.

**c** ***Judging from the above extracts, did Gärtner believe in blending inheritance?***

The phrase 'the mixing which occurs in reproduction' would indicate that he did but the second extract clearly shows that he had observed examples of segregating characters which did not ultimately blend.

**d** ***How many genetically different types of gamete can a male bee produce?***

Only one type as it is haploid and so cannot produce gametes by meiosis.

**STUDY ITEM**

## 17.31 Mendel's first experiment

*Principle*

1 Scientific progress depends on the effective communication of ideas as much as on individual talent and effort.

The terms parental type, $F_1$, $F_2$, $F_3$, backcross, genotype, monohybrid, and dihybrid are introduced in this section.

*Questions and answers*

a ***What was the ratio of yellow cotyledons to green cotyledons in the $F_2$ generation?***

3 yellow : 1 green.

b ***Does Mendel claim to have discovered dominance?***

No. He mentions that it has been observed in other cases and coins and defines the terms dominant and recessive.

c ***Are there any lessons about the international communication of scientific research to be learned from the loss of Mendel's work between 1866 and 1900?***

It is not enough that data and ideas are published. They must also be read by people with enough ability and expertise to understand them. (Mendel communicated his findings by letter to the botanist Naegeli. Naegeli had an international reputation and, had he properly understood Mendel's work, history might have been very different. Part of the reason for the total neglect of the discoveries may have been the interest generated by the Theory of Natural Selection. This directed attention away from hybridization experiments. Darwin rejected hybridization as a mechanism for producing new species.)
Lessons to be learned from the loss of Mendel's work might include:

**1** The value of international scientific conferences;
**2** The value of journals publishing abstracts;
**3** The value of an interdisciplinary approach.

Discussions in this area could include questions about the role of journals of general scientific interest such as *New Scientist*, language barriers, the impact of secrecy – whether for military or commercial reasons – and the question of whether an amateur could again make a scientific contribution as great as Mendel did.
Some reference to the enormous and explosive increase in the volume of scientific publication which has taken place in recent decades might be made. The role of publications which deal with abstracts and the increasing importance of microfilm and computer storage of data could also be mentioned. □

## STUDY ITEM
## 17.32 Analysis of monohybrid crosses

*Assumptions*

1. A knowledge of the terms $F_1$, $F_2$, $F_3$, backcross, genotype, monohybrid, and dihybrid.
2. An understanding of the term allele.
3. A knowledge of meiosis.

*Principles*

1. The fundamental 1:2:1 ratio of genotypes found in breeding experiments is the result of segregation of alleles at meiosis followed by random fertilization.
2. A homozygote is an individual the cells of which carry two identical copies of a gene. A heterozygote has two different alleles of the same gene in its cells.
3. Dominance masks the fundamental 1:2:1 ratio, converting it to a 3:1 ratio of $F_2$ phenotypes.

black female BB × brown male bb

black $F_1$ Bb

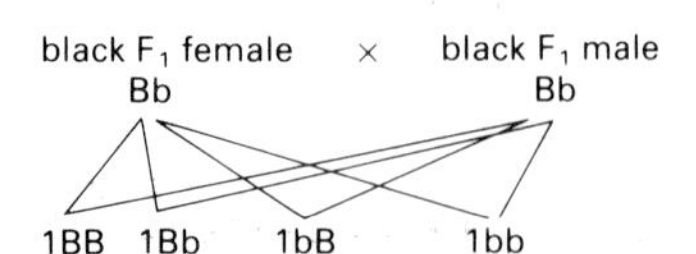


∴ 1BB : 2Bb : 1bb

**Figure 5**
A common method of deriving the genotypes of a monohybrid $F_2$, which has disadvantages.

It is considered most important that genetic ratios should be understood in terms of statistical probability. Many students have been introduced to the 1:2:1 ratio by means of a diagram such as *figure 5*.

This approach, cutting out as it does an explicit reference to ratios of gametes, should be avoided. It leads to the consciously or subconsciously held notion that pairs of gametes are seeking each other to achieve a fixed ratio and leaves students ill equipped to face a dihybrid analysis, especially if this involves linkage. The very full analysis given in table 9 of the *Study guide* can be shortened in a number of ways, for example, by writing the phenotype in each section of the matrix with the genotype and omitting the statement on random fertilization. The key to students' understanding is that they are required to make a conscious decision about the possible gametes produced by each individual and the frequency of these gametes.

*Questions and answers*

**a** ***Calculate the ratios and frequencies of the phenotypes in each $F_2$ generation and in the backcrosses.***

In the $F_2$ of both reciprocal crosses the ratios approximate to 3 grey body:1 ebony body.
In the backcross to the grey bodied parent all offspring are grey bodied. In the backcross to the ebony bodied parent there is an approximate 1:1 ratio of grey to ebony.

**b** ***Carry out a full analysis of the inheritance of the ebony body character as in tables [S]9a, b, and c.***

See table 5 which gives an analysis of the $F_1$ crossed together to give the $F_2$.
The student should state expected gamete frequencies, give a statement that random fertilization is assumed, and draw out a matrix from which genotype frequencies and ratios are deduced.

| *Parental generation* | | | |
|---|---|---|---|
| phenotype | grey | × | ebony |
| genotype | $e^+e^+$ | | ee |
| *Genotypes of parental gametes* | all eggs $e^+$ | | all sperm e |
| *$F_1$* | | | |
| phenotype | | all grey | |
| genotype | | $e^+$ e | |
| *Genotype of $F_1$ gametes* | eggs either $e^+$ or e (frequency of each 0.5) | | sperm either $e^+$ or e (frequency of each 0.5) |

Assuming that both types of sperm have an equal chance of achieving fertilization:

| | | *Frequency of $F_1$ sperm* | |
|---|---|---|---|
| | | 0.5 $e^+$ | 0.5 e |
| *Frequency of $F_1$ eggs* | 0.5 $e^+$ | 0.25 $e^+e^+$ | 0.25 $ee^+$ |
| | 0.5 e | 0.25 $e^+e$ | 0.25 ee |

∴ in $F_2$ (as $e^+$ completely dominant)

| | | | | | |
|---|---|---|---|---|---|
| Frequency of genotypes | 0.25 $e^+e^+$ | | 0.5 $e^+e$ | 0.25 ee | *Total 1* |
| Frequency of phenotypes | | 0.75 grey bodied | | 0.25 ebony bodied | *Total 1* |
| Ratio of genotypes | 1 $e^+e^+$ | | 2 $e^+e$ | 1 ee | |
| Ratio of phenotypes | | 3 grey bodied | | 1 ebony bodied | |

**Table 5**

**c** ***Carry out a similar analysis of Mendel's result as shown in table [S]8.***

The phenotype ratio in the $F_2$ is almost exactly 3 yellow: 1 green. All the $F_1$ are yellow, so yellow cotyledon is dominant to green cotyledon. An analysis exactly similar to table 5 should follow.
(This might be an opportunity to raise the question of the statistical improbability of Mendel's results. It is well known that the probability of achieving ratios as close to 3:1 as he did is low. Some have suggested that this was the result of fraud, others that he simply stopped counting progeny when he achieved his predicted result. The danger of doing this could be mentioned during practical investigations.)

**d** ***Explain this on the basis of the chromosome theory of inheritance.***

The genes responsible for all the characters so far considered are located on non-sex chromosomes (autosomes). Both sexes have a pair of chromosomes in their diploid somatic cells and produce haploid gametes by an identical process of meiosis. □

## STUDY ITEM

## 17.33 Analysis of monohybrid crosses involving sex linkage

*Principle*

1 Deviations from simple monohybrid ratios may result if a gene is located on one of the sex chromosomes and is absent from the other sex chromosome, so that the heterogametic sex is hemizygous (the term hemizygous has not been used in the *Study guide*).

*Questions and answers*

a ***Complete the analysis for other possible crosses, involving black, ginger, and tortoiseshell cats.***

The analysis should be similar in layout and content to table (S)11 on page 74 of the *Study guide*. It might be wise to ask only for a complete treatment of the cross
tortoiseshell female × black male (see table 6) or
tortoiseshell female × ginger male.
Notice that the terms $F_1$, and $F_2$ have been avoided. The cross can be regarded as a backcross to a true breeding black parental line but this has no meaning since the male could not be heterozygous.

| *Parental generation* | | | |
|---|---|---|---|
| phenotype | tortoiseshell female | × | black male |
| genotype | $b_1b_2$ | | $b_1$ Y |
| *Genotypes of parental gametes* | eggs either $b_1$ or $b_2$ (frequency of each 0.5) | | sperm either $b_1$ or Y (frequency of each 0.5) |

Assuming that fertilization is random and that all sperm carrying $b_1$ are X-chromosome-bearing cells:

| | | *Frequency of $F_1$ sperm* | |
|---|---|---|---|
| | | 0.5 $b_1$ | 0.5 Y |
| *Frequency of $F_1$ eggs* | 0.5 $b_1$ | 0.25 $b_1b_1$ | 0.25 $Yb_1$ |
| | 0.5 $b_2$ | 0.25 $b_2b_1$ | 0.25 $Yb_2$ |

∴ in the offspring

| | | | | | |
|---|---|---|---|---|---|
| Frequency of genotypes | 0.25 $b_1b_1$ | 0.25 $b_1$Y | 0.25 $b_2$Y | 0.25 $b_2b_1$ | *Total 1* |
| Frequency of phenotypes | 0.25 black female | 0.25 black male | 0.25 ginger male | 0.25 tortoiseshell female | *Total 1* |
| Ratio of genotypes | 1 $b_1b_1$ : | 1 $b_1$Y : | 1 $b_2$Y : | 1 $b_2b_1$ | |
| Ratio of phenotypes | 1 black female : | 1 black male | 1 ginger male : | 1 tortoiseshell female | |

**Table 6**

**b** ***Analyse and explain the results shown in table [S]12.***

As the majority of offspring in the $F_1$ are red eyed and, in one of the reciprocal crosses, all the $F_1$ males are of one eye colour and all the $F_1$ females the other colour, we may assume that there is sex linkage and that the allele determining red eye ($w^+$) is dominant to the allele determining white eye (w). The analysis is given in tables 7a and b.

*Parental generation*

| | | | |
|---|---|---|---|
| phenotype | red eyed female | × | white eyed male |
| genotype | $w^+w^+$ | | wY |
| *Genotypes of parental gametes* | all eggs $w^+$ | | sperm either w or Y (frequency of each 0.5) |

Assuming that fertilization is random and that Y-bearing sperm carry neither w nor $w^+$:

| | | *Frequency of parental sperm* | |
|---|---|---|---|
| | | 0.5 w | 0.5 Y |
| *Frequency of parental eggs* | $w^+$ | 0.5 $ww^+$ | 0.5 $Yw^+$ |

∴ in $F_1$

| | | | |
|---|---|---|---|
| Frequency of genotypes | 0.5 $ww^+$ | 0.5 $w^+Y$ | *Total 1* |
| Frequency of phenotypes | 0.5 red eyed female | 0.5 red eyed male | *Total 1* |
| Ratio of genotypes | 1 $ww^+$ | : 1 $w^+Y$ | |
| Ratio of phenotypes | 1 red eyed female | : 1 red eyed male | |

*$F_1$ generation*

| | | | |
|---|---|---|---|
| phenotype | red eyed female | × | red eyed male |
| genotype | $w\,w^+$ | | $w^+Y$ |
| *Genotypes of $F_1$ gametes* | eggs either w or $w^+$ (frequency of each 0.5) | | sperm either $w^+$ or Y (frequency of each 0.5) |

Assuming that fertilization is random and that Y-bearing sperm carry neither w nor $w^+$

| | | *Frequency of $F_1$ sperm* | |
|---|---|---|---|
| | | 0.5 $w^+$ | 0.5 Y |
| *Frequency of $F_1$ eggs* | 0.5 $w^+$ | 0.25 $w^+w^+$ | 0.25 Y $w^+$ |
| | 0.5 w | 0.25 $w^+w$ | 0.25 Y w |

∴ in $F_2$

| | | | | |
|---|---|---|---|---|
| Frequency of genotypes | 0.25 $w^+w^+$ : 0.25 $w^+w$ | 0.25 $w^+Y$ | : 0.25 wY | |
| Frequency of phenotypes | 0.5 red eyed female : | 0.25 red eyed male | : 0.25 white eyed male | *Total 1* |
| Ratio of genotypes | 1 $w^+w^+$ : | 1 $w^+w$ 1 w Y : | 1 wY | *Total 1* |
| Ratio of phenotypes | 2 red eyed female : | 1 red eyed male : | 1 white eyed male | |

**Table 7a**
Analysis of breeding investigation involving wild type female and white eyed male *Drosophila*.

| *Parental generation* | | |
|---|---|---|
| phenotype | white eyed female × | red eyed male |
| genotype | ww | $w^+Y$ |
| *Genotypes of parental gametes* | all eggs w | sperm either $w^+$ or Y |

Assuming that fertilization is random and that Y-bearing sperm carry neither w nor $w^+$.

| | | *Frequency of parental sperm* | |
|---|---|---|---|
| | | 0.5 $w^+$ | 0.5 Y |
| *Frequency of parental eggs* | w | 0.5 $w^+w$ | 0.5 Yw |

∴ in $F_1$

| | | | | |
|---|---|---|---|---|
| Frequency of genotypes | 0.5 $ww^+$ | | 0.5 wY | *Total 1* |
| Frequency of phenotypes | 0.5 red eyed female | | 0.5 white eyed male | *Total 1* |
| Ratio of genotypes | 1 $ww^+$ | : | 1 wY | |
| Ratio of phenotypes | 1 red eyed female | : | 1 white eyed male | |

| *$F_1$ generation* | | | |
|---|---|---|---|
| phenotype | red eyed females | × | white eyed male |
| genotype | $ww^+$ | | wY |
| *Genotypes of $F_1$ gametes* | eggs either w or $w^+$ (frequency of each 0.5) | | sperm either w or Y (frequency of each 0.5) |

Assuming that fertilization is random and that Y-bearing sperm carry neither w nor $w^+$

| | | *Frequency of $F_1$ sperm* | |
|---|---|---|---|
| | | 0.5 w | 0.5 Y |
| *Frequency of $F_1$ eggs* | 0.5 $w^+$ | 0.25 $ww^+$ | 0.25 $Yw^+$ |
| | 0.5 w | 0.25 ww | 0.25 Yw |

∴ in $F_2$

| | | | | | | | | |
|---|---|---|---|---|---|---|---|---|
| Frequency of genotypes | 0.25 $ww^+$ | | 0.25 ww | | 0.25 $w^+Y$ | | 0.25 wY | *Total 1* |
| Frequency of phenotypes | 0.25 red eyed female | | 0.25 white eyed female | | 0.25 red eyed male | | 0.25 white eyed male | *Total 2* |
| Ratio of genotypes | 1 $ww^+$ | : | 1 ww | : | $w^+Y$ | : | 1 wY | |
| Ratio of phenotypes | 1 red eyed female | : | 1 white eyed female | : | 1 red eyed male | : | 1 white eyed male | |

**Table 7b**
Analysis of reciprocal cross: white eyed female × wild type male.

c ***Explain this (that red females produce blue female chicks, not blue males) and predict the results of crossing blue females with red males.***

In pigeons, as in all birds, the female is the heterogametic sex. (The students could be reminded of this.)
A red female is of genotype RY and must pass the genetically empty Y chromosome to all her daughters. She can never transmit a wild type r

allele. All her sons must be red since R is dominant. Her daughters can be red if they inherit R from their father, or blue if they inherit r from their father.

Crosses between blue females and red males are of two possible types
rY × RR which gives all red offspring, and
rY × Rr which gives both red and blue males and females.

The above example has been chosen in preference to more familiar examples in poultry so that students will be less likely to look up the solution to the problem in other texts. The importance of sex linked crosses in commercial poultry breeding can be discussed. □

## 17.4 Dihybrid crosses

**Practical investigation. *Practical guide 5*, investigation 17C, 'Patterns of inheritance'.**

### STUDY ITEM

### 17.41 Inheritance involving two different genes in *Drosophila*

*Principles*

1 Each pair of alleles in a dihybrid situation can be considered separately.

2 The 9:3:3:1 and 1:1:1:1 ratios characteristic of dihybrid crosses without linkage show independent assortment of genes on different chromosomes.

In most texts the assumption is made that parental and recombinant gametes of a double heterozygote are produced in equal numbers, that is, that there is random assortment. There is no reason for this assumption except that Mendel made it to explain his particular results. The possibility of linkage has been assumed at the outset. The use of a test cross to measure the frequency of different types of gametes is an important principle in genetic analysis.

*Questions and answers*

**a** ***Ignoring eye colour, what is the ratio of normal winged to vestigial winged phenotypes in the $F_2$ generation?***

3 normal winged:1 vestigial winged

**b** ***Ignoring wing type, what is the ratio of red eyed to sepia eyed phenotypes in the $F_2$ generation?***

3 red eyed:1 sepia eyed

**c** ***Why are the combinations $s^+s^+$, $ss$, $v^+v^+$, $vv$, $v^+v$ and $s^+s$ not possible?***

This question is to check that the student has a sufficient grasp of the allele concept to proceed to the rest of the argument. Two types of answer are valid:

1 Fusion of any of these gametes might result in a zygote lacking a

gene completely, for example, $s^+s + ss \longrightarrow s^+sss$ which has four eye determinants and no wing determinants.

**2** Since s and $s^+$ are alleles and v and $v^+$ are alleles, they must be separated by meiosis as each pair is associated with a pair of homologous chromosomes that separate at meiosis. The combinations $s^+s$ or vv, for example, could only occur if a gamete had two copies of a chromosome, that is if non-disjunction occurred.

**1** is an argument from function, **2** from the underlying mechanism. The student should appreciate both.

**d** ***Assume that $F_1$ flies produce the four possible types of gamete with equal frequency ($\frac{1}{4}$ or 0.25 of each) and work out the frequencies and ratios expected in the $F_2$ generation on this basis.***

See table 8.

| *$F_1$ generation* | | | |
|---|---|---|---|
| genotype | $s^+s\,v^+v$ | | $s^+s\,v^+v$ |
| phenotype | normal winged red eyed | × | normal winged red eyed |
| *Genotypes of parental gametes* | *eggs* $s^+v^+$, $s^+v$, $sv^+$, and sv (frequency of each 0.25 or $\frac{1}{4}$ of each) | | *sperm* $s^+v^+$, $s^+v$, $sv^+$, and sv (frequency of each 0.25 or $\frac{1}{4}$ of each) |

Assuming that fertilization is random

| | | *Frequency of $F_1$ sperm* | | | |
|---|---|---|---|---|---|
| | | 0.25 $s^+v^+$ | 0.25 $s^+v$ | 0.25 $sv^+$ | 0.25 sv |
| *Frequency of $F_1$ eggs* | 0.25 $s^+v^+$ | $\frac{1}{16}\,s^+s^+v^+v^+$ | $\frac{1}{16}\,s^+s^+vv^+$ | $\frac{1}{16}\,ss^+\,v^+v^+$ | $\frac{1}{16}\,ss^+\,vv^+$ |
| | 0.25 $s^+v$ | $\frac{1}{16}\,s^+s^+\,v^+v$ | $\frac{1}{16}\,s^+s^+\,vv$ | $\frac{1}{16}\,ss^+\,v^+v$ | $\frac{1}{16}\,ss^+\,vv$ |
| | 0.25 $sv^+$ | $\frac{1}{16}\,s^+s\,v^+v^+$ | $\frac{1}{16}\,s^+s\,vv^+$ | $\frac{1}{16}\,ss\,v^+v^+$ | $\frac{1}{16}\,ss\,vv^+$ |
| | 0.25 sv | $\frac{1}{16}\,s^+s\,v^+v$ | $\frac{1}{16}\,s^+s\,vv$ | $\frac{1}{16}\,ss\,v^+v$ | $\frac{1}{16}\,ss\,vv$ |

∴ in $F_2$

| | | | | |
|---|---|---|---|---|
| Frequency of genotypes | $\frac{2}{16}\,s^+s^+\,vv^+$ | $\frac{2}{16}\,s^+s\,v^+v^+$ | $\frac{2}{16}\,s^+s\,vv$ | $\frac{2}{16}\,ss\,v^+v$ |
| | $\frac{4}{16}\,s^+s\,v^+v$ | $\frac{1}{16}\,s^+s^+\,vv$ | $\frac{1}{16}\,ss\,v^+v^+$ | $\frac{1}{16}\,s^+s^+\,v^+v^+$ |
| | $\frac{1}{16}\,ss\,vv$ Total 1 | | | |
| Ratio of phenotypes | 9 normal winged red eyed flies | 3 normal winged sepia eyed flies | 3 vestigial winged red eyed flies | 1 vestigial winged sepia eyed fly |

**Table 8**

**e** ***Is the prediction supported by the ratio actually found in the $F_2$ generation?***

Yes (a $\chi^2$ test could be carried out here but would be a distraction from the theme).

f ***Are the results of the test cross consistent with the hypothesis that parental and recombinant gametes are produced with equal frequency by $F_1$ individuals?***

Yes.

g ***Examine* figure [S]36 *in Chapter 16, page 43, and explain, in terms of the behaviour of chromosomes during meiosis, why four types of gamete in equal numbers are produced.***

The genes determining sepia eye and vestigial wing occur on different pairs of chromosomes. When two pairs of chromosomes are organized on the spindle at first metaphase of meiosis each pair orientates independently. This means that genes carried on different, that is, non-homologous chromosomes, show random assortment. (There is a link here with tetrad analysis in fungi – see page 58 in the *Study guide* and *Practical guide* 5, investigation 16D, 'Spore colour in *Sordaria*'.)

h ***Assume dihybrid inheritance is involved and explain the results shown in table [S]16.***

The student will need to give symbols to the alleles involved, for example,
B = allele determining black, b = allele determining fawn, H = allele determining non-hooded, h = allele determining hooded.
In the first cross, the black female is a double heterozygote (BbHh), and the male a double homozygous recessive (bbhh). There is an approximation to a 1:1:1:1 ratio.
In the second cross, two double heterozygotes have been mated (BbHh × BbHh) and one would expect a 9:3:3:1 ratio. The actual result is not very close to this because of the small numbers involved.

□

## 17.5 Autosomal linkage

**Practical investigations. *Practical guide* 5, investigation 17C, 'Patterns of inheritance' and investigation 17D, 'Linkage and linkage mapping in tomato'.**

*Assumptions*

1. A knowledge of meiosis and chiasma formation.
2. An understanding of dihybrid inheritance showing random assortment.

*Principles*

1. If genes are situated on the same chromosome they are linked.
2. This is revealed by an unexpectedly large proportion of parental types in a test cross.
3. The degree of linkage can be measured by the cross-over value.
4. Cross-over values can be used as estimates of the distance between gene loci and also to map chromosomes.

*Questions and answers*

a ***What would be the cross-over value in a case where two genes were not linked, that is, where there was independent or random assortment?***

The numbers of parental and recombinant gametes produced by the double heterozygote would be equal so the cross-over value would be 50 %.

b ***Calculate and list the cross-over values for each of the four test crosses reported in table [S]17.***

See table 9.

Test cross 1A $\frac{25 + 29}{102 + 25 + 29 + 93} \times 100 = 21.7\ \%$

Test cross 2A $\frac{0 + 0}{147 + 0 + 0 + 129} \times 100 = 0$

Test cross 1B $\frac{18 + 26}{18 + 83 + 90 + 26} \times 100 = 20.5\ \%$

Test cross 2B $\frac{0 + 0}{0 + 67 + 59 + 0} \times 100 = 0$

(One would expect the C.O.V. for the reciprocal crosses to be the same but chance must produce a certain deviation.)

**Table 9**

## STUDY ITEM

## 17.51 An association between recombination and exchanges between chromosomes

*Assumption*

1 An understanding of meiosis and of the hypothesis that chiasmata represent sites of genetic exchange.

*Principle*

1 Recombination of characters shown by parents can be associated with a visible recombination of chromosome features.

*Questions and answers*

a ***Draw and label the pair of chromosomes with associated genes for each seed type (A to E) in the table, using drawings similar to those in figure [S]56.***

A single chiasma between the two loci involved has occurred in gametes producing some of the seeds. The full reasoning is set out in *figures* 6 and 7.

b ***In which of the seed types (A to E) has no recombination taken place?***

A and E.

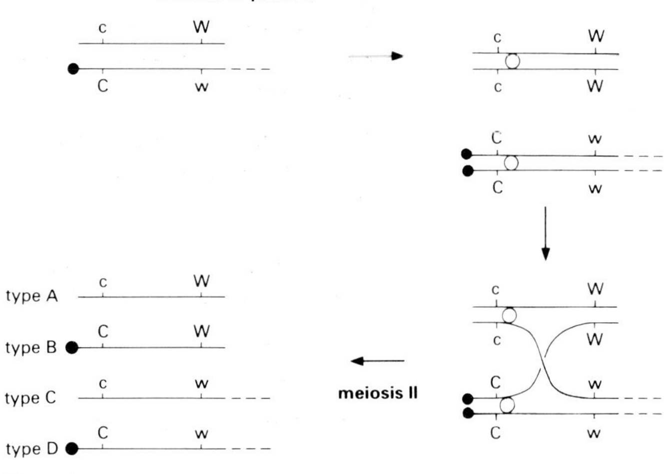


Figure 6

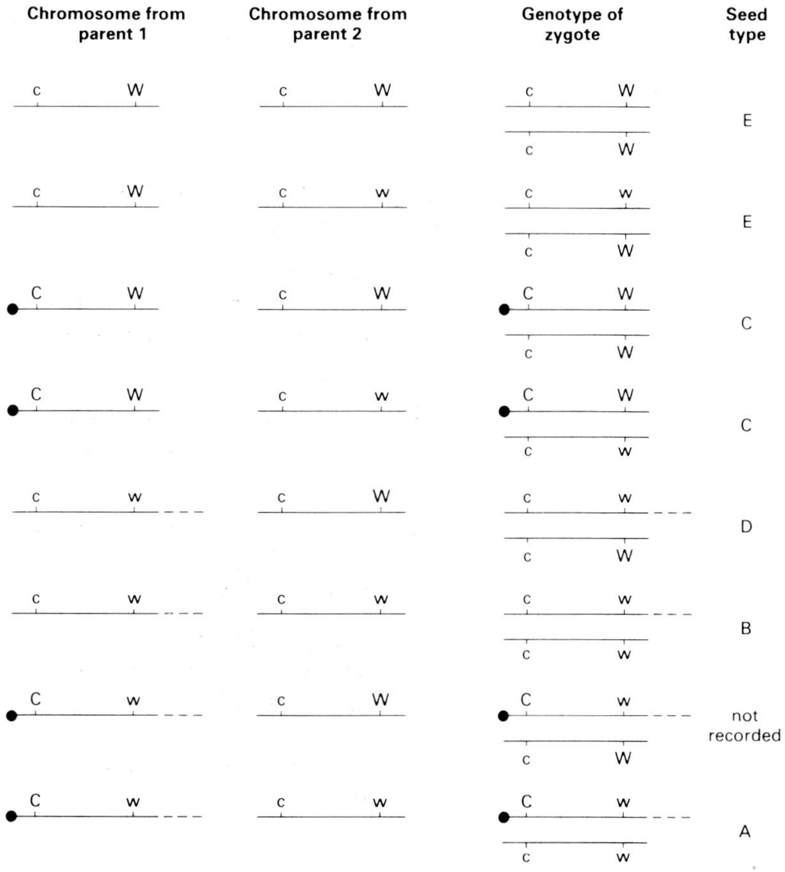

| Chromosome from parent 1 | Chromosome from parent 2 | Genotype of zygote | Seed type |
| --- | --- | --- | --- |
| c W | c W | c W / c W | E |
| c W | c w | c w / c W | E |
| ● C W | c W | ● C W / c W | C |
| ● C W | c w | ● C W / c w | C |
| c w – – – | c W | c w – – – / c W | D |
| c w – – – | c w | c w – – – / c w | B |
| ● C w – – – | c W | ● C w – – – / c W | not recorded |
| ● C w – – – | c w | ● C w – – – / c w | A |

Figure 7

c ***In those seed types where chiasmata have resulted in the separation of the knob and translocation, is there also evidence of recombination of genes?***

Yes, in all cases.

d ***Would the use of a cytologically normal parent of genotype cc ww have been justified in this experiment as a substitute for parent 2?***

Yes.

Students will probably find this short exercise quite testing unless they are very familiar with meiosis. They might be encouraged to discuss it in small groups to avoid the less able becoming discouraged, or the teacher could provide them with *figure 6*. □

## 17.6 A model for the inheritance of continuously varying characters

**Practical investigation. *Practical guide* 5, investigation 17C, 'Patterns of inheritance'.**

*Assumptions*

1 An understanding of the concept of continuous variation.
2 Ability to reason that in a trihybrid $F_2$ with no linkage one in sixty-four offspring would be triply homozygous recessive.

*Principles*

1 Different genes can have similar, additive effects.
2 If several genes, each having a small influence on the phenotype, are segregating and recombining, this will result in continuous variation and there is no straightforward means of investigating or even detecting the individual genes. (The term polygene has not been used.)
3 Statistical methods may be used to make an estimate of the extent to which variation in a population is produced by genetic differences between the individuals.
4 Selection can result in a loss of genetic variability. (This concept is developed in Chapter 6.)

*Questions and answers*

a ***What can be deduced from Johannsen's work?***

**1** The original sack of seeds was not genetically homogeneous.
**2** Selection for five generations produced genetically uniform populations.
**3** In the genetically uniform populations there was still variation, presumably produced by the environment (internal or external), of the plants producing the seeds.

b ***What was the genotype of the $F_1$ plants?***

AaBbCc.

**c** ***What would have been the genotypes of gametes produced by the $F_1$ plants?***

| | |
|---|---|
| ABC abc | – the parental combinations |
| ABC aBC<br>AbC aBc<br>ABc abC | the recombinants |

**d** ***Explain how one in sixty-four white seeds in the $F_2$ generation was to be expected from the hypothesis outlined above.***

One in eight gametes would be of genotype abc, so as $\frac{1}{8} \times \frac{1}{8} = \frac{1}{64}$, one in sixty-four of the $F_2$ generation would be of genotype aabbcc, that is triply homozygous recessive, and white.

## 17.7 Mutation

**Practical investigations. *Practical guide* 5, investigation 17E, 'Mutation in yeast' and investigation 17F, 'The effects of irradiation in plant seeds'.**

*Assumptions*

1 Familiarity with the term allele.
2 Familiarity with mitosis and meiosis.

*Principles*

1 New alleles may enter a closed population by occasional outcrossing.
2 This cannot ultimately explain the existence of alleles. They must arise by the imperfect replication of other alleles.
3 Such spontaneous mutation has been repeatedly observed in domestic animals and plants.
4 Deletion mutation, chromosome breakage, and reunion to produce inversion and translocation and point mutation are explained without reference to DNA.
5 A mutation is a random event, though the frequency of occurrence of particular mutations can vary.
6 Somatic mutation is possible.
7 Recessive point mutations have occurred generations before they become manifest in a population of a diploid species.
8 The haploid state facilitates the detection of mutation as does the X chromosome in the heterogametic sex.
9 Ionizing radiation and certain chemical substances increase mutation rates.

*Questions and answers*

a ***In* Drosophila *the X chromosome sometimes breaks and then rejoins to give an inverted segment as shown in* figure [S]60.**

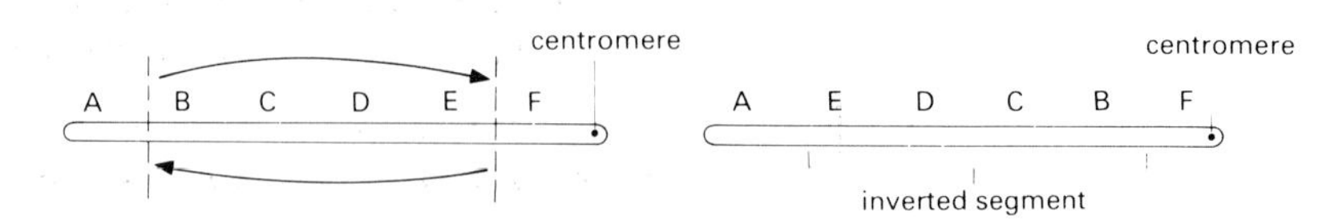


***During gamete formation in a female* Drosophila *with one normal X and one inverted X chromosome, which way would these two chromosomes be expected to pair?***

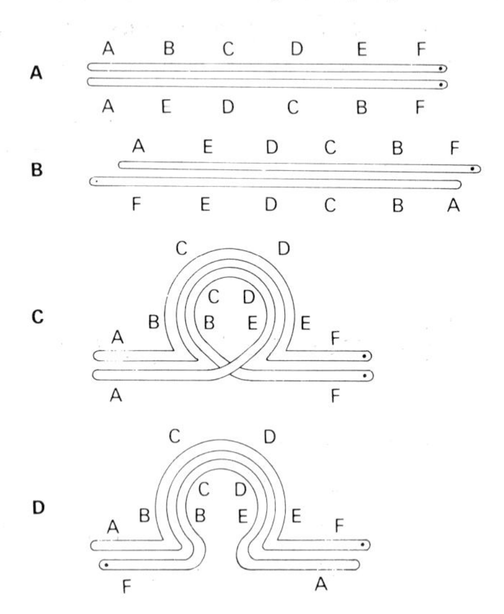


Figure 8
*Study guide* figure 60.
(*J.M.B.*)

(*J.M.B.*)

The correct choice is **C**, since this allows all genes to pair with alleles on the other chromosome. (This is also the cytologically observed situation.)

b ***Why should genes on the X chromosome be especially favourable for the detection of mutation?***

Because many gene loci on the X are absent from the Y. The allele present at any of these loci cannot be masked by a dominant allele in the heterogametic sex.

c ***Yeasts and other fungi occur in the haploid state – what other advantage do they offer in mutation research?***

Vast numbers of spores are produced in a small space so that the chance of any particular mutation being observed is fairly good.

**STUDY ITEM**

17.71 The effects of radiation

*Assumption*

1 Some understanding of radioactivity is helpful.

*Principles*

1 Radiation damages cells and organisms.
2 This damage is at least partly a consequence of mutation.
3 The relationship between dose of a mutagen and mutation rate is linear and, in theory at least, there is no safe dose.

*Questions and answers*

a ***Draw conclusions from these results.***

Radiation kills seeds and reduces the chance of survival of seedlings. The effect increases with dose received.

b ***Draw a graph to indicate what would be the relationship between dose and yield of mutations if the relationship were linear but a minimum (threshold) dose were needed to induce any mutations.***

The graph would not pass through the origin but would commence at some point on the x axis, that is, the threshold dose. □

# PART II The *Practical guide*

## INVESTIGATION 17A Describing variation

(*Study guide 17.1*, 'Introduction'; Study item 17.11 'Continuous and discontinuous variation'.)

### ITEMS NEEDED

Clover plants, collected in field
Beakers or other vessels suitable as vases for clover samples
Graph paper
Hand lens 1/group
Plastic ruler 1/1
Polythene bags and fasteners

### Part A Variation in clover

*Assumptions*

1 An appreciation of the fact that variation is produced by interaction of genotype and environment.
2 An understanding of the use of contingency tables, histograms, and the standard deviation and standard error of difference.
3 An ability to understand and use certain terms often used in descriptions of plant morphology.

*Principles*

1 Variation is conveniently classified as continuous or discontinuous, although many features of organisms show both types of variation.
2 Statistical procedures allow variation to be quantified.
3 Some variants may have advantages over others when faced with a specific environmental variable.

*Trifolium repens L.* is self-incompatible and is a highly polymorphic species. It is one of the commonest species of clover-like plants in Britain but can be confused with other legumes such as *T. fragiferum L.* and *T. hybridum L.* and with species of *Medicago* and *Lotus.* If pupils produce samples which are patently not *T. repens*, this will be a good opportunity to introduce or reinforce the species concept and the use of keys. It is important to point out that an accurate description of variation must precede the definition of species and varieties.

It would be feasible to perform the whole exercise in the field. When samples are collected they must not be crushed and cannot be stored for more than about 12 hours without changes in colour and/or wilting. A very little water in each polythene bag may prevent wilting. The leaflets have pronounced nastic movements and it could be difficult to distinguish immature and expanded leaves in the evening.

Lawns and playing fields are feasible as sources of material but waste places, neglected pastures, or road verges are better. If a suitable habitat is not accessible from the laboratory, pupils and teachers could collect material from various locations the evening before the exercise is scheduled. If significant numbers of samples are available from two or more locations, variation could possibly be correlated with habitat factors such as shade and the height of the herbage.

Pupils will probably observe variation in some of the following characters:

leaf size
petiole length
internode length
degree to which stolons branch
extent to which leaflet margins are toothed (hand lens may be needed here)
colour of flowers (white, pinkish, rarely purplish)
number of flowers in an inflorescence
colour of petioles and stolons (green–purplish)
presence or absence of a white mark or band across the upper surface of each leaflet

This last is the most obvious discontinuous variable. Plants having no mark are usually in a minority and may be difficult to find in some locations. Several alleles are known which influence the size, shape, and position of the mark. It is also influenced by the environment and appears more intense in leaves which grow in brighter conditions.

*Questions and answers*

**a** ***In these investigations a patch or clump of clover is considered to be one individual. Is this assumption justified?***

As a working approximation, yes, but it is always possible that two seedlings close together might produce a mixed clump or that a large plant might split into several clumps as a result of the death of intermediate stolons. It might be argued that each node with its attendant axillary bud and leaf was an individual. The concept of a clone could be raised here.

**b** ***Consider the list you have prepared of the ways in which* T. repens *varies and produce hypotheses to explain why some of the contrasting types you have encountered have survived in the habitats from which the plants were obtained.***

Small leaves and short petioles are likely to be advantageous if mowing or grazing pressures increase. They may also result in lower transpiration rates than large leaves with longer petioles. Large leaves with long petioles will compete more effectively for light. Anthocyanin pigments and white marks on the upper leaf surface could protect photosynthetic tissue from damage at very high light intensity but result in lower photosynthetic efficiency at low light intensity.

**c** ***Does your contingency table describing the discontinuous variation in the clover population for one character agree with the tables produced by others in the class who scored the same character? If there is substantial disagreement, try and explain this.***

Minor disagreements could be the result of observational error. Major disagreements mean either that the observers are using different criteria or that the variation is not discontinuous.

**d** ***Why, in each of a sample of stolons, were the first three leaves measured rather than a random sample gathered from each clump?***

Because a truly random method of selecting leaves is difficult to devise and the first three leaves in each stolon will be comparable. It is important to avoid bias in the selection of leaves.

**e** ***Did measurement of length of leaflet in the two clumps of clover reveal a significant difference? How could it be established whether such a difference is caused by environmental factors or by genetic differences?***

If the difference between the means of the lengths of the two samples of leaflets is equal to or greater than twice the standard error of difference, then the difference is significant. If the difference was due solely to environmental factors, new leaves produced by the cuttings taken from the two clumps and grown in a similar environment would be similar.

ITEMS NEEDED

Graph paper
Ruler 1/1

## Part B Variation in humans

*Assumptions*

1 An appreciation of the fact that variation is produced by interaction of genotype and environment.
2 An understanding of the use of contingency tables, histograms, standard deviation, and standard error of difference and of scattergrams and the coefficient of rank correlation.

*Principles*
See the notes in this *Guide* on part A of this investigation.

The exercises could be carried out in whole or in part outside formal teaching time. If this approach is adopted, it will be desirable to have some preliminary class discussion concerning suitable characters which might be variable and about procedures for choosing a group of subjects.

It may be desirable to use a fairly small group of subjects for step **2** and to extend the number observed or measured for steps **3** and **5**.

Before embarking on the exercise, the teachers will need to consider the feelings of various ethnic groups in their school or college, the possible feelings of insecurity among adolescents who consider their own phenotype to be in some way extreme and the possibility that pupils may have an adoptive relationship to their parents of which they are unaware.

Pupils will probably observe variation in some of the following characters:

hair colour and texture
eye colour
shape of eyes, ears, nose, and lips
shape of head
height
mass
ability to roll the tongue
ability to hold the hand flat with the fingers close together in two pairs and a good gap between the second and third finger

Only tongue rolling is likely to show clear discontinuous variation.

Correlation between pupil-to-pupil distance and age is likely to be slight unless young children are included in the sample.

*Questions and answers*

a ***Consider the list you have prepared of the ways in which humans vary, and produce hypotheses to explain why some of the variants you have encountered have survival value, either in modern conditions or in conditions to which the ancestors of your group may have been subjected.***

This may be an opportunity for exposing and discussing prejudices. Little is understood about the reasons for human variation though there has been much speculation. Tall, thin individuals tend to lose heat more easily than the opposite phenotypes. Dark individuals are better protected than fair people against ultra-violet light but are likely to require more vitamin D in their diet. The possibility that sexual selection has resulted in variation could be discussed.

b ***Was there any doubt about whether any of the characters vary continuously or discontinuously? If so, which characters presented difficulty?***

Eye colour is likely to cause most confusion since there is a number of genes which interact to produce a large number of phenotypes which can be diffcult to distinguish. Shape of ear lobe is another doubtful case.

c ***Do you think it probable that differences in pupil-to-pupil distance are mainly inherent or mainly determined by environment? How could this question be settled?***

There is no evidence to suggest that either genotype or environment is more important, though it is difficult to envisage how the environment could influence the character directly. A study of monozygotic and dizygotic twins would be needed to settle the question.

## INVESTIGATION
## 17B The influence of environment on development

(*Study guide* 17.2, 'The role of inheritance and environment'.)

### ITEMS NEEDED

*Drosophila melanogaster* culture, wild type 1/group
*Drosophila melanogaster* culture, yellow body 1/group
Yeast, dried

Culture medium, *Drosophila*
Ethanol
Ethoxyethane (diethyl ether)
Silver nitrate solution 0.1 % in food tubes 2/group

Beaker, for morgue 1/class
Cottonwool
Etherizer (see *Practical guide 5*, page 41) 1/group
Food tubes 2/group
Incubator at 25 °C 1/class
Labels for tubes 4/group
Microscopes, binocular or dissecting lens
Paint brush 1/group
Plastic netting (optional)
White tile 1/group

### Part A The influence of silver nitrate on fruit flies

*Assumption*

1 That students have knowledge of the inhibitory effect of some heavy metal ions on some enzymes.

*Principles*

1 A mutant gene often results in a less effective enzyme.
2 The phenotype can be changed in a similar manner by a genetic change or a change in the environment in which development takes place.

**Culture medium for *Drosophila***
Any fruit fly medium will serve – the dry 'instant' media sold by suppliers work well. Add sufficient silver nitrate to the water used to prepare the medium to achieve a concentration of about 0.1 %. The effect is roughly proportional to concentration but higher concentrations than 0.1 % markedly reduce viability and 0.2 % will be lethal. Presumably the silver ions inhibit an enzyme needed for melanin production.

Yellow body is a sex-linked mutant which results in a dilution effect; that is, there is less melanin in the cuticle. The phenocopy produced by silver nitrate is not quite the same (chlorosis produced by darkness and by mutation in seedlings) but it is sufficiently similar to allow the general principle to come across. The best possible match with yellow body might be achieved by setting up cultures with a range of silver nitrate concentrations. It is important to allow the flies to age and darken before they are scored or the untreated wild types will be pale.

It is interesting that silver nitrate has no influence on the colour of ebony bodied flies. Other mutants such as black body have not been investigated.

*Questions and answers*

a ***Explain your results in terms of genotype, phenotype, and environment.***

The phenotype is the colour of the fly. The colour can be made lighter by a change in genotype or by an environmental change, in this case, the addition of silver nitrate to the medium.

**b** ***How could both genetic variation and silver nitrate change the concentration of melanin in a fly?***

A gene, or more probably several genes, are needed to produce enzymes which synthesize melanin. The yellow body mutant presumably produces a less effective enzyme than its wild type allele. Silver in the silver nitrate may inhibit the action of a melanin-producing enzyme.

**c** ***How could you show that silver nitrate has not changed the genotype of the flies but merely altered its expression?***

Transfer the pale flies, treated with silver nitrate, to normal medium and allow them to breed. They should produce offspring similar to the wild type.

## Part B Chlorophyll production in seedlings

### ITEMS NEEDED

Seeds, tobacco or other segregating green and white (or yellow)

Agar
Sodium hypochlorite solution, 1 %
Water, distilled

Beaker, for soaking seeds 2/group
Bunsen burner 1/group
Cloth, closely woven, piece 1/group
Cotton thread
Forceps 1/group
Glass marking pen 1/group
Incubators at 25 °C, light and dark 2/class
Needles, dissecting 1/group
Petri dishes 2/group
Ruler 1/group
Scissors 1/group

**Agar** (enough for about 24 plates)
Agar, 9 g
Charcoal powder (carbon black), 5 g
Water, distilled, 600 $cm^3$
Disperse the agar in the cold water and then bring gently to the boil, stirring all the time to prevent burning, in a large beaker to prevent the agar boiling over. Allow it to boil for $\frac{1}{2}$ minute then cool, without stirring, to give the bubbles a chance to disperse. Pour gently into the Petri dishes, while still hot.
A slightly easier discrimination of seedling phenotypes may be achieved if, once the above mixture is boiling, the 5 g of charcoal powder (carbon black) are added. Add this very slowly at first since it may make the mixture boil rather vigorously.
Sterilization is not necessary if clean Petri dishes are used; only small amounts of fungal contamination will occur.
**Sodium hypochlorite solution.** See page 13.

*Assumption*
1 Understanding of the term cotyledon.

*Principles*
1 Genetic expression depends upon the presence of appropriate internal and external environmental factors.
2 Chlorophyll synthesis requires light.
3 There is no obvious phenotypic difference between the cotyledon colour of chlorotic seedlings that have resulted from light-starvation and those that are genetically deficient.

These seeds may be germinated more cheaply on filter paper thoroughly moistened with distilled water. If this method is used it is vital to seal the Petri dishes well, since evaporation will occur very rapidly and lead to desiccation and death of the seedlings. Alternatively add water in small amounts when the dishes are inspected.

It is advisable to score cotyledon colour as soon after germination as possible since the chlorotic seedlings die when still very small and all obvious trace of them is easily lost.

It would, of course, be possible to provide each group with two dishes of seedlings already germinated, to save time, but involvement in the investigation would thereby be reduced to the point where the student was a passive observer. It is recommended that students set up one investigation such as the one listed and that additional examples, such as those that follow, could be supplied to reinforce the general principles.

**Maize,** segregating 3:1, green:albino
or
**Xantha barley**, segregating 3:1, green:albino

**1** Plant surface-sterilized maize fruits in potting compost in seed trays or pots at a separation of 2–3 cm. (See page 14.)
**2** Cover the trays with newspaper and glass to keep them dark and humid until germination has occurred (about 5 days).
**3** Place one set of trays in the light and another in the dark for a further 9–10 days.
**4** Seedlings may be scored at around 14 days if kept at room temperature.

(Various other genetic ratios can be obtained, but since the object of this investigation is not the discovery of the genetics of chlorophyll production it is perhaps wise not to make the issue too confusing.)

**Tomato hypocotyl colour**

**1** Select seeds of the genotype that you wish to investigate. Fill a seed tray three-quarters full with potting compost which has been slightly dampened.
**2** Scatter the seeds on the surface of the tray and cover with a further 5 mm of compost.
**3** Keep the seed tray dark with glass and newspaper until germination has occurred and then transfer to full light.
**4** Prepare sufficient small plant pots (or plastic cups with holes pierced in the base) to take your seedlings, about 2–3 to a pot. Use nutrient-poor potting medium. Make holes in the centre of each individual pot of compost with your finger. Lift seedlings with a dinner fork, disturbing the roots as little as possible, and transplant to the small pots.
**5** Shake the small pots so that the compost settles around the roots of the seedlings. Water each pot lightly with about 10 $cm^3$ of water.
**6** After the transplanted seedlings have recovered from any root disturbance during transplanting allow all the pots to dry out until the seedlings almost begin to wilt (cotyledons will start to droop and will appear rather flaccid). Then divide the seedlings into 6 groups and treat each group in one of the following ways:

**A** Water the pot from above with 20 $cm^3$ distilled water
**B** Water the pot from above with 10 $cm^3$ distilled water
**C** Water the pot from above with 10 $cm^3$ 0.5 % sodium chloride solution
**D** Water the pot from above with 10 $cm^3$ 0.75 % sodium chloride solution

**E** Water the pot from above with 10 $cm^3$ 1.0 % sodium chloride solution
**F** Place the pot on pieces of broken porcelain in a Petri dish and pour distilled water into the pot until the excess runs through into the dish.
**7** Repeat step **6** with pots **A–E** as often as necessary (about every 2–4 weeks in winter, 1–7 days in summer). Water pot **F** daily.
**8** Record the differences in hypocotyl pigmentation when they become apparent (between 20–30 days after starting treatment). Throughout the treatment keep the plants in as high light intensity as possible and keep the temperature low, around 15 °C.

The results of this investigation should demonstrate that physiological or real drought increases anthocyanin production by the plant. Treatment **F** will also show increased pigmentation and students could work out why this is (the reasons are linked to the waterlogging which reduces soil air and hence decreases root activity and growth. Associated with this treatment may be a degree of mineral leaching which may accentuate pigment production.)

A suitable tomato seed set would be PGL 3 from Practical Plant Genetics, 18 Harsfold Road, Rustington, Sussex, BN16 2QE, which contains a number of mutants with different expressions of pigment production. Many of the anthocyanin producers give very good results with the suggested treatment.

Other examples that could be mentioned to the students are:
**1** The genetically distinct strain of sheep called Merino produces best wool when kept on a poor diet.
**2** The pink colour in flamingoes is not expressed unless their diet contains carotenoid pigments found in certain crustaceans, a fact which keepers in zoological gardens need to take into account if the birds are to keep their 'natural' colour.
**3** Rust-resistant antirrhinums only express their resistance in the presence of a rust infection.
**4** The spore colour of the fungus *Aspergillus niger* is black unless magnesium is withheld from the culture.

*Questions and answers*

a ***Explain the presence of colourless cotyledons in both treatments as fully as you can.***

This is a result of both genetic and environmental influences. In the 'dark' seedlings the pale colour is present in all seedlings and is thus a dark-induced effect. Of the 'light' seedlings, three-quarters of the total number will be green and one quarter chlorotic. This has to be a genetic effect. It can be explained if, as stated at the start, chlorophyll manufacture is an inherited ability and both parents were heterozygous for the gene controlling it. The chlorotic seedlings are homozygous recessive and cannot manufacture chlorophyll. Hence in the 'dark' seedlings, there must be a quarter of those present that are genetically unable to produce chlorophyll but these are indistinguishable from the rest.

**b** ***From your observations in this investigation, what general hypothesis can you make regarding the influence of the environment on the phenotypic expression of a genotype?***

In general, one can suggest that a genotype can only be expressed if the correct environmental conditions exist; hence
phenotype = genotype + environment.

## INVESTIGATION 17C Patterns of inheritance

(*Study guide* 17.3 'Mendel and his contemporaries', 17.4 'Dihybrid crosses', 17.5 'Autosomal linkage', and 17.6 'A model for the inheritance of continuously varying characters'.)

Teaching situations vary so widely that it is inappropriate to specify too precisely the nature of breeding experiments which should be carried out by pupils studying Advanced Level Biology. Some teachers may feel that their resources of time and space are so limited that they must confine themselves to the more highly structured investigations outlined elsewhere in *Practical guide* 5. Many others will wish to extend some or all of their students with a more open-ended investigation. Some may be seeking the sort of living visual aid that a few trays of seedlings, grown as a demonstration, can provide. Almost all teachers will be anxious to find suitable material for individual project work.

It has seemed right to explain to students some of the opportunities that are available and some of the choices which teachers will have to exercise on their behalf. The introductory section also attempts to explain some of the benefits of a longterm experiment which they may initially find rather daunting.

The time taken for an investigation may be greatly reduced if a teacher or technician or small group of pupils sets up a breeding experiment and a class is able to examine progeny and record results. The teaching packages provided by Practical Plant Genetics of Rustington, either directly or through other suppliers, lend themselves very well to this approach. Fruit fly crosses can be treated similarly; for example, a single tube of $F_1$ flies can be divided up to provide a tube of $F_2$ for each person in a class to score; students may not have seen the parents of this $F_1$.

If students are to conduct the whole experiment themselves, a great deal of thought has to be given to practical details and timing. Either a small group will need to become responsible for the care of the organisms or a rota will need to be devised.

Unless material is used as demonstration, it seems rather pointless to select very simple examples since, when results become available, they will merely confirm elementary principles, already well understood, and will seem an anti-climax. It is far better to pursue an example involving, say, linkage or epistasis, which will yield results that seem worth the effort involved in obtaining them.

## Option A Seeds and seedlings

Investigations 17D, 'Linkage and linkage mapping in tomato', and 17F, 'The effects of irradiation in plant seeds', provide details of particular demonstrations using tomatoes. Many others are available using this plant and also peas, radishes, and maize.

*Senecio vulgaris* (groundsel) provides a very nice example of incomplete dominance. Plants may have long ray florets or short ray florets, or ray florets may be completely absent. The latter is the common condition. If the population in the school grounds is entirely rayless it is well worth introducing the rayed form. Allele frequencies can be calculated and an annual record of these kept. The short rayed form is the heterozygote and if a short rayed plant, just coming into flower, is dug up and established in a pot indoors, enough seed can be collected to get a 1:2:1 ratio. The plants should be kept fairly dry as they are very subject to fungal infection when indoors. The time needed to produce flowers on a windowsill is about eight weeks.

## Option B Small mammals

It would seem wasteful to breed small mammals in a school without exploiting their potential for teaching genetics. They are especially valuable because students are unlikely to lose interest while the experiments are in progress and because, if only small numbers are available, some of the difficulties of human genetics can be simulated. The inheritance of coat colour provides beautiful examples of epistasis and it may often be possible to combine a simple monohybrid study of, for example, B (black) and b (brown) or short tail (Sd) with a study of epistasis.

Many accounts of animal house procedure suggest that small cards carrying details of pedigree should be kept attached to mammal cages. These, however, often get lost and it may be found better to keep the details in a note book or folder and label cages with indelible felt pen. Methylated spirit easily removes the label when it becomes out of date.

Purpose-built laboratory animal cages are essential. Two drinking pots should be provided for each cage to ensure a constant supply of water. (Rats drink more than mice or hamsters, while gerbils drink very little; all require water at all times.)

Avoid straw for bedding as it may be contaminated. Fresh sawdust or shavings or peat are suitable to cover the floor. The inexperienced student usually puts in too much bedding and floor litter and this generally results in water syphoning from the drinking pot after the animal piles litter over the nozzle. Gerbils are especially apt to do this. Details of colour mutants available in gerbils can be obtained from the Department of Genetics at Leeds University.

Laboratory animal pellets, such as those produced by Oxoid, are an excellent food for all small rodents. Only buy sealed bags of pellets to avoid contamination. Check that the size of pellet is suitable for a laboratory cage with wire food hopper; it must not fall through the bars. Check the protein content. Pony nuts look ideal but are not nutritionally adequate for breeding rodents.

| | Age of maturation weeks | | Recommended minimum breeding age (weeks) | Recommended maximum breeding age (months) | Average length of gestation (days) | Usual number in litter | Desirable weaning | | Normal life span (years) |
|---|---|---|---|---|---|---|---|---|---|
| | (male) | (female) | | | | | Age (days) | Mass (grams) | |
| Mouse *Mus musculus* | 6 | 6 | 8–10 | 12 | 19–21 | 8–11 | 21 | 10–12 | $1-1\frac{1}{2}$ |
| Rat *Rattus norvegicus* | 4–5 | 4–5 | 9–12 | 12–15 | 20–22 | 9–11 | 22 | 35–40 | 2–3 |
| Gerbil *Meriones unguiculatus* | 9–12 | 9–12 | 12 | 18 | 25–28 | 4–6 | 22 | 12–18 | $2\frac{1}{2}-3\frac{1}{2}$ |

**Table 10**

Table 10 provides data which may be useful in planning a breeding programme.

Gerbils are monogamous and can only be mated easily when young. A mature animal does not accept another mate and they usually fight to the death. If it is absolutely necessary to mate a gerbil to more than one partner, for example, in order to test a prediction, the new pairing must be done in an aquarium with a secure wire partition. (Two small aquaria with ends missing and a frame of double half-inch mesh wire sandwiched between make a good introduction cage.) The adult male and female to be mated are put into the two halves of the divided enclosure and kept separate for at least one week. Swop handfuls of bedding between the two halves daily. Remove the partition cautiously in the morning and leave the cage where it can be watched for the rest of the day.

Mice easily accept a succession of mates and this is a great advantage in testing predictions.

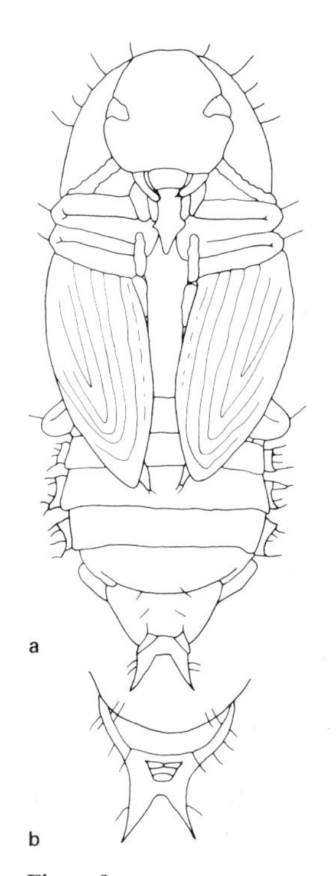


**Figure 9**
The pupa of *Tribolium castaneum*, ventral view.
**a** Female (× 17).
**b** Terminal segments of male.
*Based on Haskins, K. P.,* Using *Tribolium* for practical genetics, *Philip Harris Biological Ltd, 1975.*

## Option C Flour beetles (*Tribolium* sp.)

These animals have often been described as trouble free. They have, however, three serious disadvantages in school use.

**1** The difficulty of sexing. This is done at the pupal stage and students and teachers find it a problem (see *figure 9*).

Pairs of pupae placed together at random, one of each type to be crossed, would overcome this difficulty. Two out of four cultures would be expected to be a true pair and the cross would be expected to succeed.

**2** Populations in stock cultures often crash without warning, possibly as a result of mite infestation.

**3** The life cycle is long and, as the animals do not need attention, the experiment gets forgotten!

Stock cultures can be kept in jam jars or glass containers, covered with fine muslin or nylon, containing about 5 cm depth of food medium. A temperature between 25–30 °C and a relative humidity of the order of 65 % are ideal.

Glass tubes 10 cm × 2.5 cm make suitable containers for experimental crosses and should contain about 3–4 cm depth of food. The tubes can be sealed with foam plastic bungs or cloth.

A suitable relative humidity may be obtained by exposing a beaker with a saturated aqueous solution of sodium nitrate in the storage box.

Insects can be removed from the food medium by being sieved through a flour sieve or fine tea strainer, and may be immobilized by being chilled in a refrigerator to make them easier to handle.

The main stock should be sub-cultured at about 3-month intervals and any badly parasitized population should be destroyed.

Food medium is 9 parts wholemeal flour to 1 part of dried powdered (not pelleted) yeast. This can be sifted before use to remove coarse bran which gets in the way when the insects are being harvested. It can be heat-sterilized, though this is not essential. Pellets used to feed laying hens can be pulverized in a coffee grinder to powder and this makes a good alternative medium.

## Option D Fungi

Details of culture of the numerous fungi which can be used in genetic studies are beyond the scope of this volume.

## Option E Fruit flies (*Drosophila melanogaster*)

### ITEMS NEEDED

Use the 'Items needed' list for investigation 17B, part A (page 65), but omit silver nitrate solution. Choose appropriate varieties of *Drosophila*.

If one wishes to get significant data in a reasonable time and involve every member of a class in the experiment, these animals can give the best results. Teachers are likely to find difficulty in timing the experiment. It has often been claimed that at 25 °C the life cycle takes eleven days but, possibly because of an unreliable incubator or because of variation in the medium, it often takes longer. Different strains certainly develop at different rates and this can be very inconvenient if one is making several replicated reciprocal crosses. It is a good idea to watch the cultures one is going to use and if pupae appear in one but not the other, the more advanced culture can be moved to a cooler spot (not the refrigerator) for 24 or 48 hours to allow the slower one to catch up.

If time is pressing and a teacher is unwilling to face the uncertainty of getting enough virgins to involve the whole class in setting up a cross, it will still be very worth while to harvest just a few virgin females and set the cross up oneself with the help of a few enthusiastic students. If three healthy virgin females and some males are placed in a small tube of medium for 5 or 6 days, or until larvae are visible, they can be shifted to a fresh tube and later to yet another tube. This greatly increases the number of progeny obtained. If one wishes to move a few flies to a new tube it is best to shake them down and invert the new tube over the old one. The flies soon walk up into the new one (see *Practical guide 5*, figure 16). The less ether they get the more fertile they will be.

Mite and mould infestations are greatly increased if media are stored in the refrigerator or freezer before use. It is much better to use fresh medium. Stock cultures should be started with plenty of females, at least

15 to a 150 cm$^3$ bottle. If one is in a hurry stock can be maintained by simply tipping a good number of flies from the old bottle into the new one. It is a good idea to etherize and examine carefully at least every third transfer to ensure that each stock is pure. (Flies escape in the best regulated laboratory and may get into the wrong bottle.)

If backcrosses or test crosses are to be attempted it is important to start new cultures of the appropriate pure stock at the same time as the cross is set up.

It is a good idea to keep at least four pure cultures of each genotype one wishes to maintain and not to dispose of old cultures until larvae are observed in the new ones. Many schools find it convenient to get new stocks each year rather than to maintain their own. If one does this, one should order new stock well before the experiments are planned so that several large flourishing colonies of each genotype can be established. It is no good hoping to obtain enough virgins to satisfy a large class from the single small specimen tube the supplier will send.

Schools will be unlikely to wish to maintain more than four or five pure strains; this means 20–30 containers. Wild type flies will be required and the following mutants are hardy and very easy to recognize – vestigial wing (vg), ebony body (e), white eye (w), and sepia eye (sp). (The standard abbreviations have sometimes been further shortened in the *Study guide*.) White eye is sex-linked, and ebony and sepia eye are autosomally linked. Wing mutants other than vestigial may be hard for students to recognize.

If investigation 17B is to be attempted, yellow bodied flies will be needed. Yellow body (y) is closely linked to white eye (w) on the X chromosome. Much elegant complementation work can be carried out with the numerous eye colour mutants available.

Details of handling flies appear in the *Practical guide*. Some texts have suggested using an 'emergency etherizer' to immobilize flies which start to crawl around while being handled. This is a glass Petri dish top or bottom, with a piece of cottonwool taped inside. The cottonwool is damped with ether and the dish inverted over the moving flies. This does not work as well as sweeping the flies quickly into a tube of ether vapour. It spreads ether fumes about the laboratory and is therefore undesirable. Wild flies of several species can be trapped if a bottle of medium is left open in a sheltered spot outdoors in summer. A little sherry helps to attract them faster! They are not easy to identify but are very good for exploring the species concept. Pure-breeding colonies can be established from each female trapped and attempts can be made to cross the various lines. Interspecific competition experiments on the lines of investigation 27B, 'Interspecific competition between two *Lemna* species', make valuable projects. For example, how do two species in pure and in mixed cultures respond to variation in the water content of the medium?

Phototaxis experiments with *Drosophila*, placed in long tubes about 1 cm wide and masked along half their length, are very interesting. Vestigial-winged flies are markedly less positively phototactic than other types – a nice example of pleiotropy.

The mites that infest old fly cultures are a neglected source of interest. If an old culture is shaken hard over a Petri dish of plain agar, they fall onto it and can be observed with a dissecting microscope. They can be seen feeding on pollen or powdered yeast.

*Recipes for food media*
Many different media are successful and it is recommended that a jar of dried 'instant' medium as sold by biological suppliers is kept for use when time is short. For large scale use a fresh homemade medium is more successful and cheaper. Yeast should be added to the surface of the medium soon after it has been prepared so that it gets a good start in competition with moulds and bacteria. Plastic netting, as sold for use to shade greenhouses or cold frames is useful if placed in a tube since it allows the flies to cling and fewer get drowned (see *Practical guide 5*, figure 14). The cottonwool is better than filter paper and helps to prevent old medium from sliding down an inverted tube. Both netting and cottonwool should be pushed in after the medium has set. Neither are essential.

**'Homemade' fruit fly medium.** Place in a saucepan kept for the purpose:
Agar, 6 g
Glucose, 35 g
Methyl hydroxybenzoate (mould inhibitor), a few crystals (a pinch)
Porridge oats (Scott's Porage Oats or Quaker Oats), 72 g
Water, 600 $cm^3$
Boil until thick and pour into clean tubes or bottles, using a plastic funnel with the stem cut off. Plug tubes immediately with foam rubber. Do not boil the medium in a beaker; it often cracks suddenly and this has proved a serious hazard.

The water content of a medium can be varied. If in doubt, use less and, when larvae have appeared, use syringe and hypodermic needle to add a few drops of water to each culture if it starts to dry up seriously and shrink away from the sides of a tube.

Black treacle can be substituted for glucose in the recipe. It is, however, no better and much more trouble. Cornmeal can be substituted for porridge oats. Sucrose works but not as well as glucose.

## INVESTIGATION
## 17D Linkage and linkage mapping in tomato

(*Study guide* 17.5 'Autosomal linkage'.)

### ITEMS NEEDED

Seeds, tomato, 60–70, from either
$F_1$ *inter se*
*or*
$F_1$ backcross from PGL 9 (see note on supplier at end of this list)
Compost, Levington Universal
*or*
similar peat-based compost, 4–500 g
Water, for damping compost

Labels
Newspaper or facility for dark germination
Seed trays 210 × 150 × 60 mm with transparent lids *(continued)*

*Assumptions*
1 Ability to use the 'chi-squared test'.
2 An understanding of the techniques (for example, Punnet Square) used to predict the outcome of trihybrid crosses not showing linkage.
3 Understanding of the terms allele, locus, map unit, and linkage.

*Principles*
1 The existence of fairly close linkage (less than 50 map units apart) between loci will significantly disturb a ratio of offspring phenotypes expected on the assumption of independent segregation.
2 Very loose linkage (around 50 map units) gives results indistinguishable from independent assortment.

ITEMS NEEDED *(continued)*

*or*
glass sheets
*or*
propagator trays 2/group
Spray or watering can 1/class
Wood block for seed tray preparation

Genetics Kit PGL 9 is available from Practical Plant Genetics, 18 Harsfold Rd, Rustington, Sussex BN16 2QE. Each complete 'lesson pack' in this kit contains enough seeds to give about 2 trays of $F_1$ *inter se* and 2 trays of $F_1$ backcross.

**3** It is possible to map the position of loci on a chromosome, using the results of a test cross as determining evidence.

The three genes under investigation are the following:

**Stem colour.** Recessive allele bls, or Baby Lea syndrome (here called b), produces no anthocyanin in the stem; dominant allele Bls (here called B) does produce anthocyanin.

**Leaf colour.** Recessive allele sy, or sunny (here called s), produces leaves and cotyledons which are not the usual dark green, but which are variably yellowish and patchy; plants of genotype ss have low viability. They survive better in cool conditions; dominant Sy (here S) is phenotypically normal.

**Leaf shape.** Recessive allele sf, or solanifolia (here called f), gives a leaf which is shaped more like the potato leaf than the tomato leaf, hence solanifolia; dominant allele Sf (here called F) gives the normal 'cut' leaf of the tomato (see *Practical guide* 5, figure 19). There is complete dominance at all three loci.

It is important to ensure that seeds and seedlings are maintained at reasonably high light intensity and are not allowed to get too hot or wet.

Cool, dry, bright conditions are essential just before scoring stem colour. This can be achieved by putting the trays outside at night in a porch or in a similar situation. If trays are kept very cool and dry throughout the experiment it will take longer for the foliage leaves to develop, though there will be a lower chance of fungal attack. There may be some difficulty in calculating cross-over values if only a small number of seeds has been used. It is suggested that, to make analysis easier, two sets of seeds of the $F_2$ and of the backcross should be sown, but that only the backcross results should be analysed fully. It is not vital to work out the correct order of the three genes along the chromosome; it is more important to demonstrate the effect of linkage upon the ratio of offspring phenotypes. It will be worth while collecting the data each time this practical investigation is used so as to keep an increasing number of data from which the linkage map may be made.

*Questions and answers*

**a** ***Put into words the null hypothesis that your chi-squared test was evaluating.***

There is no significant difference between the proportions of offspring (of the different phenotypes) observed in this cross and the proportions expected, assuming that the three genes are unlinked.

**b** ***State 1 the value of chi-squared given by your results, and 2 the probability of obtaining that chi-squared value. Put into words the outcome of your chi-squared test.***

It is useful to reinforce a standard approach to the use of statistical tests and the expressions of significance associated with them. In this case, suitable wording might be; 'There is/is no significant departure from the expected results, and results as different as these are from those anticipated would be found in (p × 100) per cent of all cases examined, owing to chance effects only.'

c ***Why was it not necessary to test the results of the cross against 'expected' results which assumed that the genes were completely linked with no crossing over?***

Because it is immediately obvious that, both in the backcross and in the $F_2$, there are more than two phenotypes of offspring.

d ***Use the results of your cross to draw a linkage map of these three genes, which are all carried on chromosome III; assume that the locus of the gene for stem colour is 74 map units from the end of the chromosome. Explain how you arrived at the map that you have drawn.***

Using some typical data from the $F_1 \times P_2$ backcross, these were the backcross phenotype frequencies:

| *Stem* | *Leaf shape* | *Leaf colour* | *Frequency* |
|---|---|---|---|
| Purple | normal | green | 48 |
| Green | potato | yellowing | 38 |
| Purple | potato | green | 24 |
| Green | normal | yellowing | 29 |
| Purple | normal | yellowing | 25 |
| Green | potato | green | 21 |
| Green | normal | green | 14 |
| Purple | potato | yellowing | 13 |

These phenotype ratios allow us to make the following deductions: the parental combinations of alleles are the most frequent, being B F S and b f s, totalling 86; the allele combinations resulting from double crossovers are the least frequent, in this case b F S and B f s, totalling 27. This places the order of the loci on the chromosome as F B S. Hence the recombinations between F and B (B f S and b F s) total 53; the recombinations between B and S (B F s and b f S) total 46. To these two latter totals should be added the number of double crossovers, since the appropriate recombinations are also represented there. So the crossover values work out as follows (total number of offspring in the backcross = 212):

COV between B and F = 53 + 27 = 80/212 = 0.38 – 38%
COV between B and S = 46 + 27 = 73/212 = 0.34 – 34%

Hence the separation of F and B is 38 map units and that of B and S is 34 map units. (See figure 10.)

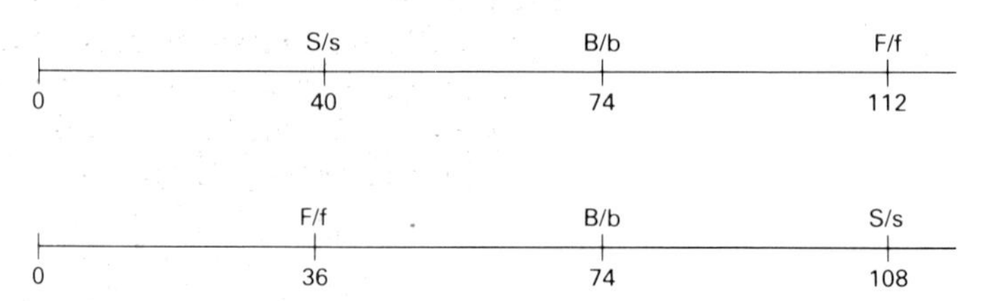


**Figure 10**
A linkage map of chromosome III to show the loci of the three genes.
The data do not allow us to distinguish between the two possibilities. Able pupils who see this clearly could be asked to plan an investigation which would allow the true map to be inferred.

e ***Suppose the investigation were to be repeated, starting with homozygous parents:***
***$P_1$ green stem, yellowing leaves, deeply lobed leaflets (bb ss FF)***
***$P_2$ purple stem, green leaves, non lobed leaflets (BB SS ff).***
***1 Would the linkage between the genes be the same?***

The loci would have the same linkage but the new parental combinations of alleles would be retained, giving a different ratio of phenotypes in the $F_2$ and backcross.
(Coupling and repulsion can be discussed here.)

***2 What cross would be needed to work out the crossover values required to map the loci?***

$F_1$ × a plant homozygous recessive at all three loci – i.e. a test cross not a backcross.

f ***Briefly outline the advantages and disadvantages for a species in possessing genes which are 1 unlinked, 2 loosely linked, and 3 tightly linked.***

Advantages
**1** The merits of independent assortment in giving rise to recurrent phenotypic variation in offspring, allowing the possibility of survival in a changing environment.
**2** There is some phenotypic variation, but a majority of offspring show parental combination of characteristics. Assuming that the parents were well-adapted for their environment, this means that the majority of offspring are also well-adapted. The small but recurrent percentage of recombinants gives them advantages over those species where no phenotypic variation is seen. These latter species would be largely asexually reproducing.
**3** The inheritance, reliably, from generation to generation of combinations of alleles which had proved to be of survival value in the past would be guaranteed by this method.

Disadvantages
**1** Maybe insufficient phenotypic stability would result in survival problems for the species if it were faced with severe competition in a relatively stable environment.
**2** This has very few disadvantages, except perhaps a slight loss of phenotypic stability which may give problems as outlined above.
**3** This has severe disadvantages if the environment is at all labile. The only way in which the problem of genetic uniformity may be reliably overcome would be by a greater degree of outbreeding and/or the establishment of a polymorphism.

## INVESTIGATION
## 17E Mutation in yeast

(*Study guide* 17.7 'Mutation'.)

### ITEMS NEEDED

Yeast dried

Agar, TTC (about 12 $cm^3$) in tube or bottle, sterile 2/group
Agar, yeast culture medium in Petri dish, sterile 2/group
Ethanol
Triphenyl tetrazolium chloride (TTC)
Water, sterile, distilled

Beakers
Bunsen burner 1/group
Forceps
Glass spreader 1/group
Heat-resistant mat
Incubator at 30 °C 1/class
Labels or marking pen
Lens, hand or microscope, binocular 1/group
McCartney bottle for sterile, distilled water 1/group
Syringe, sterile 10 $cm^3$ 1/group
Tape, adhesive
Tripod 1/group
Water bath at 40 °C 1/class

**Triphenyl tetrazolium chloride agar**
Agar powder (bacteriological grade), 1.25 g
Glucose, 0.5 g
2,3,5-triphenyl tetrazolium chloride, 0.05 g
Water, distilled, 100 $cm^3$
Autoclave for 15 minutes, dispense to bottles, and then re-autoclave.
**Yeast culture agar medium**
Agar powder (bacteriological grade), 4 g
Sucrose, 12 g
Yeast extract, 4 g
Water, distilled 500 $cm^3$
Autoclave for 15 minutes, allow to cool, pour into Petri dishes and leave to set. Then leave the plates, inverted and closed, at 35 °C, overnight to dry.

*Assumptions*

1. Familiarity with standard sterile precautions and microbiological techniques.
2. Knowledge of the life cycle of *S. cerevisiae* and familiarity with the similarities and differences between aerobiosis and anaerobiosis.

*Principles*

1. Identification of phenotypic differences between individual micro-organisms is hard, but colony-forming organisms reveal individual genetic differences in the form of differences between colonies.
2. Metabolic differences may be identified by the use of biochemical markers.
3. Sublethal mutations are not uncommon; organisms with a haploid stage in their life cycles will show such mutations readily.
4. No selection pressures are required for the initial appearance of a mutant, but these would be necessary for its continued survival in competition.

It must be assumed that each individual colony on the agar resulted from a single initial cell and one must employ the working hypothesis that the colony grew by haploid budding. The existence of haploidy in the life cycle is important in that it removes the problem of heterozygosity which would follow mutation of one of a pair of alleles in a homozygous organism and, in the case of a recessive mutation, would obscure the presence of that mutant allele. Thus we are dealing here with a single recessive mutation in a haploid organism. (See *Study guide II* page 88.)

In the initial culture agar there is nothing of demonstrable mutagenicity, so there is no reason to suppose that any observed mutations have been caused by an agent in the medium. Although TTC has been shown to induce the respiratory deficient (RD) mutation, the overlay technique used here does not allow long enough contact with the

TTC for it to have any noticeable effect. The predominance of non-mutant (respiratory-competent RC) colonies also rules out any suggestion that the medium is in any way selective for the RD mutant. So it can be argued that we are observing here the results of spontaneous mutation. It could also be argued that we are observing a polymorphism in the founding population.

The available literature suggests that the vast majority of RD mutants are governed not by a nuclear gene but by one in the mitochondrial DNA. Mitochondria are partly autonomous and partly under nuclear control. RD mutants are also called petite mutants because the colonies are smaller than the wild type.

Further (and perhaps more convincing) evidence that a spontaneous mutation has occurred is found in the existence of mixed colonies which consist chiefly of red-staining cells but which contain sectors composed entirely of white cells. (See Freeland, 1978, listed in the Bibliography for this chapter, for a fuller account.)

Practical problems are few. The most important points are as follows:

**1** Ensure that the plates are dry (free from condensation) before inoculation.
**2** Before pouring, ensure that the overlay agar has melted completely and has had sufficient time to cool to 40 °C.
**3** Pour the TTC agar slowly so that colonies are not disturbed.
Magdala red agar can be used as an alternative to TTC agar. This produces pink RC (wild type) colonies and red RD (petite) colonies.
**4** Some contaminants may produce colonies showing the mutant colour with TTC.
**5** If no mutants are apparent a further brief period of storage of the plates may reveal them.
**6** Bacteriological grade agar is more expensive but is essential for reliable results.

Magdala red agar can be poured over the visible colonies as outlined in *Practical Guide* 5, page 49, or it may be added to the yeast culture agar medium at the start. This is less trouble but it is then difficult to rule out the possibility that the dye has produced or selected mutants and not just revealed them.

*Questions and answers*

**a** ***Explain why the occurrence of white colonies can be taken to be the result of spontaneous mutation and not, for example, the influence of some environmental factor on existing mutant cells, or of contamination.***

This question has been answered in the foregoing discussion.

**b** ***Explain any observations that you made in step 12 above.***

It is hoped that students will notice that the mutant colonies are always small (petite), whereas the RC colonies are usually considerably larger. Large RC colonies may also be observed to have RD sectors within them. The first observation is best explained by suggesting that the respiratory deficiency results in a reduced rate of ATP synthesis and hence a retarded growth rate. This need not always be the case, since in the mixed colonies it is not uncommon to find that mutant sectors are growing out beyond the colony perimeter, which suggests that the RD mutants are, in such a case, growing faster than the non-mutants.

c ***Why does the technique employed here only reveal the respiratory-deficient mutants? Are there any other mutant yeast colonies present on your plates?***

Only RD mutants are revealed because the identification method only shows this type of colony. It is very unlikely that these are the only mutants; there are bound to be others where the mutation is either unexpressed in the phenotype under these conditions, or expressed in the phenotype but undetectable by any method used in this investigation.

*Alternative recipes for demonstrating RD mutants using Magdala red*
**Nagai medium**
Ammonium sulphate, 1.5 g
Glucose, 20 g
Magnesium sulphate, 1 g
Potassium dihydrogen phosphate, 1.5 g
Peptone, 1.5 g
Yeast extract, 1.5 g
Dissolve in 1 $dm^3$ of distilled water. Heat in water bath till the agar has melted; then autoclave at 102 583 N $m^{-2}$ or 15 lbf in for 15 minutes.
**Magdala red medium.** Prepare Magdala red stock solution by dissolving Magdala red in distilled water at a concentration of 0.001 g per $cm^3$. When required, sterilize 10 $cm^3$ of the stock solution in a water bath at 100 °C for 1 hour.
After autoclaving allow 1 $dm^3$ Nagai medium to cool to approximately 55 °C and then add 10 $cm^3$ Magdala red solution aseptically. Mix. Pour plates which have a dye concentration of 10 p.p.m. at a pH around 5.4.

## INVESTIGATION
## 17F The effects of irradiation in plant seeds

### ITEMS NEEDED

Seeds, tomato, from each of the four treatments, and the control from seed set PGL 17 (see note on supplier at end of this list)

Compost, Levington Universal or similar peat-based potting compost, 500 g for 1 seed tray
Water

Balance, weighing to ±10 mg 1/class
Beaker 250 $cm^3$ 1/group
Covers for pots or trays 5/group
Forceps *(continued)*

(*Study guide* 17.7 'Mutation'.)

*Assumptions*

1 An understanding of the phenomenon of incomplete dominance and its effect on phenotype ratios in a monohybrid cross.
2 An understanding of the use of histograms and graphs to display discontinuous data.
3 The ability to use the standard error of difference (SED) test and/or a correlation test.
4 Familiarity with the unit 'rad'. A rad is that amount of radiation which results in the release of $10^{-5}$ joule of energy in the absorbing tissue.

## ITEMS NEEDED *(continued)*

Graph paper or ruler 1/group
Labels 5/group
Newspaper
Paintbrushes
Scissors
Seed trays, 210 × 150 × 60 mm,
*or*
plant pots of equivalent capacity 5/group
Spray or watering can 1/class
Towel, paper
Wood blocks for seed tray preparation

Genetics Kit PGL 17 is available from Practical Plant Genetics, 18 Harsfold Road, Rustington, Sussex BN16 2QE. Each complete 'lesson pack' in this kit contains about 100 seeds for each treatment and for the control.

*Principles*

1 Radiation at sub-lethal doses will cause physiological and morphological changes in the phenotype of an organism.
2 The physiological changes are inferable from observation on the growth rate of the organism.
3 It may be argued that these phenotypic changes are the result of sub-lethal alterations to the genotype of the organisms by direct action upon its DNA. This hypothesis could be confirmed by breeding from affected seedlings. An alternative hypothesis is that cell cytoplasm has suffered damage.

The seeds in this cross, with the exception of the control seeds, have been exposed to gamma-radiation from a cobalt-60 source. Batches of seed were exposed to the radiation at a constant distance from the source for different lengths of time so that the approximate doses of radiation stated on the seed packs were received (these may vary from 20 to 80 kilorads).

The segregation of the seedlings into a 1:2:1 ratio at each radiation dose is clear, although the greater the number of seedlings used the better the fit. If the investigation is going to be used additionally to demonstrate incomplete dominance it is wise to pool all available data and to score the seedlings quite early on, since the homozygous mutant seedlings are very chlorosed and die soon after germination.

The homozygous normal plants should be removed when suggested since there tend to be very large differences between the size of these plants and the heterozygotes in all treatments. These differences are not of the same magnitude at all doses, and while this in itself could be used as another indicator of the effects of the radiation, it would serve only to mask the measured differences between the treatments in subsequent analysis. It is also important to remove the homozygotes if it is intended to grow the seedlings for any length of time, since the different size of the homozygotes, notably at low radiation doses, leads to differing degrees of competition between green and yellow plants at the different doses, and this will influence the growth of the heterozygotes.

It is wise to point out that there is no demonstrably causal relationship between the radiation and the nature of the mutation observed. Seedlings appear to show replicable mutations, such as changes in chlorophyll content, but these are phenotypic changes, and breeding experiments would be needed to demonstrate the genetic bases of the changes. We have not demonstrated that gamma-radiation causes chlorophyll content mutations more readily than other kinds of mutation. The seedlings used are already heterozygous at a locus which influences chlorophyll content and this facilitates detection of mutations at this locus. The majority of mutant alleles are recessive, so that they would not be detected in somatic tissue of a homozygous wild type plant. It is also perhaps worth reminding students that although the doses of radiation used in this investigation are high, they are markedly sub-lethal and although the mutations caused may reduce the effectiveness of the plant in competition, they are not sufficiently fundamental to lead directly and inevitably to early mortality. Lethal damage might be indicated by reduced germination at high radiation doses.

The characteristics that have been suggested for measurements have been deliberately kept simple, since they give a perfectly valid and reliable estimate of the effect of the radiation, without being impossible for the students to determine with ease. It has been found that the contribution of root mass (when grown by this method) to the whole plant mass is so slight that an entirely satisfactory investigation may be performed, using the aerial portions of the plant only. This may prove simpler for some purposes. The simplification also enables the whole investigation to be completed in a short time. Although it may be desirable in some respects to continue the work for a longer period, it is felt to be an advantage if the work gives clear results as soon as possible.

*Questions and answers*

a ***Put forward and explain a simple hypothesis to account for the occurrence of chlorosed or green patches of tissue in the leaves of a heterozygous plant.***

The simplest hypothesis is that the heterozygote (Xa–2 Xa–$2^+$) could easily be mutated to either homozygote locally and that this mutation is subsequently represented in the local phenotype. It is important for the student to realize that this is indeed a very simple hypothesis and makes many assumptions which could usefully be discussed. The green patches are more difficult to explain by an alternative hypothesis, but the chlorosed patches could also be caused by mutation at other loci than the Xa–2 locus. The concept of back mutation should be discussed.

b ***An organism which possesses cells or tissues whose genotypes differ one from another is called a mosaic. What evidence do you have that enables you to argue that some of the irradiated seedlings are mosaics? What additional evidence would give added firm support to your argument?***

If the leaves or stems on which chlorosed or green patches have developed were produced after the seeds were irradiated, it is difficult to refute the idea that they result from alterations to the genotype of the cells. However, the leaf must have begun its development before the mutation occurred or the whole leaf would be affected. It might be argued that the patch of altered coloration is the result of interference with the development of the leaf primordia (*c.f.* the limbs of children damaged by thalidomide). In the case of green patches, this seems unlikely since it is a return to normal phenotype. Additional evidence must depend on a patch of altered tissue including a bud; this would give rise to a flower which could be used to show that the condition was inheritable. Only flowers arising from mutant sectors should be used.

c ***Attempt to explain the effect of radiation dose on the germination, survival, and growth of the seeds and seedlings.***

Percentage germination, percentage survival, and growth rate of survivors are likely to be reduced at higher doses. This may be because of damage to genetic material and/or ionization in the plant cells. If only the latter had occurred, a seedling should ultimately recover and grow and reproduce normally.

# PART III Bibliography


CHAPPELL, J. B. Carolina Biology Readers No. 19, *Energetics of mitochondria.* Revd. edn. Carolina Biological Supply Company distributed by Packard Publishing, 1978.

FINCHAM, J. R. S. Carolina Biology Readers No. 2, *Genetic recombination in fungi.* 2nd edn. Carolina Biological Supply Company distributed by Packard Publishing, 1983.

FREELAND, P. W. 'Some experiments with respiratory deficient mutants of yeast (*Saccharomyces cerevisiae*).' *Journal of Biological Education,* **12**(1), 1978.

GOODENOUGH, U. *Genetics.* 3rd edn. Holt Saunders, 1984.

HORN, P. and WILKIE, D. 'Use of Magdala red for the detection of auxotrophic mutants of *Saccharomyces cerevisiae*' *Journal of Bacteriology,* **91**(3), 1966.

JINKS, J. L. Carolina Biology Readers No. 72, *Cytoplasmic inheritance.* 2nd edn. Carolina Biological Supply Company distributed by Packard Publishing, 1978.

NAGAI, S. et al. 'Advances in the study of respiratory deficient (RD) mutation in yeast and other micro-organisms', *Bacteriological Review,* **25**, 1961.

WALLACE, M. *Teaching genetics with mice.* Heinemann, 1971.

See also *Carolina tips.* Carolina Biological Supply Company, Burlington, N. Carolina 27215, USA. (A series of short leaflets published regularly.)

# CHAPTER 18 THE NATURE OF GENETIC MATERIAL

*A review of the chapter's aims and contents*

1 The concept of the gene has already been introduced and this idea is refined further in this chapter.
2 The emphasis is on the experimental evidence for the role of nucleic acids in heredity.
3 A knowledge of the structure and functions of proteins has been assumed throughout.
4 The structure of nucleic acid molecules is outlined in sufficient detail to allow an understanding of their role in the cell.
5 Some historical background to our understanding of the biochemical basis of inheritance is provided.
6 Accounts of replication, of transcription and translation, and of the genetic code are provided.

# PART I The *Study guide*

## 18.1 The search for genetic material

*Assumptions*

1 A knowledge of the terms gene and chromosome.
2 A knowledge of protein structure.
3 Some understanding of Mendelian inheritance and of cell division.

*Principles*

1 The genetic material has certain functions or properties – replication, mutability, information storage and controlled expression, which can be deduced from the observation of inheritance and development in all organisms.
2 The substance or substances of which the genetic material is composed must be able to undertake these functions and have these properties.
3 Because the structure of protein was known to be complex, and the importance of enzymes in biological systems was well understood, it was reasonable to assume, as a working hypothesis, that the genetic material was protein.

### STUDY ITEM
### 18.11 DNA in the nuclei of different kinds of cells

*Principles*

1 The amount of DNA in individual cell nuclei can be estimated.
2 The amount is constant for the somatic cells of an organism.
3 Different species have different quantities of DNA in their cells.
4 The amount of DNA in gamete cells is half that found in somatic cells.
5 If it is demonstrated that nuclei contain a definite and constant amount of DNA, this does not prove DNA to be the genetic material.

*Questions and answers*

**a** ***From the information in table [S]20, what generalization can you make about the amount of DNA per nucleus in:***
***1 the red blood cells and liver cells examined***
***2 the somatic cells and gamete cells?***

**1** For any organism the amount of DNA in the liver and red blood cell nuclei is constant.
**2** Gamete cells contain half the DNA possessed by somatic cells.

**b** ***If DNA were the genetic material, would you expect to find data similar to those given in table [S]20?***

Yes, since all somatic cells usually have the same chromosome complement and are diploid while gametes are haploid.

**c** ***If a cell divides meiotically, what would you expect to happen to the amount of:***
***1 genetic material***
***2 structural (skeletal) material of the chromosomes?***

Both types of material would be halved.

**d** ***From the data supplied by Mirsky and Ris, can we rule out the possibilities that DNA is both the genetic and the skeletal material, or just the skeletal material?***

No, we cannot rule out either possibility. If the number of chromosomes is halved at meiosis, both types of material would be halved.

**e** ***Would more DNA estimations from somatic cells and gamete cells of more kinds of animals (and plants) help to decide the role of DNA?***

No, more measurements would merely make more valid the generalization that gamete cells have half the DNA of somatic cells.

**f** ***Explain why the table contains blank spaces. Do the blanks significantly influence your conclusions?***

Blank spaces indicate that the data are unavailable or unobtainable. (A symbol could have been used instead of a blank space.) Generalizations from the data are less firmly based because of the blanks. It is a matter of personal scientific judgment whether there are sufficient data to draw conclusions. □

## 18.2 The evidence from viruses

*Principles*

1. Viruses are replicating units which can only reproduce in living cells and which lack other characteristics of life.
2. They are composed of nucleic acid and protein.
3. They are of practical importance because of their involvement in viral

diseases and their association with tumours, and because of their ability to transpose genetic material between the cells, sometimes of different species.

A brief factual account of the structure and life cycle of viruses is provided.

**STUDY ITEM**

**18.21 Labelling viral DNA**

*Assumption*

A knowledge of isotopes and of the principles of isotope labelling.

*Principle*

Labelling of the nucleic acid and protein components of viruses shows that only the nucleic acid has to enter the host cell for viral replication to occur; therefore, the nucleic acid must specify the structure of the complete virus.

*Questions and answers*

a ***What does this result show?***

That the protein coat of the bacteriophage need not enter the host cell and therefore that the DNA must carry instructions for the synthesis of the protein coat.

b ***Explain how radioactive phosphorus ($^{32}P$) could be used to make more certain the roles of DNA and protein in the infection of cells by viruses.***

If viruses are produced by bacteria grown in a medium containing radioactive phosphorus and normal sulphur, the DNA of the viruses will be labelled. If such labelled viruses were added to a bacterial culture it would be necessary for the radioactive DNA to enter the cells before they became successfully infected. The radioactivity could not be washed off the infected cells. □

## 18.3 Direct evidence for DNA as genetic material

**STUDY ITEM**

**18.31 Genetical transformation in *Pneumococci***

*Principles*

1. Bacterial cells may acquire genetic material from their environment and transmit it to their descendants (this is transformation).
2. Purified DNA from donor cells can transform recipient cells to the genotype of the donor strain.
3. Principle **2** demonstrates that the donor cell DNA must contain all the information necessary for the development of the donor cell genotype; that is, DNA is the genetic material.
4. Transformation is distinct from mutation.

*Questions and answers*

**a** ***What controls would be needed to rule out some of these alternative explanations?***

**1** Some of the 'heat-killed' smooth cells might have survived and multiplied. This is ruled out by injecting heat-killed smooth cells without living rough cells.
**2** Some of the living rough cells might have mutated to smooth. This is ruled out by injecting living rough cells without dead smooth cells.
**3** The mice could have been incubating pneumonia before the experiment started or could have acquired it by chance in the laboratory during the experiment. This is ruled out by injecting a sterile saline.

**b** ***Suggest two reasons why Griffith's result cannot be explained by mutation.***

**1** Only a small proportion of cells are transformed, but the frequency and regularity of the process are too great to be explained by mutation.
**2** Some mutation should be observable in pure cultures of rough cells.

**c** ***How could the latter hypothesis be tested?***

Produce a cell-free extract from a culture of smooth cells and see if it would transform rough cells.

**d** ***Do you think that the data in table [S]21 established that the substance extracted and isolated from the bacteria was chemically pure?***

Purity is a relative term and it is always possible that a minute trace of impurity in a so-called pure material may be of great importance (an example is the trace elements present in the water cultures of nineteenth century plant physiologists); however, within the limits of experimental error, the DNA preparations seem to be pure. A wide variety of techniques was used to assess the purity of the samples.

**e** ***What was the purpose of tabulating the ratio of nitrogen to phosphorus?***

The ratio should be constant if the sample is pure, but, more importantly, protein does not contain phosphorus, so any appreciable contamination by protein (the other possible candidate for the make-up of genetic material) would be detected. □

## 18.4 The chemical structure of DNA

*Assumptions*

1 Some knowledge of the structure and distribution of cell organelles.
2 An understanding of the principles of chromatography.

*Principle*

1 DNA consists of the subunits:
deoxyribose
phosphoric acid
purine bases – adenine and guanine
pyrimidine bases – thymine and cytosine.

## STUDY ITEM

**18.41 The relative quantities of nitrogenous bases found in various forms of DNA**

*Questions and answers*

a ***What can you say about the molar quantities of purine and pyrimidine present in any one sample of DNA?***

They are equal.

b ***What do you notice about the ratio of the number of molecules of adenine to the number of molecules of thymine in any one sample of DNA?***

There is a 1:1 ratio.

c ***Is the same relationship found between the number of molecules of guanine and cytosine in any one sample of DNA?***

Yes.

d ***Are the data in table [S]22 compatible with the idea that the four bases combine together in pairs in the DNA molecule?***

Yes.

e ***Within the limits of experimental error, are the proportions of the two types of pair constant for different tissues in the same organism? If DNA were the genetic material, would you expect this?***

The proportions of the two types of pair are constant for three different bovine tissues. This is limited evidence. One would expect the genetic material to have a constant composition in all tissues.

f ***Within the limits of experimental error, are the proportions of the pairs different for DNA extracted from different species? Would you expect this if DNA were the genetic material?***

The proportions are different for DNA from different species and one would expect this as different species would have different genes. (Actually much of the difference is due to non-transcribed regions that are not strictly 'genes'.)

g ***Does this mean that the genetic information in rat bone marrow cells and bacteriophage is the same? In answering this question it may***

***help to consider the following pairs of English words:***
***mane and name***
***cat and act***

No – the base pairs are not the genes; it is the specific sequence of these pairs in the molecule which is likely to be significant. □

## 18.5 A model of the DNA molecule

This section attempts to put the epoch-making discovery of the genetic material in a historical and human context. The way in which Watson and Crick made their discovery is outlined.

### STUDY ITEM
### 18.51 DNA replication

*Assumptions*

1 An understanding of the principles of isotope labelling.
2 A knowledge of mitosis.

*Principles*

1 If cells are provided with metabolites rich in an isotope of a normal constituent element this can be incorporated into their DNA as they grow and divide.
2 When the cells are transferred to an unlabelled medium, newly synthesized DNA will be free from label.
3 The distribution of label in the DNA after known numbers of generations is consistent with a semi-conservative hypothesis for DNA replication.
4 The incorporation of label into chromatids during mitosis is consistent with the hypothesis that they contain a single but highly condensed double helical strand of DNA.

It is mentioned that the process of DNA replication is a complex one. The spiral of the double helix must be unwound to allow pairing of free nucleotides. The details of the replication process are considered to be beyond the scope of an Advanced Level course. A good account of the replication process can be found in Szekely, M., *From DNA to protein: the transfer of genetic information*, Macmillan, 1980.

*Questions and answers*

a ***Reproduce the figure by tracing or photocopying and use the semi-conservative hypothesis to predict the position of the DNA band or bands in the experimental tubes.***

After one replication the results would indicate that each DNA molecule had one heavy and one light strand and so would lie between the heavy band and the light band. After two generations (or more) the tubes would show entirely unlabelled – that is, new – DNA, and DNA which was of intermediate density as in the first generation. (See *figure 11.*)

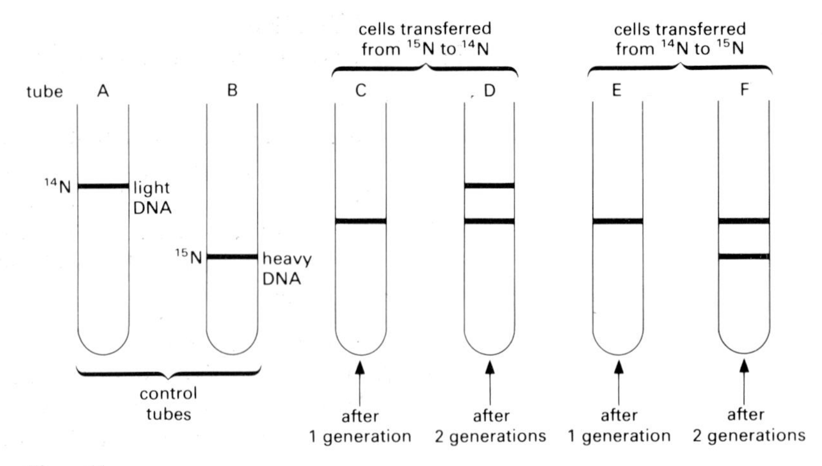


**Figure 11**
The position of DNA bands after replication of light DNA in $^{15}N$ medium, and heavy DNA in $^{14}N$ medium.

**b** ***If cells are transferred from $^{15}N$ medium to $^{14}N$ medium, how many generations would elapse before a band of DNA containing $^{15}N$ disappeared?***

The band of intermediate density should never disappear since each parental DNA strand should be conserved indefinitely, acting as a template for further synthesis with each generation. In practice, the $^{15}N$ band would be undetectable after a few generations because it would be dispersed in a very large population of unlabelled cells.

**c** ***Would one or both of the chromatids be labelled?***

Both.

**d** ***If the chromosome had replicated once in non-radioactive thymine and had been killed and fixed during the second mitosis in unlabelled medium, would one or both of the chromatids be labelled?***

One of each pair of chromatids would be labelled.

**e** ***If this experiment is to be a fair test of the semi-conservative hypothesis, how many DNA double helices must there be in a chromatid?***

One. If a chromatid contained several parallel molecules of DNA (as do polytene chromosomes in *Diptera*) then it would be expected that on replication some strands containing original, that is, labelled, material would pass to each product and both chromatids would be labelled for several generations after transfer to an unlabelled medium. □

**STUDY ITEM**

18.52 A summary of the evidence that DNA is the genetic material

*Principles*

1 A large body of evidence points to DNA as the genetic material.

2 Some of the evidence is less crucial because it would be consistent with substances other than DNA acting as the genetic material.

3 Scientific advance occurs on a number of fronts but certain ideas are crucial as they bring order to a large mass of previously unconnected knowledge.

*Questions and answers*

a ***By means of a class discussion, a group discussion, or an essay, arrange the above list of evidence in order of importance. Which pieces of evidence are only circumstantial?***

Items **3** and **4** in the list are the only direct pieces of evidence. Item **3** (transformation) is more important than **4** (the virus evidence) since it could be argued that viruses are a special case. As Study item 18.11 made clear, items **1** and **2** are consistent with DNA being either structural or genetic. Item **9** is somewhat dubious evidence since chemical mutagens are often able to react with proteins and other cell constituents as well as with DNA. It is to be hoped that students will point out that, while item **6** may be a good reason for suspecting that DNA is genetic material, it is not evidence of the same sort as the other items.

b ***The work of Watson and Crick is sometimes taken to herald the beginning of molecular biology. Is their research more crucial than that of others who are mentioned in this chapter?***

Without Chargaff's contribution, which showed that DNA has equimolar quantities of purine and pyrimidine bases, and Wilkin's X-ray diffraction data, Watson and Crick could not have undertaken their modelling. However, their work was especially important as the model provided a vastly greater stimulus to new research (and technological development) than the work of others. □

## 18.6 The breaking of the genetic code

**STUDY ITEM**

18.61 The number of bases in the genetic code

*Assumption*

An understanding of the structure of single stranded DNA as outlined in the earlier sections.

*Principles*

1 The sequence of bases in a DNA molecule can code for the order of amino acids in a polypeptide.

2 Since there are 20 amino acids in naturally occurring proteins, the minimum-sized group of bases to code for a single amino acid must be three.

3 This implies a degenerate code.

*Questions and answers*

a ***If only a single base coded for a single amino acid, how many of the latter could be coded?***

Four.

b ***Calculate the minimum-sized group of bases which could code for 20 different amino acids.***

If $n$ is the number of bases in a codon and there are 4 bases (letters), the number of possible codons ($N$) can be calculated from the formula:
$N = 4^n$
If there were two pairs of bases this would give $4^2 = 16$ codons, which is too few.

c ***1 For how many different amino acids could, in fact, the minimum-sized groups of bases code?***

$4^3$ = sixty four.

***2 If more than 20 amino acids could be coded, what might the solution to this problem be?***

Either some codons could be meaningless, or more than one codon could stand for some or all of the 20 amino acids. Some codons could stand for punctuation. These alternatives are not logically exclusive.

d ***Does this introduce a fundamental difficulty for the hypothesis that the base sequence is a code? Explain your answer.***

No – provided that some codons, or groups of codons, could act as signals to start and stop reading the genetic message. □

## STUDY ITEM

## 18.62 The site of protein synthesis

*Assumption*
An understanding of isotope, labelling and fractional centrifugation.

*Principles*
1 Protein is synthesized in organelles called ribosomes in the cytoplasm.
2 Since the DNA of the cell determines the structure of the protein synthesized, a message must pass from the DNA to the ribosomes.

*Questions and answers*

a ***Explain how figure [S]73 provides evidence for this view.***

After the labelled sulphate has been replaced by non-radioactive 'chaser', the radioactivity rapidly leaves the 70–80 S region and moves to the soluble protein. This indicates that the protein is being produced by the 70–80 S ribosomes and liberated as soluble protein.

b ***Does the DNA of cells appear directly responsible for protein synthesis?***

No. The DNA of rat liver cells is almost entirely confined to the nuclei and it is the ribosomes, rather than the nuclei, which incorporate radioactive amino acids.

c ***If the order of bases in a DNA molecule determines (codes for) the sequence of amino acids in a protein molecule, what problem is raised by the findings of Littlefield?***

How does the DNA cause the ribosomes to synthesize the protein specifically coded by it?

d ***With regard to the evidence provided by the work of Littlefield and McQuillen, are the messages DNA or protein, or are they likely to be another substance? Explain your answer.***

They are unlikely to be either as DNA remains in the nucleus and protein is synthesized in the cytoplasm where the ribosomes are found.
□ The messages are therefore some other substance.

## STUDY ITEM
## 18.63 The message linking nucleus and cytoplasm

*Assumption*
Some idea of the principles of stereochemistry.

*Principles*
1 Messenger RNA determines the structure of the proteins synthesized by ribosomes and must therefore carry information from the nucleus to the cytoplasm.
2 Artificial mRNA can be used to discover the sequence of bases that code for different amino acids.

*Questions and answers*
a ***Outline how such experiments might be carried out.***

Labelled uracil is fed to cells. This is specifically incorporated into all three types of RNA. If the cells are killed and fractionated at different times after the label has been applied to them, it can be shown to appear first in the nucleus and later in the cytoplasm.

b ***Why would rapid synthesis and breakdown of a chemical message from nucleus to cytoplasm be an advantage?***

Because it would mean that genes could produce their effect for only a short time, unless there were continuous synthesis of mRNA. Control of gene expression could be made precise.

c ***Why was it important to use mRNA from one organism and all the other constituents of the mixture from another, completely different organism?***

To show that it was the mRNA alone which determined the type of protein produced.

d ***What amino acid does the sequence UUU (or TTT in the corresponding DNA sequence) represent in the genetic code?***

Phenylalanine.

e ***Give a set of three bases that codes for the amino acid proline.***
First base, C; second base, C; third base, any of the four.

f ***Give a base substitution (mutation) which would produce a new polypeptide by substitution of a new amino acid.***

Many examples are possible. Students should notice the degenerate nature of the code and the fact that a change in the third base often produces no functional change. □

## 18.7 The synthesis of RNA

This section, and the two that follow it, are to a large extent descriptive and narrative and give a general account of the transcription and translation of the genetic code and the structure of chromatin. Much of the evidence for the structures and processes described is beyond the scope of these texts. It is important neither to expect a wealth of biochemical detail nor to allow students to regard highly schematic text book accounts as the last word on the subject.

*Principles*

1 Complementary sequences of bases in nucleic acids can be detected by molecular hybridization.
2 Transcription is the production of RNA complementary to DNA.

## 18.8 The mechanisms of protein synthesis

*Principles*

1 A ribosome consists of two subunits, each composed of RNA and protein, which associate with the mRNA to form a functional unit.
2 The genetic message is translated by means of complementary base triplets (anticodons) on the amino–acyl–tRNA units which attach to codons of mRNA.
3 The ribosome moves along the mRNA molecule with two tRNA units associated with it at any one time, one carrying the growing polypeptide and the other (the most recent to attach) carrying the next amino acid to be added to the growing chain.
4 Energy is needed for the translation process and this is provided by ATP and GTP.
5 Punctuation, provided by sequences of bases, starts and stops translation.

*Question and answer*

a ***Explain the term 'anticodon' as used in* figure [S]77.**

A triplet of bases on a strand of DNA or mRNA which stands for an amino acid is a codon. Each codon has a complementary triplet which can attach to it by base pairing. This is the anticodon.

## 18.9 The structure of chromatin

*Assumption*

A knowledge of mitosis, meiosis, and chromosome structure.

*Principles*

1 Chromosomal DNA is very long and needs to be packaged.
2 Basic histone proteins allow DNA molecules to coil into units called nucleosomes.
3 The degree of condensation of chromatin in the interphase nucleus varies. The highly condensed regions are called heterochromatin.
4 Scaffolding proteins are associated with chromatin condensation.
5 Little is understood of the relationship between DNA packaging and gene regulation.

## 18.10 What is a genome?

*Assumptions*

1 An understanding of the role of nucleic acids in inheritance.
2 An understanding of the alternation of haploid and diploid states in the life cycle of eukaryotes.
3 Some knowledge of viruses.
4 A knowledge of polyploidy.

*Principles*

1 The genome is a collective term for the genetic material of a cell.
2 The genome of a cell excludes the genetic material of autonomous or semi-autonomous structures, such as plasmids, which have genomes of their own.
3 During evolution genomes change quantitatively and qualitatively.
4 Different parts of the genome have different functions and some parts may turn out to be functionless from the point of view of the whole organism.
5 Some viruses have a genome composed of RNA rather than DNA.

# PART II The *Practical guide*

The investigations related to this chapter appear in *Practical guide 5, Inheritance.*

## INVESTIGATION 18A Testing for DNA using the Feulgen technique

(*Study guide* 18.1 'The search for genetic material'.)

**ITEMS NEEDED**

Root tips or other suitable tissue, fresh
Ethanoic-alcohol fixative
Hydrochloric acid, 1.0 mol $dm^{-3}$
Schiff's reagent (Feulgen stain)
Water, distilled

*(continued)*

**Ethanoic alcohol.** See page 13.
**Schiff's reagent** (Feulgen stain)
Basic fuchsin or magenta, 1 g
Hydrochloric acid, 1.0 mol $dm^{-3}$, 30 $cm^3$
Potassium metabisulphite, 3 g
Water, distilled, 200 $cm^3$

## ITEMS NEEDED *(continued)*

Beaker, 100 $cm^3$ 1/1
Blotting or filter paper
Measuring cylinder, 20 $cm^3$,
*or*
syringe to mix fixative 1/group
Microscope, monocular 1/1
Microscope slides and cover slips
Mounted needles 2/1
Pipettes, dropping 2/1
Watch-glasses
Water bath at 60 °C 1/class

Boil the water and pour it over the fuchsin. Stir this solution very vigorously and leave it to cool to 50 °C (this temperature is critical). Once the temperature has reached 50 °C, add the acid (*take care*) and the metabisulphite and allow it to dissolve. Then put the solution into a dark bottle and stopper firmly. Store it in this container until it has bleached to a straw colour, which should occur overnight under normal conditions. The important thing is to remove all traces of pink. Filter the solution after shaking it vigorously with 0.5 g of decolorizing charcoal. Store in a tightly stoppered, dark bottle.

*Assumptions*

1. The ability to use a microscope with high power objective.
2. The ability to recognize chromosomes and cells that are in the process of dividing.
3. Familiarity with sufficient organic chemistry to understand the meaning of the terms 'hydrolysis' and 'aldehyde'.

*Principles*

1. Chemical hydrolysis, maceration, and staining will not necessarily destroy complete cells but may react with cell components *in situ.*
2. Staining may be used as a specific histochemical test to identify the presence and location of a particular class of molecule.
3. DNA is specifically detected by its reaction with Schiff's reagent.

The method employed here is fairly standard and should cause little difficulty. It is possible to use different hydrolysis times (table 11).

| *Molarity of hydrochloric acid* | *Temperature of hydrolysis* | *Duration of hydrolysis* |
|---|---|---|
| 3.5 | 37 °C | 17 minutes |
| 6 | 20 °C | 15 minutes |

**Table 11**

The timings of the fixation and staining stages may also be varied to suit, and the following guidelines may be useful:
Fixation. Ten minutes is the absolute minimum, provided that the tissue is not very bulky. Otherwise, any duration of fixation of up to about 48 hours will be all right. Material can be stored safely in the fixative if it is kept in a deep freeze.
Staining. If the material is well hydrolysed, 15 minutes in Schiff's reagent may be quite adequate. Generally, though, the longer the material is in Schiff's reagent, the more intense is the staining, up to about 12 hours.
The method used here is quite specific for DNA and will not detect RNA. The —OH group of the second carbon $C_2$ of RNA's ribose sugar is all-important in that it is responsible for making the RNA susceptible to alkaline hydrolysis, where DNA is not. It also prevents RNA from undergoing acid hydrolysis, by somehow interfering with and stopping the removal of the bases from the molecule. The mild acid hydrolysis used here produces an apurinic acid from the DNA molecule and this acid has the available aldehyde group at the first carbon $C_1$ of the deoxyribose sugars that have lost their purine bases. There is still considerable argument over the exact mechanism of the reaction with

Schiff's reagent, but there is no doubt that the combination of the reagent and the aldehyde group results in the formation of a quinonoid ring within the complex, which naturally imparts colour.

*Question and answer*

a ***From the evidence of your preparation, what conclusions do you come to about the DNA content and distribution in the structures of 1 the nucleus and 2 the cytoplasm?***

Evidence is obtained which shows the nucleus to contain granular material at interphase, and stranded material during mitosis, which contains DNA. The stranded material closely matches the shape of the chromosomes, indicating strongly that these structures indeed contain DNA. The evidence does not suggest that they contain nothing besides DNA.
The lack of any stain in the cytoplasm supports strongly the hypothesis that little if any DNA is found there or moved to the cytoplasm from the nucleus. The fact that the cytoplasm appears to be completely structureless does not mean that it contains no structure, merely that it contains no structure dense enough or large enough to be visible without staining under the light microscope.

## INVESTIGATION
## 18B Testing for DNA and RNA using methyl green pyronin

### ITEMS NEEDED

Onion or similar bulb

Ethanol, absolute
Methyl green pyronin

Forceps
Microscope, monocular 1/group
Microscope slides and cover slips
Mounted needles 2/group
Pipettes, dropping
Scalpel 1/group
Watch-glass or small beaker 3/group

(*Study guide* 18.6 'The breaking of the genetic code'; Study item 18.63 'The message linking nucleus and cytoplasm'.)

**Methyl green pyronin**
Methyl green pyronin, 1 g
Ethanoate buffer, pH 7.3, 100 $cm^3$ (To make 1 $dm^3$, dissolve 5.45 g sodium ethanoate in 10 $cm^3$ glacial ethanoic acid, and make up volume with distilled water.)
Dissolve the stain at room temperature. It works much better if it is freshly prepared.

*Assumptions*

1 The ability to use a microscope.
2 Familiarity with the basic anatomy, as visible with a light microscope, of a vacuolated plant epidermal cell.
3 Methyl green pyronin differentiates reliably between DNA and RNA and does not appear to stain any other structure in these cells.

*Principle*
The location of a particular substance within (or as part of) a structure in a cell or organelle may reflect the role of that substance in the function of the cell or organelle.

This technique does not present many problems but the following points may be of value.
**1** Students sometimes find it inexplicably hard to locate the inner epidermal layer of a bulb leaf base, and they attempt to 'pull off' a measurably thick layer of tissue.

2 The tissue folds over onto itself rather easily on the slide during mounting, and this is a common cause of frustration.
3 Sometimes cells from the interior of the leaf base stick to the epidermis, giving two or three cell thicknesses: these should be carefully scraped away during step 6.
4 Students will probably need reminding about the size of the vacuole in these cells and the apparent paucity of the peripheral cytoplasm which appears as a very thin granular layer almost on top of the wall.
5 The results are striking, and the students should be encouraged to draw individual cells in careful detail.

*Question and answer*

a ***Using data from this and from the previous investigation, make a careful statement about the distribution of the two nucleic acids in the cell.***

DNA is found at detectable levels only in the nucleus. RNA is found more or less evenly distributed, though weakly and diffusely, in the cytoplasm. (The cytoplasm is found around the periphery of the cell and around the nucleus). RNA is also found and at far greater density within the nucleoli in the nucleus. (There is a variable number of nucleoli visible in each nucleus.)

## INVESTIGATION
## 18C Making a model to illustrate the chemical nature of a gene

### ITEMS NEEDED

'Blu-tac' or equivalent – for joining sub-units reversibly
Card, fairly stiff, 50 $cm^2$ 1/1
Modelling knife or old scalpel 1/1
Paper, thin, several sheets of each of 3 or 4 colours
Ruler 1/1
Scissors for cutting paper 1/1
Tape, adhesive

(*Study guide* 18.4 'The chemical structure of DNA'; 18.5 'A model of the DNA molecule'; 18.6 'The breaking of the genetic code'; 18.7 'The synthesis of RNA'.)

*Assumption*

1 An outline knowledge of current theory of DNA structure and biochemical function.

*Principle*

1 A simple physical model can be more instructive than diagrams when seeking to understand a complex mechanism.

Many students need no instructions to design and make a fairly satisfactory model to illustrate DNA and its functions. The instructions that are provided are based on the methods adopted by a number of students. Cardboard cartons can be used as a source of card. Figure 68, page 104 of *Study guide II* provides a satisfactory starting point. Diagrams to show the relationship of mRNA, tRNA, amino acids, peptides, and ribosomes will also be needed. (See *Study guide II*, figures 79 and 80, pages 120–1.)

If time is limited, duplicated diagrams of nucleotides can be provided for cutting out, but this will give students less opportunity for referring to diagrams and working to a sensible scale.

Discussion should make clear to the students that their models are intended only as aids to help them to visualize a process, and have little predictive value – unlike the accurate scale model of Watson and Crick. Each student can be encouraged to keep his or her model for use in revision.

## PART III Bibliography

LEHNINGER, A. L. *Biochemistry: the molecular basis of cell structure and function.* 2nd edn. Worth, 1975.

PEARSE, A. G. E. *Histochemistry: theoretical and applied.* 3rd edn. Churchill Livingstone, 1968.

STACEY, M., DERIAZ, R. E., TEECE, E. G., and WIGGINS, L. F. 'Chemistry of the Feulgen and Dische nuclear reactions', *Nature*, **157**, 1946, p. 740.

WOODS, R. A. Outline studies in biology, *Biochemical genetics.* 2nd edn. Chapman & Hall, 1980.

# CHAPTER 19 GENE ACTION

*A review of the chapter's aims and contents*

1 The one gene–one polypeptide hypothesis already introduced in this *Guide* in Chapters 15 and 18 is developed and related to metabolic pathways.
2 Mutants which prevent the synthesis of enzymes and block particular pathways are considered in yeast, *Neurospora*, humans, and tomatoes.
3 The regulation of enzyme synthesis in bacteria by the Jacob–Monod operon model is reviewed and the importance of enzyme induction for the efficient functioning of a micro-organism is discussed.
4 Students are made aware that little is known about control of gene action in higher organisms, but heterochromatin is known to be associated with inactive genes. Puffing of polytene chromosomes is a special case in which gene action can be observed and its control studied.
5 An understanding of the activity of genes at the sub-cellular level is not sufficient to explain differentiation. The development of the whole organism involves the organization of populations of cells.
6 Possible explanations for dominance and heterosis are discussed and this leads to the concepts of pleiotropy and gene interaction.
7 Epistasis is explained as a special case of gene interaction which depends on genes acting in sequence through metabolic pathways.

## PART I The *Study guide*

### 19.1 Mutant complementation in *Neurospora*

*Assumptions*

1 That transcription and translation of DNA have been studied.
2 That the terms phenotype and heterokaryon are understood.
3 That the significance of haploid and diploid states is understood.

*Principles*

1 We remain ignorant, in most cases, of the way in which gene products interact with the products of other genes and with the environment to produce the phenotype of the whole organism.
2 A minimal medium contains the least number of substances necessary for the survival of the wild type. A complete medium is a complex mixture containing many substances.
3 Beadle and Tatum devised procedures to isolate mutants, which ensured that each isolate originated in a single nucleus.
4 If two mutants are present in the same culture they may grow as the wild type as a result of complementation.

*Questions and answers*

**a** ***If a conidium contains two nuclei and a mutation occurs in one of them (or two different mutations occur, one in each nucleus) what type of mycelium would grow from the conidium?***

A heterokaryon in which the effects of a mutation might be masked.

**b** ***Which steps do each of the mutants A and B lack?***

Mutant A lacks the first step (it can use nitrite but not nitrate). Mutant B lacks the second step (it excretes nitrite which mutant A can utilize).

**c** ***If two mutants, both requiring the same substance as in* figure [S]84, *are able to complement one another, what may be deduced about the biosynthesis of the substance in the cells of the organism concerned?***

The biosynthesis requires two different genes (or the products of two different segments of DNA).

## STUDY ITEM

## 19.11 Mutations and metabolic pathways

*Assumption*

1 Familiarity with glycolysis, the Calvin cycle and other metabolic pathways, and with the use of radioactive tracers and enzyme inhibitors to verify such pathways *in vivo*.

*Principles*

1 Replica plating allows the growth of several colonies on several media to be easily compared.

2 Mutants are often unable to carry out particular steps in a reaction sequence but will grow if provided with a metabolic intermediate which follows the blocked step in the sequence.

*Questions and answers*

**a** ***Explain why four colonies have developed on minimal medium.***

They originate from cells which did not mutate and are wild type.

**b** ***How many mutants that require arginine have been produced?***

Two.

**c** ***Explain why one colony has failed to grow on minimal medium but has grown on media containing both arginine and ornithine.***

The mutant requires arginine so will not grow on minimal medium but will grow on arginine-enriched media. The gene which has mutated codes for an enzyme active in catalysing some step leading to ornithine synthesis. If provided with ornithine the mutant has the enzymes to complete the reaction sequence to arginine. (Students could be asked whether this mutant would grow if provided with citrulline or arginosuccinic acid – it would.)

**d** ***Were any of the strains used wild type?***

No, since none grew significantly on minimal medium. (Students will need to decide what is significant growth and what can be explained as slight growth resulting from arginine stored in the inoculum.)

**e** ***Classify the mutants in table [S]23 as being type A, type B, or type C.***

Strains 1, 2, and 5 are type A. Strains 3 and 4 are type B. Strain 6 is type C.

**f** ***Explain why a mutant which could grow on arginine or on ornithine but not on citrulline would be surprising.***

If a mutant could grow on ornithine, it must make arginine from ornithine via citrulline. It should therefore be able to grow on citrulline.

**g** ***Pick out from table [S]23 two* Neurospora *strains that should show complementation, and two strains which would be unlikely to complement.***

Strains 1, 2, or 5 would complement 3 or 4. Strain 6 would complement any of the others. The strains in the same groups, that is, group A (1, 2, and 5) and group B (3 and 4), would not complement. □

## 19.2 Operators, repressors, and promoters

*Assumptions*

1. A knowledge of transcription and translation of the genetic code.
2. Some knowledge of the absorption and catabolism of sugars may be helpful but is not essential.

*Principles*

1. Small prokaryote cells have only room for a limited number of enzyme molecules.
2. Economy demands strict control over synthesis of enzymes and metabolites.
3. Enzyme induction is explained as a means whereby an enzyme or group of enzymes is produced only when its presence is an advantage.
4. Some groups of genes are controlled by operators and regulators. These groups are called operons.
5. Control of gene action in higher eukaryotes is likely to be different from the operon system.
6. A promoter is a site to which a DNA-dependent RNA polymerase becomes attached.

The concepts introduced here are important in biotechnology (see Chapter 21).

*Questions and answers*

**a** ***Explain how the Jacob and Monod hypothesis accounts for both types of mutation at both loci.***

Mutation at the r locus may produce an inactive repressor (giving a constitutive phenotype), or a repressor unable to interact with lactose (giving a permanently switched-off phenotype). Mutation at the operator locus may give an operator region unable to interact with the repressor (giving a constitutive phenotype), or damage to the operator region may prevent transcription (giving a permanently switched-off phenotype).

**b** ***Suggest how the Jacob and Monod hypothesis could be adapted to explain the repression of enzyme synthesis by a product of their action.***

A repressor could be produced which, instead of being inactivated by the substrate, is activated by the product of the metabolic pathway. In the presence of the amino acid in the medium, excess amino acid starts to accumulate in the cell. This forms a complex with repressor molecules which interact with the operator and switch off the whole operon. If the level of amino acid falls, the complex comes apart and the relevant structural genes are activated – a nice example of homeostasis. Teachers might at this point wish to review other mechanisms of control of metabolic pathways and of enzyme activity. mRNA is a labile substance and its rapid turnover presumably prevents the accumulation of excess protein by translation. Many metabolic pathways are subject to end-product (allosteric) inhibition. The accumulating produce interacts directly with an enzyme early in the reaction sequence and so slows its own production (more feedback!). End-product inhibition normally occurs when low levels of product are present and allows fine control of the pathway. The operon is only switched off by higher concentrations of product. It is important for students to understand that metabolic pathways often branch, so that a single substrate may be the raw material for two alternative reaction sequences which may be in competition.

**c** ***Explain the possible value of constitutive mutants of bacteria in industrial processes.***

The bacterium is likely to be cultured in order to produce an enzyme, a group of enzymes, or a metabolite. A constitutive mutant is likely to produce a higher yield. It is important to remember that in nature, and sometimes in an industrial fermentation, the mutant will fail because it is not competitive.

## STUDY ITEM

## 19.21 An experiment demonstrating changes in the ability of yeast to ferment galactose

*Assumptions*

**1** That students understand that yeast can respire anaerobically, with the production of carbon dioxide.
**2** That students have previous experience of simple respirometers.
**3** That they have some knowledge of mutation and enzyme induction.

*Principles*

1 Carbon dioxide output can be used as a measure of the rate of fermentation and hence of the levels of enzymes present.

2 Organisms are often unable to produce enzymes needed to utilize a substrate but will acquire the ability to use the substrate by enzyme induction.

*Questions and answers*

a ***Why were the respirometers flushed with nitrogen gas?***

In air the cells could carry out aerobic respiration and absorb oxygen. Nitrogen forces all the cells to undertake anaerobic respiration and questions about which type of respiration is occurring need not be raised.

b ***List assumptions which must be made before the data can be interpreted.***

Each respirometer contains comparable numbers of yeast cells.
Similar concentrations of glucose and galactose are present in the solutions both during the 24-hour pretreatment and during the experiment.
Factors such as temperature and mineral nutrition are similar in all treatments.

c ***Display the data graphically.***

Time should be on the horizontal axis and cumulative gas production on the vertical axis. Points should be shown using four different colours (or using crosses, dots, triangles, and squares). Four lines of best fit should be drawn and labelled.

d ***What indication is there that the thermobarometers have failed adequately to reflect temperature fluctuations in the experimental respirometers?***

Smooth linear increases in gas volume are not shown, and there is contraction in gas volume on three occasions, indicating that the respirometer may have cooled.

e ***Explain how the results supported both these hypotheses.***

Twenty-four hours of treatment with galactose have resulted in an ability of the yeast cells to ferment galactose. This is shown by comparing respirometers 2 and 4. Either hypothesis accounts for the difference between the results of these two respirometers.

f ***Suggest further experiments which might allow one or other of the hypotheses to be disproved.***

**1** The cultures could be repeatedly sampled during the pretreatment period and cell population and ability to ferment galactose compared. This might indicate that very many cells were acquiring the ability to ferment galactose at the same time – disproving the mutation hypothesis.

**2** Equal volumes of yeast suspension, grown for many cell generations without galactose, could be diluted and plated on both glucose and galactose agars. If the induction hypothesis were correct, one would expect that comparable numbers of colonies would grow but that the development would be slower on the galactose agar.

**3** If the mutation hypothesis were correct a few colonies only would develop on galactose agar, these being spontaneous mutants which had been selected.

g ***Does treatment with galactose affect the ability of yeast to use glucose?***

No. (Some students may suggest that the ability to use glucose is enhanced, but the difference between respirometers 1 and 3 is not significant.) □

## 19.3 Control of transcription in higher organisms

*Assumption*

1 Familiarity with the concepts concerning differentiation outlined in Chapter 15.

*Principles*

1 The nature of protein regulators of gene action in higher organisms is unknown, but inactive genetic material is condensed and packaged by protein.

2 Packaging may be a result of the genes being inactive, rather than a cause.

*Question and answer*

a ***Give two reasons why this evidence is inconclusive.***

**1** The two X chromosomes cannot be visually distinguished so it is not possible to identify the paternal X chromosome as the Barr body in a second patch.

**2** The evidence provided does not rule out the possibility that it is the condensed chromosome (the Barr body) which produces the local coat colour.

### Gene activity in polytene chromosomes

*Assumptions*

1 Familiarity with chromosome structure and behaviour during meiosis and mitosis.

2 Knowledge of autoradiography.

*Principles*

1 Most transcription of DNA occurs in the interphase nucleus where the chromosomes are too diffuse to be visualized. The polytene chromosomes of *Diptera* provide an opportunity to observe large aggregations of genes all working together.

2 This allows us to confirm that some regions of a chromosome are transcriptionally active while others are switched off or repressed.
3 Gene activity can be correlated with developmental events and stimulated by hormones but we do not understand how control is exercised over which genes become active.

The function of the polytene state has not been stressed in this section. Cells with polytene chromosomes tend to be secretory cells, and the production of many copies of the DNA presumably allows rapid production of mRNA from several identical DNA strands. The situation can be compared with the polyploidy found in adult mammalian liver cells.

*Question and answer*

b ***How could this prediction be tested experimentally?***

If a pulse of labelled uracil was provided for the tissue to incorporate into mRNA, this mRNA would be located on the chromosome puffs by autoradiography (see Chapter 18 of this *Guide*).

## 19.4 Cell differentiation as a population phenomenon

*Assumption*

1 That students are familiar with the possibility that animal tissues can be cultured.

*Principles*

1 The phenotype of the whole organism depends partly on the enzymes and other materials it can synthesize, and partly on the range of cell types it contains. It also depends on the degree and type of organization of cells and tissues.
2 Gene action must be understood in terms of the rate at which different genes or groups of genes are transcribed and translated – but development cannot be viewed narrowly in these terms.

## 19.5 Genes that influence metabolic reactions in humans

*Assumptions*

1 Familiarity with monohybrid inheritance.
2 Familiarity with the one gene–one polypeptide hypothesis.
3 Knowledge of the term essential amino acid.
4 Some knowledge of eye structure is helpful.

*Principles*

1 Mammals have more limited abilities to produce amino acids than do most microbes, but several can be synthesized by interconversion.
2 Excess amino acids in the diet must be deaminated and broken down by metabolic pathways.
3 Melanin is a metabolic product for which phenylalanine or tyrosine are the raw materials.
4 Higher eukaryotes illustrate the relationship between genes and enzymes just as micro-organisms do.

*Questions and answers*

**a** ***What is the genotype of each of the parents of a baby who excretes phenylpyruvic acid?***

Father is Pp, as is the mother. (Mutation of P to p is extremely rare, so this is unlikely to explain the situation.)

**b** ***If these same parents choose to have a second baby, what is the probability that it will be affected?***

One in four or 0.25 or 25 %.

**c** ***In* figure [S]93 *tyrosine is shown as a product of phenylalanine. Explain why pp individuals are not albino in phenotype.***

Tyrosine is also available from the diet.

## STUDY ITEM

## 19.51 The influence of thiamine on mutant tomatoes: an analysis of an experiment

*Assumption*

1 Knowledge of standard deviation.

*Principles*

1 If a mutant is unable to produce an important metabolite, it will die as soon as stores of the metabolite are exhausted.
2 Since the metabolite affects development in several ways, the mutant allele is pleiotropic (see section 19.6).
3 The metabolite will reverse the effects of the mutant allele. The degree of reversal is proportional to concentration until a saturation value is achieved.

This exercise could be carried out as a practical demonstration or a class or group experiment since thiamine-less tomato seeds and thiamine are readily available from suppliers.

The example provides a good opportunity to discuss the role of vitamins in metabolism. Thiamine is a vitamin for normal humans but not for normal tomatoes since they can synthesize it. Could the requirement for vitamin $B_1$ in vertebrates have arisen because it was usually present in excess in the diet so that the genes needed for its synthesis rarely had an advantage over alleles which resulted in lack of synthetic ability?

In mammals and birds, vitamin $B_1$ deficiency results in an accumulation of pyruvic acid in the body. Thiamine is used to synthesize acetyl-coenzyme A and deficiency partially blocks the link between glycolysis and the TCA cycle. Pyruvate in the blood produces the syndrome beri-beri which was investigated by Eijkman in chickens. Many students will have heard of Eijkman and beri-beri from 16+ courses.

*Questions and answers*

a ***Do the results shown in figures [S]94 and [S]95 indicate that the experiment was adequately controlled?***

Yes. The results were compared with normal phenotype.

b ***From your study of the graph in figure [S]94 what can you say about the relationships between the concentration of thiamine used, the formation of chlorophyll, and the growth of the plants as measured by fresh mass?***

Both fresh mass and chlorophyll formation are increased by increasing concentrations of thiamine up to about 100 (*N.B.* not 10) p.p.m. of thiamine. There is some suggestion that the effect of 900 p.p.m. of thiamine is inhibiting, particularly on fresh mass.

c ***Why were the concentrations given as the square roots of parts per million?***

To condense the horizontal axis. (It might be useful to discuss with the class whether the log of concentration could have been used).

d ***Why were the amounts of chlorophyll and the fresh mass given as percentages of the normal genotype?***

The percentage of untreated controls helps to reduce the impact of chance fluctuation in environmental variables such as temperature. It also shows that the mutant phenotype can be completely eliminated by an optimum concentration of thiamine.

e ***What was the point of measuring chlorophyll content as well as fresh mass?***

Slow growth and chlorosis are both prominent features of the phenotypes. (Students may speculate that low chlorophyll content is a direct cause of slow growth; there is no evidence for this in the data.)

f ***What statistical techniques would you use to assess the significance of the results?***

The standard deviation of each set of results would be shown on figure [S]95 as a vertical line passing through each point.

g ***How do you account for the fresh mass increase in the untreated mutant up to about the fourth week?***

By thiamine stored in the cotyledons.

h ***Thiamine-deficient mutant tomato seedlings grow normally if sprayed with a pyrimidine. Another substance, thiazole, is implicated in thiamine synthesis, but no growth occurs if mutant plants are sprayed only with this. In which enzyme-controlled step in the pathway that follows is there most likely to be a metabolic block?***

precursor 1 $\xrightarrow{A}$ pyrimidine $\searrow^{B}$

precursor 3 + precursor 4 $\longrightarrow$ thiamine

precursor 2 $\xrightarrow[D]{}$ thiazole $\nearrow_{C}$

☐ Step A.

**Practical investigation. *Practical guide* 5, investigation 19A, 'Dwarfism in peas'.**

## 19.6 Gene expression in heterozygotes

*Assumptions*

1 That the haploid and diploid states are well understood.

*Principles*

1 The heterozygous state can only be manifest if a diploid stage of the life cycle can be influenced by a gene and the phenotype observed in the diploid.

2 Deletion mutants would be expected to be recessive unless there were a dosage effect.

3 A heterozygote may contain the product of both the alleles, even when this does not apparently influence the obvious manifestations of the phenotype.

4 Heterozygotes sometimes show heterosis and this can be explained in a number of ways.

*Question and answer*

a ***Are hypotheses 1 and 2 as stated above mutually exclusive?***

No. One hypothesis could be true for some genes and the other hypothesis true for others.

**Practical investigation. *Practical guide* 5, investigation 19C, 'Gene expression in round and wrinkled peas (*Pisum sativum*)'.**

## 19.7 Pleiotropy

*Assumption*

1 An understanding of linkage.

*Principles*

1 The term gene can be used with a number of shades of meaning.

2 In earlier sections gene action has been considered in terms of the one gene–one polypeptide hypothesis. The case of sickle-cell anaemia illustrates very clearly that even where we understand very precisely the change in a polypeptide brought about by a single mutation, we cannot predict all the effects this may have on the phenotype of the whole organism.

*Question and answer*

a **_Design an experimental procedure which would demonstrate this._**

Flies can be shaken into a long glass tube which is plugged at both ends with sponge rubber. Half of the tube is blacked out with black paper. The flies in the unmasked end of the tube are counted at one-minute intervals after which the tube is disturbed and its position moved. The mask is shifted to the previously unmasked end. Wild type flies would be expected to accumulate in the illuminated part of the tube. Vestigial flies remain more randomly distributed. The tube must be too narrow for the flies to fly. An alternative method would be to offer flies a choice in a Y or T maze, one limb of which would be darkened.

## 19.8 Inheritance of comb shape in poultry

*Assumptions*

1 Familiarity with the 9:3:3:1 and 1:1:1:1 ratios as indication of dihybrid inheritance and independent assortment.

*Principles*

1 Several gene loci interact to influence the development of a complex structure like a comb.
2 Mutations at these loci can interact to produce a variety of phenotypes.

*Questions and answers*

a **_From this information write down all the genotypes which result in_**
**_1 the rose phenotype_**
**_2 the pea phenotype_**
**_3 the walnut phenotype._**

Rose – RRpp or Rrpp
Pea – rrPP or rrPp
Walnut – RRPP or RrPP or RRPp or RrPp.

b **_What do these data tell us about the way genes act?_**

The gene loci P/p and R/r show independent assortment. They must be carried on different chromosomes. They must act together to produce the comb since four comb shape phenotypes result from the combinations of the alleles of each gene.

c **_What does this tell us about gene action?_**

Many loci besides the pea and rose loci influence comb development, but only at these loci are there alleles which switch development in such a way as to produce discontinuous variation.

d **_Does this indicate that a single gene determines the development of the hands of humans or the paws of cats?_**

No. Many different genes must be involved, but the polydactylism gene has an allele which produces distinctive discontinuous variation.

## 19.9 Epistasis

*Assumptions*

1. That the allele concept is familiar.
2. That the one gene–one polypeptide hypothesis is understood.

*Principles*

1. In higher organisms, a metabolic block introduced by a mutation is usually lethal if homozygous. Pathways leading to the production of certain pigments are less likely to be lethal and are convenient to study.
2. Genes controlling stages early in a pathway are epistatic to those controlling later stages.
3. Dominance is a relationship between alleles but epistasis is a relationship between non-allelic genes influencing the same character.

Teachers may wish to consider examples involving a branched pathway or alleles which do not cause a complete block and result in a dilute phenotype (such as pink-eyed dilution in mice).

*Questions and answers*

**a** ***If an individual is of genotype QQRRssTT, can it produce either the red or the yellow pigment?***

No, since the third stage is blocked.

**b** ***What would be the colour of an individual of genotype QQRRSStt if allele t produced an ineffective enzyme?***

Yellow, since the final stage is blocked and the yellow metabolite 3 accumulates.

**c** ***What would be the colour of a doubly heterozygous individual QQRRSsTt?***

Red, assuming T is dominant to t and S dominant to s. Each stage in the pathway has a wild type allele controlling it.

**d** ***If a mutation at locus Q blocks the pathway would mutation at loci R, S, or T make a difference to the phenotype?***

No. All three genes control stages later than Q and the enzymes they produce have no substrate on which to work.

**e** ***To which genes is R epistatic?***

Genes S and T.

**Practical investigation. *Practical guide 5*, investigation 19B, 'The biochemistry of cyanogenesis in *Trifolium repens* (white clover)'.**

## STUDY ITEM

## 19.91 An unexpected result from breeding sweet peas

*Assumption*

1 An initial understanding of dihybrid inheritance and of the principles on which epistasis rests, as set out in section 19.9 of the *Study guide.*

*Principle*

1 Two different genotypes may be unable to produce a character because different stages in the process by which the character develops are blocked by mutation. A cross between the genotypes produces an $F_1$ in which all the stages are functional.

Pre-Mendelian biologists, including Darwin, were much fascinated by this phenomenon. It was called 'throwing back' or 'reversion to type'. Darwin used the recovery of the wild type among the progeny of highly inbred strains of pigeons, such as white fantails and pouters, as evidence that the strains had been selected by humans from wild ancestors.

*Questions and answers*

a ***Put forward a hypothesis to explain this result.***

One white variety is of genotype aaBB, the other of genotype AAbb. The cross produces the $F_1$ genotype AaBb. Both A and B are needed to complete the metabolic pathway producing flower pigment. Selfing of the $F_1$ produces a 9:7 ratio of coloured flowers to white flowers. The chi-squared test shows this to be in close agreement with the result obtained.

b ***Suggest a cross which would test your hypothesis and predict the outcome.***

Several valid answers are possible. For example, 1 in 7 of the white flowered $F_2$ would be a homozygous double mutant aabb and would yield white flowers when back-crossed with either parental type.

c ***Why is a result of this type important in commercial plant breeding?***

It cannot be assumed that because different varieties are phenotypically similar, they can be safely intercrossed and will produce offspring of the same phenotype as the parents. On the other hand, new phenotypes may unexpectedly arise if crosses are deliberately made, and some of these may be desirable to establish new varieties. □

## STUDY ITEM

## 19.92 Breeding brown mice from black and white parents

*Assumptions*

1 An initial understanding of dihybrid inheritance and of the principles on which epistasis rests as set out in section 19.9 of the *Study guide.*

2 A familiarity with the chi-squared test.

*Principle*

**1** Two different genotypes may be unable to produce a character because different stages in the process by which the character develops are blocked by mutation. A cross between the genotypes produces an $F_1$ in which all the stages are functional.

*Questions and answers*

**a** ***What are the possible genotypes of the two parents?***

There are two possibilities, shown in table 12.

| *Hypothesis* | *Black parent* | *Albino parent* |
|---|---|---|
| 1 | $c^+cb^+b$ | $ccb^+b$ |
| 2 | $c^+cb^+b$ | $ccbb$ |

**Table 12**

**b** ***Work out the expected number of offspring of each phenotype on the basis of each hypothesis you suggested in* a.**

See tables 13 and 14.

**Hypothesis 1**

| black | × | albino |
|---|---|---|
| $c^+cb^+b$ | | $ccb^+b$ |
| gametes $c^+b^+$, $c^+b$, $cb^+$, and $cb$ | | $cb^+$ and $cb$ |
| (frequency of each is $\frac{1}{4}$) | | (frequency of each is $\frac{1}{2}$) |

| | | *Gametes from albino parent* | |
|---|---|---|---|
| | | $\frac{1}{2}cb^+$ | $\frac{1}{2}cb$ |
| *Gametes from black parent* | $\frac{1}{4}c^+b^+$ | $\frac{1}{8}cc^+b^+b^+$ black | $\frac{1}{8}cc^+bb^+$ black |
| | $\frac{1}{4}c^+b$ | $\frac{1}{8}cc^+b^+b$ black | $\frac{1}{8}cc^+bb$ brown |
| | $\frac{1}{4}cb^+$ | $\frac{1}{8}ccb^+b^+$ albino | $\frac{1}{8}ccbb^+$ albino |
| | $\frac{1}{4}cb$ | $\frac{1}{8}ccb^+b$ albino | $\frac{1}{8}ccbb$ albino |

Ratio of phenotypes: 4 albino:3 black:1 brown.
Expected numbers: 47 albino, 35 black, 12 brown (to the nearest whole number).

**Table 13**

**Hypothesis 2**

| black | × | albino |
|---|---|---|
| $c^+cb^+b$ | | cc bb |
| gametes $c^+b^+$, $c^+b$, $cb^+$, and cb | | cb |
| (frequency of each is $\frac{1}{4}$) | | (frequency is 1) |

| | | *Gamete of albino parent* |
|---|---|---|
| | | cb |
| *Gametes from black parent* | $\frac{1}{4}c^+b^+$ | $\frac{1}{4}cc^+ bb^+$ black |
| | $\frac{1}{4}c^+b$ | $\frac{1}{4}cc^+$ bb brown |
| | $\frac{1}{4}cb^+$ | $\frac{1}{4}cc\,bb^+$ albino |
| | $\frac{1}{4}$cb | $\frac{1}{4}$cc bb albino |

Ratio of phenotypes: 2 albino: 1 black: 1 brown.
Expected numbers: 47 albino:24 black:24 brown (to the nearest whole number).

**Table 14**

c ***Calculate the value of chi-squared ($\chi^2$) for each hypothesis.***

Hypothesis 1 $\chi^2 = 0.364$
Hypothesis 2 $\chi^2 = 15.47$

d ***Using probability tables, determine whether any of the hypotheses can be rejected as unlikely.***

There are two degrees of freedom.
For hypothesis 1, p is between 0.75 and 0.9.
For hypothesis 2, p is much less than 0.01.
The second hypothesis can be rejected as unlikely.

e ***Suggest crosses which would test the hypotheses experimentally.***

Backcross one of the brown offspring to its albino parent. If any black young were produced, hypothesis 2 could finally be rejected and hypothesis 1 accepted. □

# PART II The *Practical guide*

The investigations related to this chapter appear in *Practical guide 5, Inheritance.*

## INVESTIGATION
## 19A Dwarfism in peas

### ITEMS NEEDED

Seeds, pea round and wrinkled, for example, Meteor (dwarf) and Pilot (tall)
Compost, potting

Gibberellic acid (GA) solution 30 ppm, 100 $cm^3$
Indol-3-yl ethanoic (acetic) acid solution 30 ppm, 100 $cm^3$
Sodium hypochlorite solution, 1 %
Water, distilled

Atomizer or refillable aerosol spray 3/class
Beakers 2/group
Cloth, small piece 1/group
Cotton thread
Forceps
Labels or marker pens
Pipettes, dropping 3/group
Plant pots or plastic cups with drainage holes cut in base
Ruler or tape measure 1/group
Sticks to support pea plants

(*Study guide* 19.3 'Control of transcription in higher organisms' – see 'Gene activity in polytene chromosomes'; and 19.5 'Genes that influence metabolic reactions in humans – see Study item 19.51 'The influence of thiamine on mutant tomatoes; an analysis of an experiment'.)

**Gibberellic acid solution**
2 $cm^3$ ethanol
0.1 g gibberellic acid (GA)
1 $dm^3$ water, distilled
Dissolve the GA in the ethanol. Add this slowly to half of the water and then make up to 1 $dm^3$. This gives a 100 p.p.m. solution. Dilute 30 $cm^3$ of this with 70 $cm^3$ of distilled water to give a working solution of 30 p.p.m. The 100 p.p.m. solution can be kept in a refrigerator for about 10 days at 1–4 °C.

**Indol-3-yl ethanoic (acetic) acid solution**
2 $cm^3$ ethanol
0.1 g indol-3-yl ethanoic (acetic) acid (IAA)
1 $dm^3$ water, distilled
Dissolve the IAA in the ethanol and add this slowly to 900 $cm^3$ of the water, stirring continuously. Warm this solution to 80 °C and keep at this temperature for 5 minutes. Make up the volume to 1 $dm^3$. This gives a 100 p.p.m. solution. Dilute 30 $cm^3$ of this with 70 $cm^3$ of distilled water to give a working solution of 30 p.p.m. The 100 p.p.m. solution can be kept in a refrigerator for about 10 days at 1–4 °C.

*Assumptions*

1 Uptake of each growth substance is possible through the epidermis and is similar in both varieties of plant.
2 That students are familiar with the concept that several genes may influence the same character, each gene acting by a different mechanism.
3 An understanding of the terms node and bract.

*Principles*

1 Dwarfism is under genetic control.
2 The effect of the homozygous recessive allele producing dwarfism can be reversed by an exogenously applied growth substance.

Germination of peas is a little unpredictable. Good results are usually obtained if the seed is well sterilized and soaked in very shallow clean water – with just one layer of seeds barely covered – so that anaerobic conditions cannot develop. Damaged seeds should be picked out before soaking as they are sources of infection. Once soaked and planted the compost must be well drained and kept just damp. If in doubt, let it dry out just a little more! Pots are superior to trays as they can be sorted into conveniently sized groups for treatment, however patchy the germination, and because they can be distributed round the class for measuring. It may be a good idea to start a few spare seeds in pure sawdust or vermiculite as transplants. The tall plants will need staking so stakes must of course be provided for them. Twigs from a hedge are ideal. They should be firmly in place before sowing the seeds as they will damage roots if pushed in later.

The pots should ideally be placed on a bright windowsill and will need to be watered with a measured volume of water, such as a 25 $cm^3$ specimen tube full each. Positions can be rotated to overcome the influence of light or temperature differences. Ten plants of each variety for each treatment give significant data.

Hormone treatment should be started as soon as possible. (It may be assumed that if a seed has not germinated after ten days that it never will.) Spraying twice per week produces excellent results after about three weeks. Ideally, a new batch of hormone solutions should be prepared twice during the experiment. To compare hormone treatments quantitatively, spraying is useless and drops of a defined size must be applied to the apex after carefully separating the protecting bracts.

Once the plants have been measured they can be planted in the greenhouse and their seed harvested for the following year. Sutton's purple podded variety is a good tall one as an alternative to Pilot. Dwarf varieties are common.

*Questions and answers*

**a** ***In what way precisely do the untreated (control) plants of each variety differ in phenotype?***

The most significant difference is in the elongation of the internodes.

**b** ***What effect or effects on growth have the two plant hormones used had on each variety?***

IAA may have some very slight effect but it is unlikely to be significant. The most significant effect will have been that gibberellic acid has produced elongation of the internodes of the dwarf variety without greatly changing the number of leaves.

**c** ***Can we assume that the allele t blocks the synthesis of one or both of the hormones?***

Many students will jump to the conclusion that t blocks the synthesis of gibberellin. It is unlikely that the plant could survive a complete blockage of the pathway producing gibberellin. An alternative hypothesis is that the mechanism in the plant which breaks down gibberellin has been enhanced by homozygous t. It can be pointed out to students that several dwarfing loci are known in maize and that different types of dwarf respond to hormones in different ways showing that the genetic control of growth is a very complex matter.

**d** ***Suggest other investigations which might be performed to further investigate the situation revealed by your results.***

Investigations which could be performed by the students themselves include:

**1** Trying both gibberellic acid and IAA together.
**2** Varying the quantity of gibberellic acid applied.
**3** Varying the age at which treatment was started and terminated.
**4** Seeing if one large dose was equivalent in effect to two or more smaller doses.
**5** Finding what was the optimum time interval between applications of hormone.
**6** Seeing if all dwarf pea varieties responded in a similar manner to a standardized application of gibberellin.

**e** ***It might be supposed that farmers would lose yield by growing dwarf varieties of plants such as peas or wheat. In fact the yield of seeds from a field of dwarf and of tall plants is similar or the dwarf crop yield may be greater. Do your observations suggest a reason for this?***

The leaf areas of a tall pea plant and a dwarf one of the same age are similar. The tall one may expend more energy in stem elongation and therefore have a smaller leaf area. The differences are almost entirely due to internodal length. Tall plants only have an advantage if there is competition for light in a mixed population. A stand of tall plants are much more likely to suffer wind damage. (Tall peas need very extensive supports such as sticks or wires and cannot be mechanically harvested.)

## INVESTIGATION
## 19B The biochemistry of cyanogenesis in *Trifolium repens* (white clover)

### ITEMS NEEDED

Clover plants of all four genotypes

Methyl benzene (toluene) (caution: dispense from a fume cupboard, avoid breathing vapour or contact with skin and eyes)
Sodium picrate papers
Water, distilled

Autoclave or pressure cooker 1/class
Forceps
Glass rods
Incubator at 25–35 °C 1/class
Labels or glass marking pens
Pipettes, dropping
Tubes, autoclavable
Tubes, specimen with stoppers

(*Study guide* 19.1 'Mutant complementation in *Neurospora*' – see Study item 19.11 'Mutations and metabolic pathways'; and 19.6 'Gene expression in heterozygotes'.)

**Sodium picrate papers**
The papers may be purchased ready prepared. The papers should be handled with care since they are poisonous. They are useless if they get damp or remain exposed to light.

*Assumptions*

**1** A knowledge of dihybrid inheritance.
**2** An understanding of enzymes and their properties.

*Principles*

**1** Biochemical characters are under genetic control – support for the 1-gene–1-polypeptide hypothesis.
**2** Genetic polymorphism exists in natural populations and this does not necessarily involve visible traits.

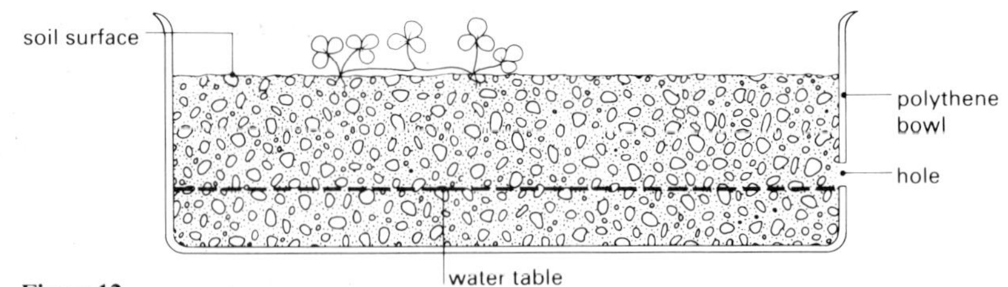


**Figure 12**
A satisfactory method of maintaining large clones of clover.

White clover grows quite well indoors if kept in a strong light but is then very prone to aphid attack. A highly satisfactory method of maintaining large clones (also useful for investigation 17A, 'Describing variation') is to bore a hole in the side of a plastic washing up bowl with a very hot cork borer. The hole should be about 1 cm in diameter and about a quarter of the way up the side of the bowl from the bottom. Fill the bowl three quarters full of soil or compost and plant rooted stolons of the plant one wishes to cultivate. The stolons are encouraged by the high rim of the container to creep around rooting inside to form a dense sward. Drought is minimized by the water table provided below the hole in the side of the container. (See *figure 12.*) The containers can be numbered with indelible felt pen in several places and kept in a sheltered spot outside until required. They should be weeded from time to time and watered in dry weather. Some nitrogen-free Knops solution or other nutrient solution is helpful in summer. Take care that seedlings of an unwanted genotype of clover do not become established by repotting once a year. Maintain at least four clones (G-E-, G-ee, ggE-, and ggee). (If they can be distinguished by white marks on the leaf or other features so much the better.) Wild clones can easily be established by cuttings in small pots indoors and transplanted into a polythene bowl if desired.

The story of cyanogenesis in *Trifolium* is not yet understood in full. Perhaps the easiest account to consult is that of Pusey in Darlington and Bradshaw (1966). *T. repens* is an allotetraploid behaving as a diploid. The use of the symbols G, g, E, and e for the genotypes in the explanation of the results rather oversimplifies the case, but helps to make it comprehensible. Several genes appear to be involved in the production of the substrate. Assuming a simple genetic model with two loci and two alleles at each locus, a teaching group of ten students can easily obtain enough data to estimate the allele frequency at the glucoside locus (G/g). Each can test three or four clumps using procedure A. They must be encouraged to spread out and not to use clumps growing close together or they may sample the same clone twice. *T. repens* has a multi-allelic incompatibility mechanism and is so highly outbreeding that cultivated populations in nearby pastures may influence the allele frequency.

It may be desirable to ask the students to record the extent to which the leaves of plants they collect are nibbled by animals and to tabulate this data alongside the frequency of the three cyanogenesis phenotypes. It will rapidly emerge that strong cyanogenesis does not confer perfect protection.

Leaves of the cherry laurel are so highly cyanogenic that they can be crushed in a bottle and used to kill butterflies. A small portion of the leaf has a spectacular effect on picrate paper. Bracken shows a polymorphism for the ability to produce cyanide analagous to that of clover. The whole question of the function of secondary metabolites in plants could be linked to this investigation and investigation 20B.

*Questions and answers*

**a** ***Describe carefully the function of each of the tubes you set up in part B, procedure 2.***

The tube containing only the autoclaved leaves tested the effectiveness of that procedure and confirmed that the enzyme had been denatured. The tube with only the leaves under investigation would confirm that, under these conditions of incubation, no cyanide release could occur without the addition of substrate. The control tube should have been a tube containing the methylbenzene (toluene) drop and picrate paper but no clover leaves. This was necessary for accurate colour comparison after incubation. Incubation conditions may conceivably have affected the colour of the paper. The fourth tube, containing an autoclaved, strongly cyanogenic leaf and a leaf under test, should have brought together enzyme and substrate and caused the release of cyanide if the tested leaf contained the enzyme.

**b** ***What genotypes would be possessed by:***
*1 **the strongly cyanogenic plants***
*2 **the weakly cyanogenic plants***
*3 **the acyanogenic plants?***

1 GGEE or GgEE or GGEe or GgEe
2 GGee or Ggee
3 ggee or ggEe or ggEE.

**c** ***Explain the results of part B of this investigation in terms of the allelic pairs G/g and E/e.***

The strongly cyanogenic leaves that were autoclaved had their enzyme destroyed and became functionally G-ee. The acyanogenic plant under test was of the enzyme-containing phenotype or lacked the enzyme. (It had already been shown to be gg by procedure A.) If the enzyme was present it would act on the glucoside from the autoclaved leaf and release cyanide. The test can thus distinguish ggee plants from ggEe and ggEE plants. (It is worth pointing out that only breeding could distinguish the homozygotes from the heterozygotes.)

**d** ***Methylbenzene (toluene) is a good solvent for lipids. It is effective in promoting the release of HCN from the strongly cyanogenic plants. Put forward a hypothesis to connect these two facts.***

The enzyme is separated from the glucoside by phospholipid membranes. (We know of no work that has compared the effectiveness of methylbenzene on strongly and weakly cyanogenic plants. This would form the basis of a project. Methylbenzene might not affect the weakly cyanogenic leaf, unless the escape of the glucoside from some membrane-bound structure in itself promoted hydrolysis.)

## INVESTIGATION

## 19C Gene expression in round and wrinkled peas *(Pisum sativum)*

### ITEMS NEEDED

Seeds, pea, round and wrinkled both dry and soaked

Agar, glucose-1-phosphate
Iodine in potassium iodide solution
Water, distilled

Balance 1/class
Beakers, large 2/group
Beakers, small 4/group
Centrifuge and centrifuge tubes 1/class
Electric grinder 1/class
Eyepiece graticule and stage micrometer, (optional)
Glass marking pen 1/group
Glass rod 1/group
Incubator at 30–35 °C 1/class
Microscope, monocular 1/group
Microscope slides and cover slips
Paper towel or blotting paper
Petri dish containing glucose-1-phosphate agar 2/group
Pipettes, dropping 3/group
Scalpel or razor blade 1/group
Test-tubes 2/group
Tile, white 1/group

(*Study guide* 19.1 'Mutant complementation in *Neurospora*' – see 'Mutations and metabolic pathways; and 19.9 'Epistasis'.)

**Glucose-1-phosphate agar**
4 g agar
1 g glucose-1-phosphate
200 cm$^3$ water, distilled
Boil the agar until it dissolves, then add the glucose-1-phosphate and stir thoroughly before pouring to a depth of no more than 2 mm in petri dishes. Sufficient for 30 petri dishes.
**Iodine in potassium iodide solution.** – See page 18.

*Pea varieties*
It is interesting that the majority of cultivars are wrinkled because the immature peas, that is, the ones one eats, are softer and sweeter than the round types. (Differences in sugar content of the mature dry seeds are marginal.) Round varieties are more frost-hardy so that early varieties are often round. A seed supplier's catalogue, for example, Suttons, will specify whether a variety is round or wrinkled. Round varieties include Meteor and Pilot. Wrinkled varieties include Little Marvel, Gradus, and Chieftain.

*Assumption*
1 Familiarity with the use of microscope and graticule.

*Principles*
1 Gene expression is not necessarily simple. Different characters may be influenced by a single gene, that is, pleiotropy may occur – though this term has not been used in the *Practical guide.*
2 It may be possible to link the different modes of expression of a gene to one basic biochemical mechanism.

To save time, the teacher or technician can weigh out dry seeds and put them to soak one or two days before the lesson, or a student could be asked to do this. The initial mass and type of pea can be recorded on each beaker and the final weighing done at the start of the practical session.

Some students find the starch grains difficult to distinguish or become confused by the appearance of the occasional complete cell packed with grains. Most find the exercise straightforward if they can be persuaded to take only a very small scraping of cotyledon. Round peas have smooth oval starch grains which tend to be larger than those of wrinkled peas. Wrinkled pea starch grains often look like a batch of scones and break up into small angular fragments (see *figure 13.*)

Interpretation of the measurements may be confused by the definition of a grain. Is a wrinkled grain a whole batch of fragments or do the angular bits constitute grains? Grains can be chosen at random by moving the slide slightly, with the eye away from the microscope, and measuring those that are under the graticule after this random movement.

The peas must be ground several times in a good coffee grinder. Care

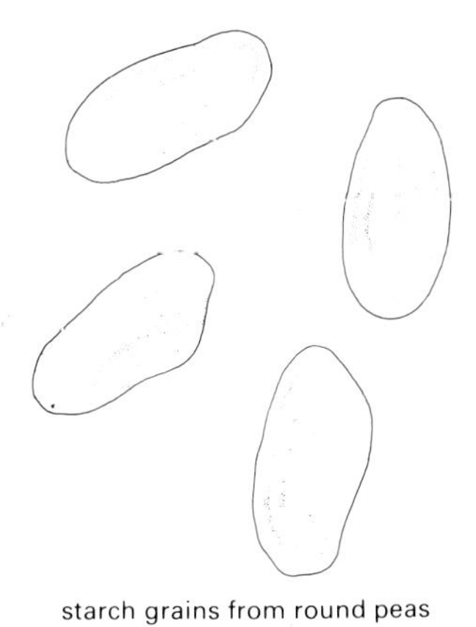

starch grains from round peas

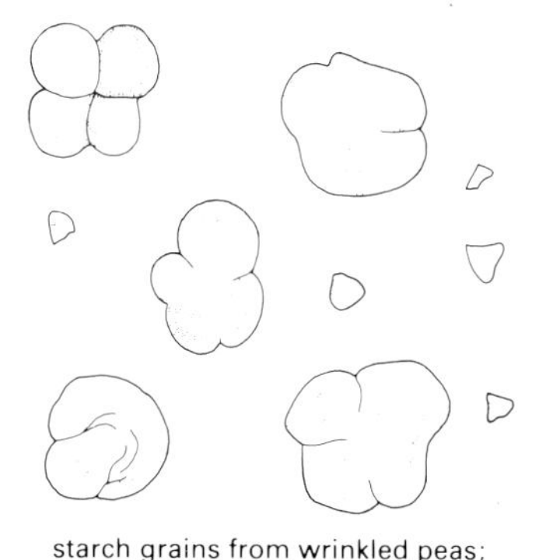

starch grains from wrinkled peas; *note angular fragments*

**Figure 13**
Starch grains in round and wrinkled peas.

should be taken to ensure that the motor is not burnt out and that the two varieties are ground to the same extent (the round peas are noticeably harder). Differences in colour of the pea flour may be observed, these are due to the yellow cotyledon/green cotyledon locus and should be ignored. The flour from wrinkled peas make a much more viscous suspension than does round pea flour and this should be pointed out to students as it correlates well with the other characters being observed. Preparing the enzyme extracts takes some time and students doing this will only be able to take a quick glance down a microscope to confirm the starch grain differences. Teachers may wish to grind the peas before the lesson.

When placed on the glucose-phosphate agar, the extract from wrinkled peas produces starch in the agar more rapidly than does the extract from round peas (provided all factors, including the time taken to prepare the two extracts, have been controlled). When stained with iodine solution the colour of the starch produced by wrinkled extracts is often more purple than that produced by round extracts.

The time taken for the extracts to act is variable and it would be a good idea to try one plate after only ten minutes incubation and to speed up the whole schedule of testing if the extracts are very active.

It is interesting that the less active enzyme seems to be produced by the dominant allele. Some students may raise this, and it can be pointed out that we are testing the diffusibility of the enzyme as well as its activity. As pea seeds have no endosperm, it is the cotyledons which are diploid which determine all the characters observed in the investigation. The testa is also diploid but of maternal genotype.

As an alternative to petri dishes of glucose-1-phosphate agar, it is possible to use small test-tubes with a little of the agar set at the bottom. Pea seed extract is pipetted onto this agar and washed away after a suitable time. Iodine solution is then added to the tube and the intensity of the purple colour in the agar recorded. This technique avoids the problem of the extract spreading out over the surface of the petri dish but test-tubes are tedious to fill with agar. Using small tubes is much more economical on materials, however, than using petri dishes.

*Questions and answers*

**a** ***Suggest how the difference in water uptake by the two seed types may be correlated with the appearance of the dry seed.***

Most students will anticipate that the wrinkled peas will absorb more water because of their greater surface/volume ratio or because there seems to be more room for it if the testa is stretched as the seed swells. When the flour or meal produced by grinding the two varieties is compared it is clear that the wrinkled tissue has a far greater water absorbing capacity which has nothing to do, directly, with the seed shape. Small fragmented starch grains are presumably more active in imbibing water than large, smooth, dense ones. Could the wrinkled appearance of the recessive phenotype be because it has lost more water while the seed was ripening?

b ***In what ways are the starch grains of the two seeds different? How could the differences help explain the differences between the appearances of the round and the wrinkled seeds?***

The starch grains of the wrinkled seeds are smaller and more irregular than those of the round seeds. (See *figure 13*.) It is likely that the wrinkled seeds' grains would pack together better as the seed loses water, allowing greater shrinkage of the seed. It is also probable that the starch grains absorb different amounts of water, mass for mass. (A loss in mass study of ripening seeds of two types would make a good topic for a project.)

c ***Would you be justified, on the evidence available to you, in saying that the differences in the structure of the starch grains and in the overall appearance of the seeds are due to differences in enzyme activity?***

No, but this hypothesis is consistent with the facts and seems likely. It could be verified by a detection of the enzyme differences in the young embryo, before the other characters made their appearance and by an investigation of the levels of the different enzymes in the heterozygote (which resembles the round type).

d ***What evidence would you need in order to suggest that the three phenotypic differences between round and wrinkled pea seeds were the result of the action of two different alleles of a single gene rather than a chance combination of independent traits?***

Strong evidence supporting this hypothesis would be given if the three characters always occurred together and were shown, by experiment to be inherited as a unit and never separated by genetic recombination, even if millions of progeny of heterozygotes were examined.

It is important not to stress a proper scientific caution in the interpretation of very limited data to such an extent that students fail to see the point of this investigation. It does demonstrate pleiotropy and the validity of the one gene–one polypeptide hypothesis.

# PART III Bibliography

DADAY, H. 'Gene frequencies in wild populations of *Trifolium repens*', *Heredity*, **8**, 1954, pp. 61–78.

DARLINGTON, C. D. and BRADSHAW, A. D. (Eds) *Teaching genetics.* Oliver & Boyd, 1966.

FINCHAM, J. R. S. *Microbial and molecular genetics.* 2nd edn. Hodder & Stoughton, 1976.

# CHAPTER 20 POPULATION GENETICS AND SELECTION

*A review of the chapter's aims and contents*

1 If two or more alleles exist at any locus their frequency in a mixed population is likely to change with time.
2 The Hardy–Weinberg model is explained and illustrated by reference to hypothetical and real human polymorphisms. Its use is to show that allele frequencies in ideal populations are stable and unaffected by dominance.
3 The idea of inbreeding as opposed to outbreeding is introduced and some examples of inbreeding and outbreeding organisms are mentioned.
4 The consequences of inbreeding and outbreeding for the level of heterozygosity in a population are discussed and the concept of a breeding system is explained.
5 The meaning of the term selection is explained and the operation of selection is illustrated by a number of cases of polymorphism. The possibility that some polymorphisms may be the result of chance (genetic drift) is mentioned.
6 Electrophoresis and other biochemical techniques can detect polymorphism and are sometimes able to distinguish homozygotes and heterozygotes.
7 Some of the principles and methods used in the artificial selection of animals and plants are explained. The possibility of positive eugenics is considered and the need for the conservation of genetic diversity explained.
8 The species concept is briefly reviewed and the need for reproductive isolation for the development of new varieties and species explained. Geographical, ecological, and behavioural isolation are briefly considered.

# PART I The *Study guide*

## 20.1 Changes in population of *Drosophila*

*Assumptions*

1 Familiarity with the terms point mutation and allele.
2 Some practical experience of handling and breeding fruit flies.
(Data of the sort discussed in this exercise are fairly easy to acquire and practical details are given in *Practical guide* 7, investigation 27A.)

*Principles*

1 A population homozygous for a mutation such as vestigial wing will thrive in pure culture, but as soon as the wild type allele is introduced it increases in frequency while the mutant declines. This is because the

probability of a mutant individual surviving and/or successfully reproducing is less than for a wild type individual.

2. There are several possible explanations of this differential success which are not mutually exclusive.
3. Different mutant phenotypes are disadvantageous for different reasons.
4. Not all mutants are disadvantageous.
5. The frequency of the mutant at the start of an experiment has little influence on the final outcome of selection.

*Questions and answers*

a ***Explain why the rapid reduction in the proportion of vestigial-winged flies cannot be the result of the dominance of the wild type allele.***

After the first generation, crosses would occur between heterozygotes to give the recessive homozygote. This means that the recessive phenotype would not disappear even if we made the unlikely assumption that mating between two homozygous mutants would never occur.

b ***Roughly how many generations have elapsed during the experiment?***

The duration of the experiment was about 22 months.

$$\text{The number of generations} \approx \frac{22 \times 28}{21} \approx 30$$

(If an organism with a life cycle of similar duration to humans had been used, this would mean an experiment lasting about 700 years.)

c ***Discuss and evaluate these hypotheses, suggesting experiments which would test some of them.***

**Hypothesis 4** is easily tested by setting up control population cages, under the same conditions, with only homozygous flies. The hypothesis seems unlikely since vestigial-winged flies are easy to culture.
**Hypothesis 5** has often been suggested as an explanation for the elimination of mutants. It cannot be valid unless it can be shown that attacks on mutants are more frequent or more damaging than could be predicted by chance. Fruit flies do not attack each other, so in this case, the explanation is invalid.
**Hypothesis 2** seems to be at least part of the explanation for the lack of success of the vestigial-winged flies. It could be tested by putting homozygous virgin females with two males, one of each type, in small food tubes for a short period, say 12 to 15 hours. The males would then be removed and the $F_1$ and $F_2$ offspring scored.
The other hypotheses are all likely, but more difficult to test. Female fertility can be assessed by placing a female in a Petri dish of banana agar darkened by carbon black and counting with a binocular microscope the eggs she lays. (She may continue to lay for three

weeks.) Interactions between larvae of two genotypes can be arranged by transferring them from pure cultures to a fresh food tube with a needle, but this changes the conditions of culture and newly hatched larvae are difficult to move in sufficient number. The decline in the proportion of vestigial-winged flies is unlikely to be the result of chance unless the proportion was very small. Chance might account for the minor fluctuations shown by the graphs.

There is a rich vein of project work to be exploited as a follow up to this section.

## 20.2 The Hardy–Weinberg model

*Assumptions*

1 An understanding of Mendelian inheritance.
2 An understanding of the principle that the probability of two events occurring together is the product of the probabilities of the events occurring alone.

*Principles*

1 If there are two alleles of a gene in a population, each with a frequency of 0.5, this frequency will be maintained in each subsequent generation. This is a special case of the Hardy–Weinberg equilibrium in which the frequencies of the two alleles are p and q.
2 A number of assumptions must be true if a population is to be in Hardy–Weinberg equilibrium. As these are unlikely to be true for natural populations the equilibrium will only apply to an ideal population. It must be applied to natural populations with great caution.

**Practical investigation. *Practical guide* 5, investigation 20A, 'Models of a gene pool'.**

The assumptions listed in the *Study guide* could be summarized by phrases such as 'no selection' and 'random mating'. Such phrases have been deliberately omitted. A series of phrases which can be memorized should be avoided, at least until some understanding has been achieved.

Students should appreciate that the assumptions are made with reference to a particular gene locus – all populations are subject to selection and mating is never random, but certain loci may have no influence on mating choice and may influence survival very little.

A lack of understanding of percentages often presents difficulty when substituting values into the Hardy–Weinberg equations. For example, if the recessive homozygote frequency is 45%

$$q^2 = 45\%$$

$$q^2 = 0.45 = \frac{45}{100}$$

$$q = \sqrt{0.45}$$

**STUDY ITEM**

## 20.21 Allele frequencies in human populations

*Principles*

1 The Hardy–Weinberg model can be used to estimate human allele frequencies, but this is inaccurate and should only be attempted when no direct method is available.

2 When the frequency of the recessive allele becomes low, most individuals carrying it are heterozygotes.

3 The model can be adapted for use when several alleles are present in the population.

*Questions and answers*

a ***What proportions of the population would be expected to have the genotypes RR, Rr, and rr, assuming the population is in Hardy–Weinberg equilibrium?***

The homozygous recessives (rr) are the 36 per cent who could not roll their tongues.

$\therefore q^2 = 0.36$

$\therefore q = \sqrt{0.36} = 0.6$

$p = 1 - 0.6 = 0.4$

The homozygous dominants (RR) have the frequency $p^2 = 0.4^2 = 0.16$
$\therefore$ 16 per cent are of the genotype RR
The heterozygotes (Rr) have the frequency $2pq = 2 \times 0.4 \times 0.6 = 0.48$
$\therefore$ 48 per cent are of the genotype Rr.

b ***If 4 per cent of the population is albino, what is the percentage of albinos in the population likely to be in 100 years' time?***

4 per cent.

c ***Give an important extra assumption that you have had to make in order to answer b. Do not question the facts you were given about the imaginary population but consider the special circumstances which must be fulfilled if the Hardy–Weinberg model is to be applicable.***

A number of valid answers are possible
**1** The factors which have produced the observed genotype frequency, for example, lack of selection, remain unchanged.
**2** There is no immigration into the island of peoples with a different genotype frequency.
**3** The mutation rate is negligible.
**4** Mating is not merely without reference to skin colour but is random. A tendency to marry relatives, for instance cousins, would cause a departure from the Hardy–Weinberg equilibrium. (This last point is unlikely to be mentioned by students unless they have thought about inbreeding and outbreeding and should perhaps be put to one side until this topic has been covered.)

d ***Complete table [S]29 for the imaginary population. It shows several frequencies of the albino allele.***

| *Frequency of the recessive allele* c | *Frequency of albinos* (cc) | *Frequency of heterozygotes* (Cc) |
|---|---|---|
| 0.9 (90 %) | 0.81 (81 %) | 0.18 (18 %) |
| 0.4 (40 %) | 0.16 (16 %) | 0.48 (48 %) |
| 0.1 (10 %) | 0.01 (1 %) | 0.18 (18 %) |
| 0.0001 (0.01 %) | $1 \times 10^{-8}$ (1 in 100 million) | 0.0002 (2 in 10 thousand) |

**Table 15**
Genotype frequencies in a human population

It would be a good use of computer facilities to write a programme which would compute diploid genotype frequencies from the frequency of the recessive allele. Two curves could be generated on screen, one showing frequency of recessive homozygotes against frequency of the recessive allele, the other showing frequency of heterozygotes against frequency of the recessive allele. Students should realize from the table that the rarer the recessive allele the lower is the ratio of homozygous recessives to heterozygotes.

The point can be made in class discussion that because we have thousands of gene loci each one of us is probably heterozygous for at least one rare recessive allele and will transmit this to half our children. Many anthropologists have suggested that this provides a function for the incest taboo since inbreeding increases the frequency of disadvantaged homozygotes. This is true of groups such as Anglo–Saxons which have been outbreeders. In some societies which have practised a greater degree of inbreeding in the past, for example Hindus who have arranged marriages within a particular caste for many generations, disadvantageous recessive mutants are less common. This is because inbreeding has increased the proportion of the recessive alleles which are present in homozygotes rather than heterozygotes, and it is the homozygotes that are selected against. People in Britain of Indian origin have lower frequencies of alleles for cystic fibrosis (an inherited degenerative condition affecting the pancreas and other glandular tissue) and phenylketonuria (see section 19.5) than do Anglo–Saxons.

**e** ***Does the frequency of the genotypes in the population depend on which of the alleles is dominant?***

No. When the frequency of either allele is 0.9, the other allele is 0.1 and the frequency of heterozygotes is 0.18. The frequency of genotypes depends on the truth of the assumptions stated and on nothing else.

**f** ***Tabulate all the diploid genotypes which are possible for these alleles and the corresponding phenotypes (blood groups).***

| *Genotypes* | *Blood group* |
|---|---|
| $I^AI^A$, $I^Ai$ | A |
| $I^BI^B$, $I^Bi$ | B |
| $I^AI^B$ | AB |
| ii | 0 |

**Table 16**
Blood group genotypes and phenotypes.

g ***Calculate the frequency of the blood groups A, B, AB, and O in the English population. Assume it to be an ideal population in Hardy–Weinberg equilibrium.***

Frequency of group A $= 0.26^2 + (0.26 \times 0.68 \times 2) = 0.42$ (42 %)
Frequency of group B $= 0.06^2 + (0.06 \times 0.68 \times 2) = 0.09$ (9 %)
Frequency of group AB $= 0.26 \times 0.06 \times 2 = 0.03$ (3 %)
Frequency of group O $= 0.68^2 = 0.46$ (46 %)
The calculations are to two significant figures. The subgroups $A_1$ and $A_2$ have been lumped together in these calculations.

h ***Are all of the assumptions that must be true for a population to be in Hardy–Weinberg equilibrium likely to be true for any real human population?***

No. Such factors as assortative mating, migration, selection, and small population size are likely to alter allele frequencies. (One must remember that only a few generations ago in Britain the population was much smaller and less mobile than it is now.) □

## 20.3 Inbreeding and outbreeding

*Assumption*

1 An understanding of the Hardy–Weinberg model.

*Principles*

1 Random mating is rare in natural populations, partly because assortative mating may occur but also for the more important reason that the effective choice of mates is restricted in various ways.
2 Different types of animals and plants have a variety of inbreeding and outbreeding mechanisms.
3 Inbreeding and outbreeding are extreme states in a continuous spectrum.

*Questions and answers*

a ***What disadvantages might a plant which is self-incompatible suffer in reproduction?***

If individuals are well spaced and the species has a low population density, cross-pollination may fail to occur and sexual reproduction will then be prevented. (Mention can be made of the need to grow two compatible varieties of cherries in the same garden to get a good fruit set).

b ***Are humans inbreeders or outbreeders?***

Self-fertilization is not possible and mating between close relatives is prevented by incest taboos. Large communities under modern conditions are highly outbreeding, for example, the Anglo-Saxon populations of Britain and the USA. Many human cultures either habitually inbreed or have done so until recently. In such cultures marriage between more distant relatives, for example cousins, is common. Inbreeding may be forced by small population size in isolated communities. Social class and religious practices may restrict the choice of marriage partner and so encourage inbreeding as, for example, in the Hindu caste system.

c ***What economic, cultural, and technological changes during the past 100 years may have changed the degree of outbreeding or inbreeding shown by the British population?***

Greater mobility of population, partly as a result of motorized transport, has increased the effective choice of mates. Migrations resulting from new employment opportunities have brought previously isolated groups into reproductive contact. A breakdown of the class structure has increased the choice of mates. Cousin marriage has become rarer. All these circumstances make the population more outbreeding.

## The consequences of inbreeding and outbreeding

*Assumption*

1 An understanding of the significance of chiasmata and linkage.

*Principles*

1 Inbreeding decreases the frequency of heterozygotes and increases the frequency of homozygotes.
2 Meiosis in a heterozygote can produce new gene combinations.
3 New combinations may have a greater chance of survival in a changing or fluctuating environment, but existing combinations are likely to be better adapted to a stable environment.
4 Linkage tends to preserve existing combinations of alleles at linked loci.
5 *Hymenoptera* and *Diptera* have special features which ensure a balance between the conservation of existing gene combinations and the generation of new combinations.
6 The mechanisms which control the degree of heterozygosity and the frequency of recombination in a species are collectively referred to as its breeding system.
7 Individuals able to exchange genetic material share a common gene pool.

*Questions and answers*

d ***Under what circumstances would this be an advantage?***

When the environment is likely to change little and the existing genotype adapts the organisms effectively for survival.

e ***When would it be a disadvantage?***

If the environment were changing or fluctuating or very patchy so that a variety of genotypes would be at an advantage at different times and in different locations within the habitat.

f ***How many haploid combinations could be produced by meiosis in an individual heterozygous at ten unlinked loci?***

Each additional locus where there is heterozygosity doubles the number of possible combinations. $2^{10} = 1024$.

g ***Does asexual reproduction change the variability of a species?***

Since both homozygotes and heterozygotes can reproduce asexually (as a result of mitosis) the average level of heterozygosity may not be changed by asexual reproduction. If large clones are produced, the choice of genetically different mates will be altered when the population is reproducing sexually. (The cycle of sexual and asexual reproduction shown by aphids could be mentioned in discussing answers to this question.)

## 20.4 Selection and genetic drift

*Principles*

1 Selection is a general term used to describe any set of circumstances which result in some genotypes being more successful in reproduction than others.

2 Increased chances of individual survival may often result in selection but it is reproductive success which is all-important.

3 Selection may lead to changes in allele frequency or it may stabilize existing allele frequencies.

*Questions and answers*

a ***In which hypothetical case is selection encouraging variation?***

Case 2.

b ***In which case is selection tending to maintain phenotypic stability?***

Case 1.

c ***Will this process necessarily lead to stable frequencies of genotypes?***

Not if the modal phenotype is the phenotype of heterozygotes, while the homozygous types at the extremes of the distribution are disadvantaged.

d ***In which case is the gene pool being changed unidirectionally?***

Case 3.

### STUDY ITEM

### 20.41 Interaction between strains of clover

*Principle*

1 A phenotype may confer advantages in competition for a resource (in this case light) and this will result in selection if the variation is inherited.

*Questions and answers*

a ***In what ways do the plants of the two strains differ in form under these conditions?***

Yarloop has longer petioles and larger leaves.

b ***What might be the consequences if plants of the two strains were grown close together?***

Yarloop might shade Tallarook.

c ***What were the advantages and disadvantages of thinning to a desired density rather than sowing seed at that density?***

If seedlings are thinned out some that remain may be damaged. The alternative is to sow an exact number of seeds and transplant seedlings to make up numbers if any seeds fail to germinate. This would be tedious and cause greater disturbance than thinning.

d ***What was the combined dry mass of both varieties when grown in mixed culture after 34 days? How does this combined mass compare with that of each variety when grown alone for the same time?***

Dry mass of pure Yarloop culture $= 350\,\mathrm{g\,m^{-2}}$
Dry mass of pure Tallarook culture $= 275\,\mathrm{g\,m^{-2}}$
Dry mass of the mixed culture $= 250 + 50 = 300\,\mathrm{g\,m^{-2}}$

The combined dry mass after 34 days is intermediate between that of pure Yarloop and pure Tallarook. (As growth proceeds the combined mass approaches that of pure Yarloop more and more closely.)

e ***Assuming no plants had died after 34 days, was the mean dry mass of each Yarloop plant greater in the mixed or in the pure culture? Explain your answer.***

In the mixed culture. The total Yarloop mass in the pure culture is greater, but there are twice as many plants.

f ***From* figure [S]105, *what was the relative light intensity 12 cm above the ground in the mixed culture after 62 days? Does this indicate that Tallarook is shaded in the mixed culture?***

The relative light intensity is zero. The graphs indicate that full daylight is between 8 and 10 cm above the ground in pure Tallarook cultures, in other words, this is the height of their upper leaves. This indicates that Tallarook will be completely shaded in mixed cultures, unless it has responded by growing taller in mixed culture than it does in pure culture.

g ***Can it be assumed that similar selection would operate under field conditions?***

No. Tallarook might be more resistant to drought or survive intense grazing pressure more easily. □

## 20.5 Polymorphism

*Assumptions*

1 That the concepts of selection and gene pool are understood.

2 A knowledge of the concept of heterosis is helpful (see *Study Guide II*, page 145).

*Principles*

1 Polymorphism is very common in natural populations.

2 Complex interaction of selective forces produces polymorphism.

3 The greater fitness of heterozygotes when compared with either homozygote is often responsible for polymorphism.

4 Polymorphism allows the intensity of selection to be observed.

There are a very large number of interesting examples of polymorphism. Practical investigation 20B could be used to introduce or reinforce the topic. Other examples which might be used for practical investigations or observed during field work include:
The ray-floret polymorphism of *Senecis vulgaris* referred to on page 70; the grey colour–orange colour polymorphism of woodlice, *e.g. Porcellio scaber*; and heterostyly in *Primula vulgaris.*

*Cepaea* populations may be observed at first hand in many parts of Britain and museum collections of the shells are useful. Practical investigation 17A mentions other polymorphisms in humans and white clover.

*Questions and answers*

**a** ***What precautions would need to be taken when marking and releasing the snails?***

The mark should be in a position which did not change the appearance of the shell. It should be harmless and the snails should be released at random.

**b** ***Under what circumstances might the ability to absorb radiant energy be an advantage and when might it be a disadvantage?***

In a diurnal animal, a form which absorbed radiant heat would warm up and become active earlier in the morning. In the case of *Cepaea* pale forms may reflect sunlight during hot weather and so are protected from overheating. (This effect may play a part in some cases of melanism.)

**c** ***The locus controlling the presence or absence of bands in* Cepaea *is closely linked to the locus controlling the background colour of the shell. How is this linkage an advantage to a population of snails?***

Favourable combinations of colour and banding which arise by recombination will persist and can be selected.

**d** ***Malaria has been eliminated in many parts of the World where it was once common. What will happen to the sickle-cell anaemia polymorphism in an area where malaria has ceased to be common?***

The $Hb^s$ allele will confer a disadvantage and will be reduced to a frequency where elimination is balanced by new mutation. (The existence of the allele in populations of African origin in the New World can be discussed.)

**Practical investigation. *Practical guide 5*, investigation 20B, 'Cyanogenesis and selection in *Trifolium repens* (white clover)'.**

**STUDY ITEM**

20.51 Polymorphism in the peppered moth (*Biston betularia*)

*Questions and answers*

a ***In which areas is* typica *most frequent?***

The west.

b ***Explain this distribution of* typica.**

The prevailing winds carry pollution from west to east, and there is little industry in the south-west peninsula.

c ***What ratio of offspring would be expected in the $F_2$ of a cross between a homozygous* carbonaria *and a homozygous* typica*?***

3 *carbonaria*: 1 *typica*.

d ***If such selection took place in the wild, what would be expected to happen to the* typica *allele frequency?***

It would decrease.

e ***What inference can be drawn from table [S]30?***

The morphs are selected by predation. Those that do not match their background are taken.

f ***Does the photograph confirm that predation is an important influence in maintaining the polymorphism of the peppered moth?***

Yes.

g ***Suppose 99 per cent of moths in an area are* typica*, 1 per cent are* carbonaria*, and the* insularia *allele is absent from the local population. Calculate the expected frequency of homozygous and heterozygous* carbonaria.**

Frequency of *typica* allele

$$= \sqrt{0.99}$$
$$= 0.995$$
$$p + q = 1$$
$$\therefore p + 0.995 = 1$$
$$\therefore p = 0.005$$
$$\therefore p^2 = 0.000\,025$$ = frequency of *carbonaria* homozygotes

$2pq = 2 \times 0.99 \times 0.01 = 0.02$ = frequency of heterozygotes.

**h** ***Is it likely that* carbonaria *individuals captured at a period before this form became common would be homozygotes?***

No.

**i** ***What changes in the distribution of industrial melanic forms would be expected as a result of these developments?***

The frequency of melanic forms would be expected to decline in Britain and increase in Norway.

☐

**STUDY ITEM**

**20.52** Evolution of Warfarin resistance

*Principle*

**1** When the elimination of a pest is attempted this results in new selective forces which may result in the evolution of the pest.

*Questions and answers*

**a** ***Explain how a small Warfarin-resistant population might have developed initially?***

Mutation.

**b** ***How would the fact that resistance is determined by an allele dominant over the pre-existing allele affect the increase in numbers and the spread of the resistant population?***

All individuals carrying the allele will show the phenotype so the numbers will increase more rapidly and the population of resistant rats spread more rapidly when compared to a hypothetical case where the resistant allele was recessive. Rates of selection for dominant and recessive alleles will show the greatest difference when the mutant is rare, especially in outbreeding populations.

**c** ***Account for the rate of spread of resistance to Warfarin between 1967 and 1970.***

The rate is uniform and is probably determined by the distance that rats may be likely to travel as they disperse.

**d** ***Studies have shown that resistant rats rarely constitute more than 50 per cent of the population. Put forward some hypotheses to explain this.***

**1** Heterozygotes are fitter than resistant homozygotes.
**2** The resistance to the poison is outweighed by the disadvantage of needing more vitamin K and this limits the frequency of the resistance allele.
**3** Poison is used only intermittently. It eliminates recessive homozygotes but these reappear as a result of breeding by resistant heterozygotes.

e ***Explain the likely implications of this genetic situation for the selection of the population in the central area of the map.***

The use of Warfarin might well decline and this would result in a decrease in frequency of the resistant allele. If Warfarin use continued, there might be the possibility of selection of resistant forms which did not require such a large amount of vitamin K as others. □

### Alcohol dehydrogenase in *Drosophila* – polymorphism at the molecular level

*Principles*

1 Electrophoresis can be used to identify gene products, even when masked by a dominant allele in a heterozygote.
2 Such techniques allow allele frequencies to be determined without the use of the Hardy–Weinberg model.
3 The same techniques are important in the detection of disadvantageous human mutants.

*Questions and answers*

e ***Examine* figure [S]110 *and explain why a heterozygote results in three bands and a homozygote in only one band.***

Since the enzyme forms dimers, a heterozygote can produce three types of dimer: S + S, S + F, and F + F. A homozygote can produce only one dimer, either S + S or F + F.

f ***Count the number of flies of each phenotype shown in* figure [S]110 *and calculate the allele frequency. Is the sample large enough for the frequencies to provide estimates of the true allele frequencies?***

The student must remember that each homozygote has two identical alleles.

$$\text{Frequency of A} = \frac{18}{32} \times 100 = 56.3\ \%$$

$$\text{Frequency of B} = \frac{14}{32} \times 100 = 43.7\ \%$$

The sample is too small for the frequencies to be reliable estimates.

g ***Why might the ability to produce two slightly different types of alcohol dehydrogenase be of advantage:***
***1 for an individual fly***
***2 for a fly population?***

Each version of the enzyme may work most efficiently under slightly different conditions.
**1** A heterozygous fly might therefore operate efficiently under a wider range of conditions than either homozygote.
**2** A polymorphic population would be able to adapt to changes in the environment.

**h** ***What is the probability of such parents producing a healthy child?***

The parents are proven heterozygotes since they have already produced a recessive homozygote. The chance of producing a healthy child is 0.75 (75 per cent) but one would expect two-thirds of the healthy offspring to be carriers.

## 20.6 Artificial selection

*Assumptions*

1. An understanding of the inheritance of continuous variation.
2. An understanding of the genetic consequences of inbreeding.

*Principles*

1. Domestication changes the environment of previously wild organisms and this results in selection of genotypes which were previously disadvantaged.
2. Artificial selection is not very different in its operation or result from natural selection.
3. Selection may be consciously carried out by the farmer or it may be the by-product of agricultural practice.
4. Very large changes in phenotype have been produced by many generations of artificial selection.
5. Artificial insemination and the storage of semen have provided new opportunities for efficient, selective breeding because progeny testing on a large scale can take place.
6. It may be important to prevent inbreeding during a programme of livestock improvement if variability is to be maintained.

*Questions and answers*

**a** ***Why would a geneticist regard this as undesirable?***

It would lead to inbreeding. The next generation would be much more closely related than before, and the gene pool would decrease in size. Rare disadvantageous recessive mutants might be carried by one or more of the bulls and these would become common in future generations.

**b** ***Ignoring aesthetic or moral objections, what reservations might a geneticist have about this idea?***

The same objections apply as those given in answer **a** above. In addition, there is uncertainty about the extent to which human qualities such as good sportsmanship vary as a result of inheritance rather than as a result of upbringing. It is not possible to produce a universally agreed list of qualities which would be desirable in those men to be selected as semen donors.

**STUDY ITEM**

## 20.61 Inheritance of resistance to tobacco mosaic virus (TMV)

*Assumptions*

1 An elementary knowledge of viruses.
2 An understanding of multiple alleles.

*Principles*

1 Breeding for disease resistance is an important aim of plant breeders.
2 Mutation of a pathogen can overcome resistance. Precautions should be taken to minimize the chance of pathogen evolution.

*Questions and answers*

**a** ***Why were cuttings of the plants tested for resistance rather than the plants themselves?***

Cuttings grow as a result of mitosis and are the same genotype as the parent plant. Testing a cutting for resistance avoids harming the plant itself if it turns out to be susceptible.

**b** ***Explain this result and give the genotypes of the two parents.***

Plant B must carry TM–$2^2$ since it is resistant to both virus strains. Plant A must be the homozygous recessive tm–2 tm–2 because both virus strains infect it.
Half the offspring carry TM–$2^2$ as they are resistant to both virus strains. The remaining offspring must carry TM–2 since they resist only strain 0. TM–2 must have come from plant B.
Plant A is of genotype tm–2 tm–2. Plant B is of genotype TM–2 TM–$2^2$.

**c** ***If resistance of the type discussed above was first discovered in a wild tomato species of no commercial value, suggest how a plant breeder might incorporate the newly discovered character into an existing commercial variety.***

The breeder could produce an $F_1$ hybrid and backcross to the commercial variety. Resistant plants showing as many good qualities as possible would be selected from the backcross progeny and then backcrossed again. Repeated backcrossing (usually at least five times) with similar selection would eventually result in a strain similar to the original commercial variety but incorporating the allele for resistance. As this is dominant, two selfing generations would need to follow the backcrossing to select a family true-breeding for resistance.

**d** ***Explain the reason for this precaution.***

If conditions favour multiplication and spread of the pathogen, the chance that a mutant pathogen able to overcome the resistance of the host will spread is increased should such a mutant appear.

e ***Explain why it is desirable for a plant to be resistant at more than one locus.***

Each resistance gene probably operates by a different mechanism. The chance of a pathogen undergoing two mutations at the same time, so that it can overcome the double resistance, is very small. □

### Conservation of gene pools

*Principles*

1. Selective breeding has been highly successful, but it has resulted in modern varieties having an impoverished gene pool with little variability to provide raw material for future improvements.
2. Traditional breeds of stock and varieties of crop plant have been developed in particular regions and cultivated for long periods of time.
3. Each variety has a unique gene pool from which valuable qualities may be selected and incorporated into modern varieties.
4. Some traditional varieties compare favourably with modern breeds but have become nearly extinct through the vagaries of fashion. Such varieties may be improved by selection and increase in popularity in the future.
5. Apart from their ornamental value, traditional varieties may be sources of valuable alleles which can be incorporated into commercial strains. They may become valuable as an alternative to existing commercial strains, perhaps under changed agricultural conditions.

(It should be emphasized to students that mutation, either spontaneous or induced, is unlikely to make good this deficiency, partly because most mutation is deleterious, but also because desirable qualities often depend on combinations of particular alleles of several genes.)

It may be desirable to use the topic of $F_1$ hybrids to revise sex linkage by referring to the use of sex-linked genes for plumage colour in the poultry industry.

c ***What is likely to be the result of maintaining a small population of a rare breed of animal or plant (or a population of a rare wild species such as the mountain gorilla) for many generations?***

Unless great care is taken, inbreeding will result in loss of variability and genetic drift may result in undesirable characters becoming fixed in the population.

## 20.7 Reproductive isolation

*Assumptions*

1. That students are aware of the existence of chromosome mutations and the disturbance in meiosis which may result in hybrids between two karyotypes.
2. That students understand continuous and discontinuous variation and also the concept that many loci each have a small and cumulative influence in producing continuous variation.

*Principles*

1 A species is often defined as a group which will not breed successfully with other groups. Very distinctive varieties may be able to mate successfully while apparently very similar forms may be unable to mate.

2 When reproductive behaviour is unknown, species must be defined by morphological or other criteria.

3 Distinct varieties, whether domesticated or occurring in the wild, must have a unique gene pool which is separated from other gene pools by some isolation mechanism.

4 The most obvious isolation mechanism is geographical.

5 If geographical isolation breaks down:

**1** Forms may merge and develop a common gene pool.

**2** Differences in structure, behaviour, or karyotype may have arisen which result in reproductive isolation.

**3** Hybridization may be possible, but very disadvantageous, in which case there will be selection pressure for the evolution of isolation mechanisms.

The important example of the Galápagos Islands has been mentioned, but has not been discussed in detail since it has been well reviewed in many student texts.

## STUDY ITEM

## 20.71 An isolated population of house mice

*Principles*

1 An isolated population will evolve independently of other populations either as a result of selection or gene drift.

2 Extreme habitats may result in very large selection pressures.

3 Islands have empty niches that poorly adapted species may colonize in the absence of competition.

*Questions and answers*

**a** ***Assuming dominance, how would the number of heterozygotes be determined?***

By test crosses to known recessive homozygotes, or by electrophoresis.

**b** ***Were the mainland and island populations both polymorphic?***

Yes. Allele frequencies are too high to be maintained by mutation.

**c** ***Advance two possible hypotheses to explain why the number of heterozygotes observed in the autumn population on Skokholm was less than expected.***

**1** They suffer a selective disadvantage during the summer.

**2** The population is small and the number is less than expected as a result of chance (gene drift).

**d** ***Suggest two possible reasons for the different frequency of the mutant allele in the Skokholm mice compared with the mainland mice.***

**1** Chance (genetic drift).
**2** The mutant has a greater selective advantage on the island.

**e** ***Suggest why house mice were able to survive in the absence of human habitation on Skokholm but were unable to do so on St Kilda.***

Both islands offer rather unfavourable habitats. On St Kilda competition for food (and possibly nest sites) from the field mice would explain the extinction of the house mice. The 'small omnivorous rodent niche' was full on St Kilda but empty on Skokholm. □

## STUDY ITEM

### 20.72 Isolation in North American crickets

*Principles*

**1** Closely related species in the same area occupy different habitats.
**2** Differences in courtship behaviour may result in or reinforce isolation.

*Questions and answers*

**a** ***Is the northern race of* G. fultoni *likely to compete with* G. vernalis*? Explain your answer.***

No, because they live in different habitats.

**b** ***Is the southern race of* G. fultoni *likely to compete with* G. vernalis*? Explain your answer.***

No. The two forms have a similar habitat preference but are geographically separate.

**c** ***Use the information provided above to suggest two reasons why hybrids are rare in the field.***

**1** They occur in different habitats and so will only meet at the margins of woods and fields.
**2** The females may only respond to the song of males of their own species.

**d** ***Does this information support your hypothesis produced in answer to c?***

Yes, since it suggests that species that are likely to meet have distinctive songs.

**e** ***What lesson should this scattergraph provide for taxonomists?***

That the behaviour and ecology of organisms should be considered when they are classified into species and not merely their morphology. □

# PART II The *Practical guide*

The investigations related to this chapter appear in *Practical guide 5, Inheritance*.

## INVESTIGATION 20A Models of a gene pool

### ITEMS NEEDED

Beads or equivalent of one colour 500/group
Beads or equivalent of a second colour 500/group
Containers, e.g. plastic beakers or boxes 6/group
(2 to store beads and 4 for the model)
Labels
Microcomputer and appropriate software (optional)

(*Study guide* 20.2 'The Hardy–Weinberg model'.)

*Assumption*

1 A knowledge of monohybrid inheritance, meiosis, and $\chi^2$ test.

*Principles*

1 The frequencies of genotypes in future generations may be predicted by a physical or a computer model, provided that certain assumptions form part of the model.
2 The model helps to demonstrate the importance of chance, and hence of population size, in the maintenance of allele frequencies.
3 It can be used to predict the effects of population size and selection in natural populations.

The instructions have deliberately been kept fairly simple so that the model could be used at a fairly early stage in the students' exploration of population genetics. It might be used as a lead into the Hardy–Weinberg equation or as a reinforcement after this equation had been studied.

The ideas of a breeding system and of a gene pool should come across from the investigations, but these ideas will need to be further refined and reinforced – see *Study guide II*, sections 20.2, 'The Hardy–Weinberg model' and 20.3, 'Inbreeding and outbreeding'.

All but the most mathematically gifted will benefit from some work with beads before they use a computer model – this must not be allowed to become a magic toy. Indications are given in the *Practical guide* of ways in which both types of model could be developed. Mutation, immigration, and other factors could of course be investigated also. The bead model becomes very tedious after more than about three generations and if a computer is not available, it would be wise to divide up the class in advance so that some groups are working with large populations and others with small ones.

It would be quite expensive to provide 1000 beads for each individual in a large class. A good alternative is to use dried peas or haricot beans that have been dyed with suitable stains such as Sainsbury's red food colouring and methylene blue. They swell slightly in the dyeing process so it is unwise to use dyed and undyed as the colour contrast – there must be no systematic variation in size or shape of the two colours of object used or bias will result. Large trays, for example cut-down cardboard cartons, may be useful to help sort out the colours between generations.

A variety of computer simulations is available. The more complex simulations should perhaps be avoided at first until the concept of gene pool and selection are thoroughly understood.

Changes in the hardware and software in common use are likely to occur, so specific references to software have been avoided. Fairly broad hints are given of what might be expected of good software. The possibilities are limitless, e.g. models which allow three or more alleles to segregate or the pressure of selection to change in a cyclical manner–simulating seasonal change.

*Questions and answers*
The questions have been deliberately interspersed in the procedure to avoid the task becoming mechanical and boredom setting in.

**a** ***Why would you expect the pairs of beads to be in the ratio of 1 red, red (AA): 2 red, blue (Aa): 1 blue, blue (aa)?***

Because the frequency of A and a gametes are equal, that is, 0.5 or $\frac{1}{2}$ of each, the chance of two red beads being chosen is $0.5 \times 0.5$, that is, 0.25 or $\frac{1}{4}$. There is a similar chance of two blue beads being chosen. The chance of red meeting blue is also $0.5 \times 0.5$, that is, 0.25, but there is a similar chance of blue meeting red. These two events are indistinguishable so that the total chance of getting one red and one blue bead, that is, a heterozygote, is 0.5 or $\frac{1}{2}$. This gives a 1:2:1 ratio of red red, red blue, and blue blue.

**b** ***Why would you be surprised if the actual result was exactly 1:2:1?***

Because chance is determining the selection of each pair of beads (just as it determines the fusion of particular pairs of gametes in 'real' reproduction).

**c** ***Use the $\chi^2$ test to see if your actual result departs significantly from the 1:2:1 ratio expected.***

There are clearly two degrees of freedom and the calculation should be very straightforward. If a computer is available it might be best to wait until the data of the whole class were recorded and then compute $\chi^2$ for every set of results. Individuals or pairs of students could type in each result as it becomes available and the $\chi^2$ program run when everyone had finished.

**d** ***Does the ratio of diploid genotypes in generation 2 differ significantly from a ratio of 1 red, red (AA): 2 red, blue (Aa): 1 blue, blue (aa)?***

One would not expect it to as the allele frequencies are each about 0.5 and the population fairly large.

**e** ***Considering the results as a whole, in how many cases did a result depart significantly from a 1:2:1 ratio?***

If several students were at work one might expect some departures as a result of mistakes in counting or through chance.

**f** ***Does a significant departure mean that bias has been introduced***

***(for example, by failing to mix the beads properly, or by miscounting) or would some significant departures occur by chance?***

This is a good test of the students' understanding of the concept of probability. If the experiment is repeated many times improbable events, that is, significant departures, are bound to be encountered.

g ***In the model, we have used a fairly large population (100 individuals and 800 gametes). Would the ratios produced have been very different if we had varied the total size of the population between certain limits, for instance 50 and 200, as it moved from one generation to the next?***

No – unless populations became so small that chance deviations became very significant. The population is created by withdrawing 'individuals' from the pool of gametes container. The ratio of genotypes expected depends not on the ultimate size of the population but on the probability of getting certain combinations of beads at each 'fertilization'.

h ***Provided that homozygotes always produced gametes of one colour and heterozygotes produced equal numbers of the two colours, would the number of gametes produced by each individual in the model have affected the ratio?***

No – as long as there is no selection, that is, the total number of gametes (of any colour) contributed by each individual is the same. (Many bead models allow each individual to produce just 2 gametes. A really sophisticated model would allow the number of gametes to vary within wide limits by chance and the ratio would still be unaffected. Only if selection caused one type to produce systematically fewer gametes than another would the frequency of alleles start to shift away from that in the founding population.)

## INVESTIGATION
## 20B Cyanogenesis and selection in *Trifolium repens* (white clover)

### ITEMS NEEDED

Clover plants, 15 cyanogenic and 15 acyanogenic
Young plants growing individually are required (see PG5 page 71) About 30/group
Slugs or snails (not recently fed) 10+/group

Sodium picrate papers 4/group

Forceps 1/group
Plastic bag, large (or several large plastic boxes with lids) 1/group
Refrigerator 1/class
Scissors 1/group
Spirit based felt pen 1/group
Thermometers 2/class
*(continued)*

(*Study guide* 20.5 'Polymorphism'.)

**Sodium picrate papers.** See page 117.

*Assumptions*

1 That students are already familiar with the genetics and biochemistry of cyanogenesis (see investigation 19B).
2 Some previous experience of the use of choice chambers is an advantage.

*Principles*

1 Animals may act as selective agents favouring the evolution of plants which are distasteful.
2 Frost can act as a selective agent.
3 A polymorphism may be the result of a balance between two opposed forces of selection which favour different genotypes.

## ITEMS NEEDED *(continued)*

Tubes, specimen, large, with stopper 4/group
Tubes, specimen, small (or small test tubes) 4/group

For part A, the teacher may wish to establish the clover cuttings well in advance of the experiment and present the pots to the class ready for use. If three inch plastic flowerpots are used and the stolons regularly trimmed, the plants can survive for years – but they must be well watered. Details of a procedure for the mass cultivation of clover clones are given on page 118. It is much easier to grow fresh cuttings each year from such clones. Clover takes a good time to become established from seed and it would be necessary to test each plant raised from seed to be sure of its degree of cyanogenesis.

The grey slug *Agriolimax reticulatus*, which is common in gardens and produces a milky mucus, is satisfactory. Other species of mollusc may be tried. Significant differences in the preference shown by different species have been discovered in student project work. There has been considerable controversy among research workers about the extent of selective predation. Hungry slugs, starved for 48 hours in a box with moist soil, may eat a good deal but are they likely to be as selective? Results obtained in this work are often tantalizing – the mollusc may be a Darwinian or it may not! It should be quite clear to pupils that they are not simply confirming a well understood scientific phenomenon – that full time research workers have obtained contradictory results in this type of experiment and that the possibilities for project work are extensive. It is puzzling that the petioles always lack the ability to produce cyanide since if they are cut through, the plant is presumably more disadvantaged than if the leaf is partly eaten.

Dead or yellowing or wilting leaves are likely to be acyanogenic whatever their genotype so that experiments with harvested leaves are unlikely to give evidence for selection unless the slugs feed before the leaves start to die.

Students have been asked to record leaflets rather than leaves since this increases the number of choices recorded and slightly reduces the problem of equating a slightly nibbled leaf with a totally consumed one.

It must not be forgotten, when discussing this work, that the secondary metabolism involved in cyanogenesis may have some unknown function unrelated to selection by frost or slugs. It is also important to realize that both influences might be far more important as selective agents at the seedling stage of the life cycle, when the plant is vulnerable, but difficult to test with picrate paper without destroying it.

For students who find mathematics difficult, it may be easier to arrange for there to be equal numbers of the two types of leaves at the start. It is then easier to work out the expected number of nibbled leaflets, on the null hypothesis that chance alone determined selection.

Part B of the investigation should offer few problems if everything is kept dry and the only water is a small volume at the bottom of the inner tube. Assessment of frost damage to the leaves is highly subjective and treatment of whole seedlings or runners might be tried since it would then be possible to grow them and measure growth rate and survival. There would not be time for this in a class experiment but it could be a project. Frosted cyanogenic leaves look darker and more translucent than acyanogenic ones and this can be taken as evidence of damage. Presumably frost produces ice crystals that disrupt cell membranes and bring glucoside and enzyme together. The cyanide released might increase the damage to the cells.

*Questions and answers*

a ***Is the degree of cyanogenesis the only way in which the leaves presented for choice by the molluscs varied?***

It is likely that they varied in age, surface texture, and even flavour and this variation might not have been randomly distributed between the plants offered to the molluscs.

b ***Would the results of the experiment be more or less meaningful if the molluscs were unusually hungry at the start?***

Very hungry molluscs would eat more so might give a more clear cut result but they might also become less selective and so give a misleading result.

c ***If other food plants had also been available, as they almost certainly would have been in a natural situation, how might this have affected the results?***

We do not know what the particular species of slug or snail used in the experiment would prefer to feed on in the field. Many types of mollusc might never feed on clover, either cyanogenic or acyanogenic, if offered a preferred alternative.

d ***What further investigations of predation of the different types of clover might be worth while?***

**1** Trying different species of molluscs or other invertebrates under standardized conditions.
**2** Trying slightly cyanogenic (G-ee) plants in a choice situation as well as G-E- and gg-- ones used already.
**3** Carrying out a field survey to see if cyanogenic leaves are nibbled less in the field. (Data on this may already be available from a previous investigation.)
**4** Seeing if there is a positive correlation between high populations of animals that prefer acyanogenic plants and high frequencies of the G and E alleles determining a cyanogenic phenotype.

e ***What do the results suggest about the effect of low temperatures on the leaves of the two sorts of clover plant?***

There is no question that temperatures sufficiently low to freeze the leaves, rather than just chill them, cause the release of cyanide. This may be correlated with greater visible damage caused by the freezing process.

f ***Look carefully at* figure [P]27. *To what extent do your results suggest a reason for the data represented by the figure?***

It gets colder as one gets higher in a mountainous region. Frosts would be more severe at high altitude and this would select against cyanogenic types.

g ***Imagine that there is a mixed population of cyanogenic and acyanogenic clovers growing at around 700 m in a mountainous region. Explain the effects that predation and climatic conditions could have upon the evolution of the clover population in that region.***

Frosts might select positively the acyanogenic phenotypes while predation might select positively the cyanogenic phenotypes. The result would be a polymorphism in which the alleles G/g and E/e were held at definite frequencies in a dynamic equilibrium.

## PART III Bibliography

BISHOP, J. A. and COOK, L. M. (Ed.) *Genetic consequences of man-made change.* Academic Press, 1981. (A good review of industrial melanism.)
CLARKE, C. A. Studies in Biology No. 20, *Human genetics and medicine.* 2nd revd edn. Edward Arnold, 1977.
DRITSCHILO, W. et al 'Herbivorous insects colonising Cyanogenic and Acyanogenic *Trifolium repens*'. *Heredity*, **42**, 1979, pp. 49–56.
EBLING, F. J. (ed.) Symposium of the Institute of Biology No. 22, *Racial variation in Man.* 1975.
MANWELL, C. and BAKER, C. M. *Molecular biology and the origin of species.* Sidgwick and Jackson, 1970. (A good account of protein polymorphism.)

# CHAPTER 21 THE PRINCIPLES AND APPLICATIONS OF BIOTECHNOLOGY

*A review of the chapter's aims and contents*

1 This chapter defines biotechnology and describes how biological principles and concepts can be applied in a commercial environment.
2 The emphasis is on the crucial role played by recent developments in recombinant DNA techniques and in the upsurge of interest in biotechnology.
3 The application of mutants, considered in Chapter 19, to fermentation processes is considered.
4 Specific examples of fermentation products are discussed, covering the organisms involved and the uses of the products.
5 The techniques involved in genetic manipulation are explained. A knowledge of the structure and function of nucleic acids has been assumed.
6 The application of enzymes (including their purification and immobilization) to industrial and analytical processes is described.
7 The production of monoclonal antibodies and their use in biomedical applications is discussed.
8 A summary of possible future developments in biotechnology is provided.

# PART I The *Study guide*

## 21.1 What is biotechnology?

*Principles*

1 Biotechnology (of which there are many definitions), is defined as the application of biological organisms, systems, or processes to industry.
2 The subject is multidisciplinary, diverse, and in a rapid phase of development.
3 The upsurge of financial interest in biotechnology is due to a number of factors such as advances in basic biological research and the oil crisis.

## 21.2 Fermentation

*Assumption*

1 A basic knowledge of the biochemistry of fermentation. This has probably been encountered in the conversion of glucose to ethanol and carbon dioxide in the alcoholic fermentation by yeasts, or in the anaerobic conversion of glucose (or glycogen) to lactic acid in the muscles of animals.

*Principles*

1 Humans have exploited the process of fermentation, without understanding it, for thousands of years in the preparation of food and drink.
2 Pasteur's work enabled microbiology to be established as an exact science, and led to the development of new fermentation products and techniques.
3 Aseptic fermentation was a very important development in the fermentation industry, making possible the large scale production of antibiotics and other important new products.
4 Improvements in the performance of the micro-organisms used in industrial fermentation processes can be achieved by mutation and selection.
5 The fermentation industry was revolutionized by the development of the antibiotic industry. Antibiotics are complex compounds synthesized and secreted by some micro-organisms which are selectively toxic to other micro-organisms. A number of important therapeutic agents have been found since the discovery of penicillin, in addition to the development of semi-synthetic antibiotics.
6 The industrial production of other important commodities, such as amino acids and polysaccharides, is also carried out by fermentation processes.
7 The use of microbial cells for human and animal food (single cell protein). In theory the process is attractive as being an efficient conversion of a variety of substrates, including wastes, into high quality protein. There are, however, political and social obstacles to be overcome.
8 The recent energy crisis has led to renewed interest in microbially derived fuels. One example is the use of ethanol obtained from the fermentation of sugar cane to extend petrol for motor fuel; another example is the use of methane, produced by the anaerobic fermentation of sewage and agricultural and industrial wastes.

## STUDY ITEM

### 21.21 Lysine production by 'overproducing' strains of bacteria

*Principle*

One mechanism of metabolic control is that of feedback inhibition, where the products of an enzyme-catalysed reaction accumulate and inhibit the activity of that enzyme. In order to create 'overproducing' strains of bacteria, this feedback inhibition must be overcome.

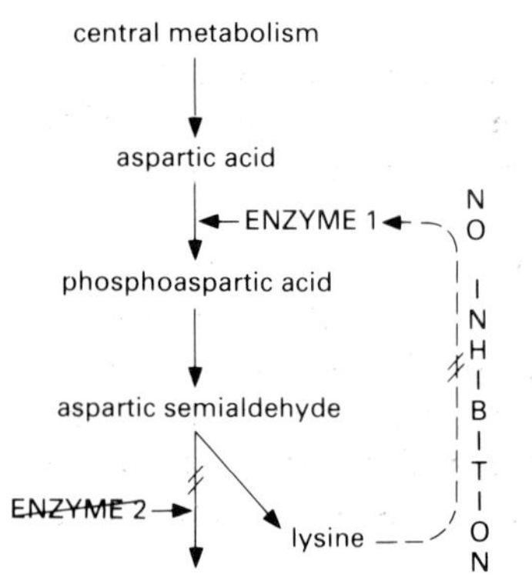


**Figure 14**
A possible mechanism for the increased production of lysine by inactivation of enzyme 2.

*Question and answer*

**a** ***With reference to the pathway shown here, suggest possible mechanisms for this increased production of lysine.***

**1** The mutation inactivates enzyme 2, preventing threonine production and thereby removing the feedback inhibition, since both lysine and threonine are required for this inhibition. (Figure 14)
**2** The mutation alters enzyme 1 in such a way that it still acts as a catalyst, but is no longer sensitive to lysine even when present in high concentrations. (Figure 15) □

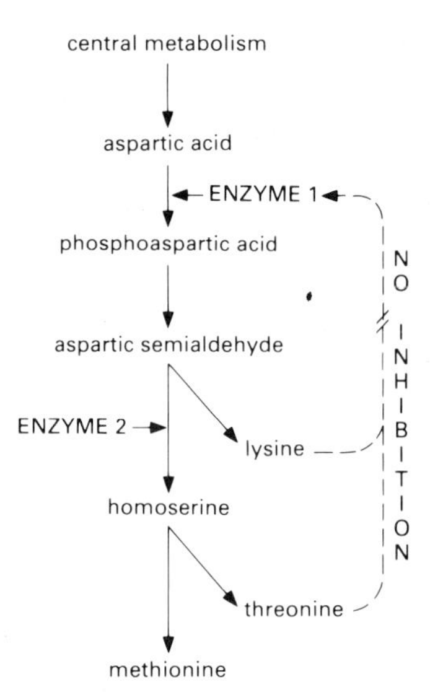


**Figure 15**
A possible mechanism for the increased production of lysine by an alteration in enzyme 1.

## STUDY ITEM
## 21.22 Selective toxicity of antibiotics

*Principle*
Antibiotics are selectively toxic to some infecting organisms but relatively harmless to the host. The basis of this selectivity is differences in cellular structure, such as the presence of a cell wall in bacteria which is absent in mammalian cells.

*Questions and answers*

**a** ***There are several major types of pathogenic agents – viruses, bacteria, fungi, and protozoa. From your knowledge of the cellular structure of these organisms and of the mammalian host cells, say which types will be most and least susceptible to selectively toxic antibiotics.***

The most profound biochemical differences for selective toxicity are found when bacteria are the target organisms. The prokaryotic cell is relatively susceptible to selective antibiotic attack. Eukaryotes are more difficult to destroy selectively or to inhibit as their cells are similar in structure and biochemistry to those of the host organism. There are a number of effective drugs for treatment of diseases caused by eukaryotes but in many of these the mode of action is unknown. The fungi, as eukaryotic organisms, are also difficult to combat with antibiotics, but some antifungal agents are available which attack the cell membranes (which contain sterols, unlike mammalian cells) or cell walls.

Students are asked to refer to table (S)33 when answering questions **b–g**.

b ***Why is penicillin toxic to Gram-positive bacteria and not mammalian cells?***

Mammalian cells lack a cell wall and so are unaffected by penicillin. The degradation of the bacterial cell wall eventually leads to the disruption of the cell membrane.

c ***Penicillin is ineffective against Gram-negative bacteria. What can you deduce from this?***

That the cell walls of Gram-positive and Gram-negative bacteria are different. The Gram-positive wall contains a polymer called peptidoglycan, the synthesis of which is inhibited by penicillin. The cell walls of Gram-negative bacteria do not contain peptidoglycan, and are unaffected.

d ***Penicillin and Polyoxin D are both known to inhibit cell wall growth, but are effective on different organisms. What can you deduce about the cell walls of Gram-positive bacteria and filamentous fungi?***

That the cell walls of Gram-positive and filamentous fungi are different. The main component of the fungal cell wall is chitin, a polysaccharide, which does not appear in bacteria. Similarly, there is no peptidoglycan in the fungal cell wall.

e ***Rifampicin is selectively toxic towards some bacteria but does not harm mammalian cells. Can you suggest a reason for this?***

Many bacterial enzymes differ from their counterparts in mammalian systems. Bacterial RNA polymerase is one such enzyme, having a different amino acid sequence and structure. Rifampicin selectively inhibits this enzyme but cannot inhibit the mammalian RNA polymerase.

f ***Tetracycline and chloramphenicol both inhibit bacterial ribosomes. Why are they ineffective against eukaryotic ribosomes?***

The ribosomes of prokaryotic and eukaryotic organisms have different sizes, masses, and composition. These two antibiotics specifically inhibit bacterial ribosomes, probably by binding to certain proteins on the ribosome. These proteins are not present in eukaryotic ribosomes.

g ***Ketoconazole inhibits the synthesis of sterols (large insoluble molecules) in fungal cell walls. What can you say about the presence of sterols in mammalian or bacterial cell membranes?***

There are no significant amounts of sterols in mammalian or bacterial cell membranes. □

## 21.3 Genetic engineering

*Principles*

1 Humans have used selective breeding to obtain desired characteristics in animals for many years.
2 Genetic information can now be transferred from one micro-organism to another using plasmids or phages as vectors, enabling the direct transfer of foreign genes for specific purposes.
3 The four stages in gene manipulation are
 1 Replication – the formation of DNA fragments containing the desired gene.
 2 Splicing – the insertion of the DNA fragments into vectors.
 3 Transformation – the introduction of the vectors into the organism.
 4 Selection – the selection of the newly transformed organisms.
4 Genetic engineering is now used in the production of important compounds such as insulin, somatostatin, and interferon, which are difficult and expensive to extract from mammalian sources. The mammalian genes coding for these polypeptides are inserted into the plasmid pBR322 which is used to transform the bacterium *E. coli*. This is then cultured, using fermentation techniques.

### STUDY ITEM

### 21.31 The selection of a transformed organism

*Principle*

1 Plasmid pBR322 is a commonly used vector. It carries genes for resistance to the antibiotics tetracycline and ampicillin, and therefore can confer a selectable phenotype on the transformed bacteria.

*Question and answer*

a ***Which of the newly formed plasmids (A, B, and C) would allow bacteria containing it to grow on media containing***
***1 ampicillin?***
***2 tetracycline?***
***3 amylase?***

**1** A, B and C, as all the plasmids contain the gene for resistance to ampicillin.
**2** A, as only plasmid A contains the gene for resistance to tetracycline.
□ **3** C, as only C contains the gene for the enzyme amylase.

## 21.4 Enzymes

*Principles*

1 The nature and function of enzymes and their application to industrial processes and biochemical analysis.
2 Enzymes may be obtained in a pure form from micro-organisms.
3 For some industrial processes, it is preferable to immobilize the enzyme rather than use it in a soluble form.
4 Industrial applications of enzymes includes their use in biological

washing powders, starch and sugar processing, the food industry, textiles, and leather treatment.

5 Enzymes are widely used analytically: compounds are detected and measured by adding the appropriate enzyme and measuring the product formed.

6 The analytical application of immobilized enzymes has given rise to the biosensor, of which the most common example is the enzyme electrode. In this, the immobilized enzyme is held close to an electrochemical detector, which measures the concentration of a product or reactant of the enzyme-catalysed reaction. Three examples of glucose electrodes are illustrated.

## 21.5 Immunology

*Assumption*

1 Some knowledge of the immune system and antigens and antibodies.

*Principles*

1 Antibodies are of great use in the diagnosis and treatment of disease, and also analytically for the detection and measurement of minute quantities of chemicals.

2 Obtaining antibodies used to be a difficult and expensive process, and it was impossible to obtain a single, pure antibody.

3 Antibodies can now be produced relatively simply and in a pure form, using the technique of monoclonal antibody production.

## 21.6 Future trends in biotechnology

*Principles*

1 The possible future developments in biotechnology are numerous.

2 Genetic manipulation will enable further progress in health care through the production of new therapeutic compounds and new vaccines. A more longterm possibility is the alteration of the human genome to correct genetic defects.

(The teacher may wish to conduct a discussion on the social consequences and implications of performing recombinant DNA experiments on humans.)

3 Veterinary science is likely to benefit from the advances made in medical biotechnology.

4 Genetic manipulation offers great potential in the improvement of plant crops, for example in the insertion of genes coding for essential amino acids into the genome of cereal plants which often lack these amino acids.

5 As fossil fuels diminish, biotechnology will become increasingly important in the provision of alternative fuel supplies.

6 The use of enzymes in industry will continue to increase as economic forces change and techniques improve. Not only are new enzymes being discovered, but the biotechnologist can now use genetic manipulation to produce 'tailor made' enzymes for specific industrial processes.

7 One of the most speculative areas of biotechnology is the suggestion that proteins could be used in 'molecular electronics' in order to develop the 'biochip'.

## PART II Relevant practical work

There are no Practical investigations accompanying this chapter, although the investigations involving enzymes and fermentation in *Practical guide 2* are relevant.

## Bibliography

HOUWINK, E. H. *A realistic view on biotechnology*. Proc. of the 3rd European Congress on Biotechnology, Sept. 1984. DECHEMA, 1984.

*Commercial biotechnology: an international analysis*. Washington DC US Congress Office of Technology Assessment, OTA-BA-218, Jan. 1984.

RUSSELL, G. E. (Ed.) *Biotechnology and genetic engineering reviews*, Vol. **1**, 1984. Intercept Ltd., Newcastle upon Tyne, 1984.

PYE, E. K. and WINGARD JNR., L. B. (eds.) *Enzyme engineering*. Plenum Press, New York and London, 1974.

SCIENTIFIC AMERICAN BOOK *Industrial microbiology and the advent of genetic engineering*. W. H. Freeman & Co., 1981.

WISEMAN, A. (ed.) *Principles of biotechnology*. Surrey University Press, 1983.

# CHAPTER 22 METHODS OF REPRODUCTION

*A review of the chapter's aims and contents*

1. The aim of this chapter is to stimulate and direct discussion of the essential features of the mechanisms and consequences of reproduction and their inter-relationships with other aspects of biology.
2. In addition to outlining the essential features of a number of reproductive life cycles, consideration is given to the role of reproduction in relation to the maintenance and evolution of populations, to the relative benefits and disadvantages of sexual, asexual, and intermediate types of reproduction, and to the genetic and ecological implications of the variations which occur in sexual cycles and in the methods of asexual propagation.
3. Sexual cycles differ in respect of the position in the sequence of the two essential events, meiosis and fertilization of gametes, in relation to the mitotic activity which produces the multicellular stages that characterize all but the most primitive organisms.
4. Stages in the evolution of basic sexual cycles (from haplontic to diplontic) are revealed by a survey of the wide variety of life cycles which still exists in the plant kingdom. The significance of these evolutionary changes, leading to the predominance of the diplophase developed mitotically from the zygote, is considered.
5. The most advanced, diplontic, cycles are mainly found in animals, including all vertebrates, and the most elaborate, or at least the best studied, control mechanism for such a cycle is in the higher vertebrates, in particular humans. The reproductive cycle and its control in humans is discussed in some detail.
6. A central feature of sexual reproduction is the production of gametes, their transfer between individuals, and their subsequent fusion. The events in these processes are described for higher plants and for mammals, especially humans.
7. The movement of genetic information through the population is determined by the breeding system. Together with the events of meiosis, this determines the level of heterozygosity and thus of variation amongst the progeny.
8. The level of variation is itself an adaptive feature and different levels, and thus the mechanisms to achieve them, are favoured in different environments.
9. Some of the mechanisms which determine breeding systems are introduced, demonstrating that events in reproduction are inseparable from genetical, ecological, and evolutionary considerations.
10. The final step in successful reproduction is the establishment of the progeny as independent individuals, and embryo protection and nutrition in higher plants and vertebrates are briefly reviewed in order to illustrate two answers to the same essential needs.

*Assumptions*

1. An understanding of mitotic and meiotic cell division.
2. A knowledge of introductory genetics.

# PART I The *Study guide*

## 22.1 Why organisms reproduce

*Principles*

1. Reproduction is the production of new individuals.
2. All cells are capable of reproduction in unicellular organisms.
3. Multicellular organisms show specialization for reproduction and possess finite life spans.
4. Environmental factors affect the numbers in a population which survive to reproduce.
5. The two main forms of reproduction are sexual and asexual.

*Question and answer*

a ***How do scientists estimate the ages of such trees?***

The standard technique is to take a core out of the bole (trunk) of the tree and count the growth rings, on the assumption that there is one growing season in a year and that there is an increase in girth each year. In practice, environmental factors, such as climatic variation or attacks by parasites, may affect the growth in any year. The growth rings tend to vary and to form a pattern which is common to the trees in the same area, because of shared environmental experiences. By relating changes in recent growth rings to known environmental changes, it is possible to suggest what climatic changes occurred in the more distant past. In addition, it is possible to date long dead timber, which has been preserved in dry conditions. By overlapping the band patterns of living and dead trees, scientists have been able to extend the ring pattern back another 1000 years or more before the origins of the oldest living specimens. This is the same technique as is used in archaeology for dating timbers of buildings. The pattern of the growth rings reveals not only how long a tree took to grow, but when, or between which years, it grew.

## 22.2 Asexual reproduction

**Practical investigation. *Practical guide 6*, investigation 22A, 'Reproduction in fungi'.**

*Principles*

1. Asexual reproduction depends upon the retention, during development, of cells with totipotency.
2. Vegetative asexual reproduction occurs when new individuals are generated from somatic tissues of structures which are never involved in sexual reproduction.
3. The other main type of asexual reproduction involves the formation of embryos and some structures normally involved in sexual reproduction, but without meiosis or fertilization.

*Question and answer*

a **In such a case, how would you define an individual plant?**

A definition is not strictly possible. The clump represents a clonal colony of grass plants, rather than individuals. In time the original plant may die, or the connections with the asexually produced tillers may be broken, resulting in new individuals. This is a useful question to relate to ecological sampling, for when you are surveying the plants in an area, such as grass, you cannot use 'individuals' as a unit of measurement. The potentially independent and structurally complete units are called 'ramets' – the coining of a special term being in itself recognition of the unsuitability of 'individual' or 'plant'.

## STUDY ITEM
## 22.21 Asexual reproduction

*Principles*

1 Asexual reproduction occurs naturally, in addition to, and sometimes in place of, sexual reproduction.

2 Asexual reproduction depends entirely upon mitotic cell division which, apart from mutations, replicates exactly the genetic information in each cell. Progeny of asexual reproduction are therefore normally genetically identical to the parent and each other, forming a clone. The only source of variation is by mutation, which occurs at a low frequency, but accumulates in long-lived clones.

3 Asexual reproduction by vegetative means requires fewer resources and produces mature progeny more rapidly than sexual reproduction, but permits little or no dispersal.

4 Asexual reproduction is often exploited when it occurs naturally in crop plants, and it can be induced artificially in many other crops in order to propagate rapidly individuals with desirable genotypes to form uniform populations.

*Questions and answers*

a ***Consider the examples of asexual reproduction illustrated in* figures [S] 142 *and* [S] 143. *What do all examples of asexual reproduction have in common?***

They all have cells capable of dividing mitotically and producing more cells with the potential to form a complete new individual.

b ***What are the advantages and disadvantages of asexual reproduction in the wild?***

The advantages include:

**1** The ability to propagate rapidly a successful genotype in order to exploit the immediate environment fully and to compete successfully with other plants.

**2** The ability to reproduce plants which have been rendered sexually sterile by polyploidy, hybridity, or mutation, thus allowing survival, at least for a while.

The main disadvantages are:
**1** The inability to adapt and evolve by the production of new genotypes by recombinations.
**2** The gradual accumulation of harmful mutations with no means of eliminating them.

**c** ***Some plants reproduce asexually by vegetative means, as shown in the examples in* figure [S] 142. *Others, such as* Citrus, *reproduce asexually by means of seeds with asexual embryos. What are the advantages and disadvantages of these two different types of asexual reproduction?***

Vegetative reproduction has the disadvantage of little or no dispersal of the progeny which therefore competes with the parent. Asexual seeds have the advantage of normal seed/fruit dispersal. However, as they are formed by modification of the sexual cycle and involve most of the developmental processes of flowering and fruiting, they require more energy resources.

**d** ***Some domesticated crop plants are propagated for commercial purposes by means of vegetative propagation which occurs naturally in the species. Name three examples, and describe how they are propagated.***

Familiar examples include:
**1** Strawberry (*Fragaria x ananassa*). Runners (horizontal stems) spreading on the surface of the soil produce daughter plants at alternate nodes when these are in contact with the soil. Once established, the plantlets can be separated from the parent plant.
**2** Potato (*Solanum tuberosum*). Many stem tubers are formed underground on each plant. The next season each tuber forms roots and shoots, from axillary buds or eyes, and produces a new plant. Each separate tuber, or piece of a tuber, can therefore be sown the following year as a so-called 'seed' potato. 'Seed' potatoes are mostly produced in Scotland where there are fewer disease-carrying aphids. Most UK potato crops are grown in England. Many important tropical 'root' crops (so-called even though some, like the potato, are stems) are propagated in the same way, including taro, cassava, and sweet potato.
**3** Onion (*Allium cepa*). Most forms are propagated from seed, but the shallot, like the ornamental bulb species such as tulip and daffodil, is usually propagated from bulb 'offsets' produced at the side of the parent bulb from axillary buds.
**4** Bramble (*Rubus fructicosus*). The bramble reproduces by root 'suckers' and by layering. Shoots form from adventitious buds arising on roots and subsequently produce their own root system and eventually become independent plants, rather like the strawberry, but from roots instead of shoots. Raspberries (*Rubus idaeus*) reproduce in the same way. Layering occurs when a shoot arches over until the tip touches the ground. The shoot then turns to grow upwards, roots form where it touches the soil, and in time an independent new plant is

formed. Frequently seen in hedgerows, this form of reproduction can be encouraged by pegging shoot tips to the ground and covering with soil.

e ***Many other crop and ornamental plants do not normally reproduce vegetatively, but are propagated asexually by horticultural techniques. Name as many of these techniques as you can, in each case giving an example of a species which is commercially propagated by that means. What do all these techniques have in common?***

The main techniques are:
**1** Hard and soft-wood stem cuttings. Many species produce adventitious roots, with or without the aid of growth substance treatment, from the cut base of the excised shoots, which can then grow as a new individual. Examples which will do this readily, even in water alone, include Busy Lizzie (*Impatiens*), mint (*Mentha*), currants (*Ribes*), and willow (*Salix*).
**2** Root cuttings. Some plants, including horseradish (*Armoracia*), chicory (*Chicorum*), and *Acanthus*, propagate very readily by producing adventitious buds from root cuttings (excised segments of roots). Anyone who has tried to eliminate dandelions by simply cutting off the top of the tap root will have experienced this.
**3** Leaf/petiole cuttings. African violets (*Saint paulia*) regenerate new plantlets from adventitious buds produced on the cut petiole of leaves placed in soil. *Begonia* produces them from the edges of cuts made in the main veins of leaves pegged down on the soil.
**4** Grafting. Many slow growing species, including most fruit and ornamental trees, such as apple (*Malus*) and cherry (*Prunus*), are propagated by grafting an excised shoot (scion) onto a stock of a related variety or species. Each parent tree can produce a large number of suitable shoots which, with a supply of root stocks often of a wild species raised from seed, can produce a large number of clonal progeny. The position of the callused union of stock and scion can frequently be recognized even in old trees planted in orchards and gardens and along suburban roads.
**5** Budding. Roses (*Rosa* cultivars) are propagated by budding, a technique similar to grafting except that only one axillary bud on a small amount of underlying tissue is inserted under the bark of the stock.
**6** Layering. Some slow-growing woody species can be propagated most rapidly by artificial layering techniques, involving the production of a root system while still partially attached to, and supported by, the parent plant. This is induced by cutting part-way through the stem and wedging open the cut with, say, a pebble. Then, either the cut region is pegged to the soil, as with *Rhododendron*, or, if this is not possible, it is surrounded with damp moss and polythene (air-layering), a technique used with the rubber plant (*Ficus elastica*) and *Magnolia*.

These techniques all involve the removal of tissues from the intended parent plant which are capable of regenerating missing organs to produce a new complete individual. In many cases, one or more shoot meristems are removed and induced to produce, or provided with, a root system. In the other methods, adventitious shoot meristems and then plantlets are induced on roots or leaves. In recent years this approach has been transferred from the potting shed to the laboratory, and, in many species, buds and whole plantlets are induced *in vitro* in meristem, callus, and cell cultures derived from a variety of organs (see also section 23.9).

**f** ***More recently, some plants have been successfully propagated* in vitro *from meristems, tissue, or cell cultures. In some cases, several identical plants have been regenerated from a single cell isolated from a differentiated tissue. What does this tell you about these plant cells?***

It shows that a simple plant cell, perhaps all plant cells, can retain during differentiation all the genetic information of the zygote and embryo. Also, cells derived from it by mitotic division inherit the retained ability to express this information and thus regenerate all cell and tissue types of a complete plant. This ability is known as 'totipotency'.

**g** ***Why, and under what circumstances, are plants commercially propagated by asexual rather than sexual means?***

An asexual method is used when it is cheaper or when it is the only available means of maintaining the desired cultivar. It is thus particularly important when:

**1** natural or induced asexual reproduction is easily used on a large scale, as with potato and strawberry;

**2** a plant is slow to mature from seed, as with many shrubs and trees;

**3** sexual reproduction fails due to sterility, as with hybrid lilies, 'double-flowered' forms of many species, and plants lacking suitable climatic conditions, pollinating agents, or, if self-sterile, other compatible individuals;

**4** the desired character is not transmissible sexually, as with decorative white and green chimeras of *Pelargonium* and *Chlorophytum*.

**h** ***Asexual reproduction occurs in vertebrates, including humans, but only rarely and at only one stage of development. When does it occur and why does it not occur at other stages?***

It occurs only at an early stage of embryo development, to produce identical offspring. In humans it is usually after some development has taken place (see section 23.2). It presumably does not occur at later stages because, under natural conditions, totipotency is not retained. □

## STUDY ITEM

## 22.22 The life cycle of aphids

*Principles*

1 Parthenogenesis provides a means of rapid population increase (when conditions are relatively constant).

2 The presence of both sexual and asexual phases in an annual cycle allows variation to occur and selection to act, while at the same time enabling successful genotypes to replicate rapidly to fully exploit the environment.

3 Adverse winter conditions are spent as dormant eggs.

*Questions and answers*

a ***What is the major advantage of reproduction by parthenogenesis in the black bean aphid's life cycle?***

Parthenogenesis is an economical way of producing large numbers of identical offspring in a short time, thus enabling an ephemeral, uniform, near optimum, environment to be exploited.

b ***Some species of aphid reproduce only by parthenogenesis. What might be the disadvantage of such a life cycle?***

It reduces the genetic variation in the species. This is no disadvantage, apart from the accumulation of mutations, under stable conditions, but it would be in the longer term if the environment changed.

c ***How would you try to prove that an insect, such as* Aphis fabae, *was reproducing parthenogenetically?***

It would be necessary to isolate individual eggs, grow isolated individual insects on the host plant, and determine whether they laid eggs which hatched into the next generation.

d ***What factors might cause sexually reproducing forms to appear in September?***

Fall in temperature or shortening of the day length are two likely factors. In this particular species, both factors have been found to promote the production of sexual forms.

e ***After a wingless aphid population has been living on a host plant for some time, winged forms begin to appear. What factors might promote the production of winged forms?***

Various factors have been shown to affect the development of winged forms: quality of the host plant, overcrowding, temperature, day length, attendance by ants. Mature plants, dense populations, low temperatures, shorter day length, and absence of ant attendance, have all been shown, in different species of aphid, to promote the production of winged forms.

f ***How might knowledge of the aphid's life cycle help in planning control measures?***

There are two periods when aphids migrate. A primary migration from spindle trees occurs between mid May and early June. The secondary migration, bean aphids living on the crop plant, takes place in July. It has been found that beans sown in February and March are very attractive plants to the primary migrants, but not to the secondary – probably because the plants are too mature by July. A single chemical treatment of the crop at the end of the primary migration gives good control for the rest of the growing season. Crops sown in April are attractive to both primary and secondary migrants and so one chemical treatment is unlikely to give good control.
It has been found that aphid attacks can be reduced by dense planting of the beans. This apparently works because densely grown plants are individually less nutritious and mature more quickly than those in a low density cultivation. When aphids are abundant, dense crops suffer less damage than sparsely sown crops. If only moderate numbers of aphids are present it is likely that dense crops will not need chemical treatment, whereas sparse crops almost certainly will need treatment.
It might be thought that removing spindle trees in the vicinity would prove an effective control measure, as such trees provide the winter hosts for the bean aphid. However this is not a practical solution, because aphids migrate long distances, partly by their own flight but also blown by air currents. □

*Questions and answers*

**b** ***Why do you think that there may be such a disadvantage?***

Because no environment remains stable for ever. A change in the environment is likely to put a clonally produced organism at a selective disadvantage. In addition, the gradual accumulation of deleterious mutations with no mechanism for 'cleaning-up' may lead to loss of vigour and eventually to death. (See also Study item 22.31.)

**c** ***Why might it be, at least in the short term, a definite advantage?***

Because most mechanisms of clonal propagation use less resources than sexual reproduction. Therefore more resources can be used in vegetative growth to exploit the habitat to the full. In addition, vegetative reproduction is often capable of producing progeny which mature faster than seedlings and give a quicker 'turn-over' of asexual generations.

## 22.3 Sexual reproduction

*Principles*

1. Sexual reproduction involves the fusion of two gametic cells.
2. All sexual cycles show an alternation of fertilization with a meiotic division together with many mitotic divisions.
3. In most organisms the two gametes can be distinguished by both structure and behaviour.
4. In some organisms the zygote formed by fertilization is retained within the parent's body and develops into an embryo protected by parental tissues.

*Question and answer*

a **_What different meanings do biologists give to the term 'egg'? Which is the correct definition?_**

The term can mean: an oocyte; an unfertilized ovum or egg cell; a zygote; a blastocyst, as, for example when talking of the 'egg implanting in the uterine wall'; an ovum/zygote/embryo with a shell and membranes, as with a bird's egg. Correctly the term should be used to refer to an ovum, or female gamete.

## STUDY ITEM
### 22.31 Genetic variation

*Principles*

1 For every organism there is an optimum level of recombination and potential variation in the next generation.

2 This optimum level is always lower than the random variation of all gene loci, which would immediately dismantle every successful combination of alleles as it arose.

3 The optimum level varies with species and with current circumstances. For some organisms there may be an advantage, at least in the short term, in restricting variation or even eliminating it altogether.

4 Recombination is restricted in all eukaryotes by the location of more than one locus (usually many loci) on each chromosome. It is further restricted in some by reduced numbers of chromosomes per genome, by reduced numbers and restricted distribution of chiasmata, and by restriction of the independent movement of bivalents through structural heterozygosity.

*Questions and answers*

a **_Under what conditions would a high degree of genetic variation be disadvantageous?_**

Generalizations are difficult and occasionally misleading because of the many interacting factors which may have an influence, but completely random recombination of all loci, if it were possible in eukaryotes, would probably be disadvantageous under all conditions. It would dismantle all allele combinations and reduce the proportion of offspring which has a genotype sufficiently similar to the successful parent to allow establishment in the same microhabitat. Furthermore, the highest levels of recombination which can be achieved by sexual reproduction would also be disadvantageous under any conditions favouring conservation of parental genotypes (see **b** below). However, a high degree of variation would be found when severe inter-specific competition in a very heterogeneous habitat eliminates most juveniles regardless of genotype and allows only some of the most closely adapted to survive.

b **_Under what conditions could little or no genetic variation be advantageous?_**

Restricted variation is likely to be advantageous:

**1** When juvenile survival and establishment are infrequent due to extreme environmental conditions. For example, in many plants of desert or tundra, many seedlings with the closely adapted and specialized genotype of the successful parent inherited, more or less, unchanged, will have to be produced to ensure that one will find the microhabitat in which it can establish.
**2** In a pioneering species colonizing a recently disturbed open site with uniform conditions and no competition, at least over a limited range and time span. Plants which can conserve the successful genotype of the initial colonizing individual while reproducing rapidly and in large numbers will exploit the site fully and have an advantage over competitors.

**c** ***By what means is the variation generated by recombination at meiosis restricted in all eukaryotic organisms?***

The fact that genes are grouped on chromosomes leads to linkage, that is, less than random recombination between loci. Despite the effects of chiasmata and random assortment of bivalents, alleles of neighbouring genes on the same chromosome do not recombine at random. Random chromosome rearrangement followed by selection can therefore result in particular and advantageous assemblages of functionally selected alleles being inherited intact.

**d** ***Discuss means by which variation is further restricted in some species. Give examples where possible.***

Recombination is further restricted by:
**1** Low chromosomal number reducing the effect of random assortment of bivalents and increasing the size of linkage groups; for example in *Crepis capillaris* and *Crocus balansae* (n = 3).
**2** Low frequency of chiasmata, often coupled with localized distribution to produce large groups of loci with 100 per cent linkage in each arm; for example in *Tradescantia* spp. (distal localization) and *Tulbaghia* spp. (proximal localization).
**3** Structural heterozygosity for chromosome interchanges preventing random assortment of bivalents; for example in the boat lily, or oyster plant, *Rhoeo spathacea* (2n = 12), where multiple interchanges have resulted in two large linkage groups, each involving a complete gametic complement of six chromosomes.

It is not intended that this question be given to the students without prior discussion. In fact it might be best used directly by the teacher to extend the students' understanding of the genetics involved. □

*Question and answer*

**b** ***What might be the benefits of sexual reproduction?***

Sexual reproduction is usually associated, at least sporadically, with cross-fertilization and thus with recombination. Recombination generates new genotypes at a frequency which is orders of magnitude greater than that of new genotypes created by mutation. This is, in essence, the advantage of sexual reproduction.

However, the answer is not so simple as that; the advantage cannot be simply and only the generation of new genotypes which allow adaptation to changing conditions, important though this will have been during an organism's evolutionary history. For most organisms the time scales are wrong. Climatic changes are very slow relative to the life cycles of even the longest-lived organisms. Thus survival of the progeny will depend on producing genotypes similar to the parent and equally adapted to the same conditions. Long-term evolutionary survival requires changes in the genotype, and thus sexual reproduction; short term survival – without which there is no long-term survival and evolution – is more likely to require a relatively stable genotype.

So why is the resource-expensive mechanism of sexual reproduction maintained? There cannot be selection for survival value in an evolutionary future, for there is no way in which the environment can select for that. It is conceivable that, in general, only those organisms which have retained sexual reproduction have survived through time until now and thus most organisms we now see reproduce sexually. There may have been insufficient time for a less resource-expensive mechanism to evolve. However this does not explain why sexual reproduction has not been lost in the past. There has been time for all organisms to lose it. So it raises the question of possible short-term advantages of sexual reproduction to explain its survival, despite its cost, until it can confer evolutionary benefits. Such possible short-term advantages are much debated and are difficult to prove. Possibilities include:

**1** By generating heterozygosity in an outbreeding population, sexual reproduction confers effective protection against, for example, elimination of plants by herbivores and other predators (polymorphism for chemical defence, etc.).

**2** Interaction between genetically different individuals in a mixed population results in greater vigour, with an optimum at a particular proportion (frequency-dependent selection.)

**3** By recombination followed by selection of gametes and zygotes, sexual reproduction can 'flush out' deleterious mutations between one generation and the next. Further, it can reassemble the original pre-mutation genotype as well as new and non-deleterious combinations. Progeny with the parental genotype can then be selected for by the environment, if that has remained unchanged. Thus, in this case, sexual reproduction followed by stabilizing selection actually maintains an unchanged genotype from one generation to the next – the reverse of the popular image of sexual reproduction.

**4** The unique events of meiosis and/or gametogenesis and fertilization might well include a process of intracellular repair and 'rejuvenation' to eliminate damage accumulated during the lifecycle of repeated cell divisions. These processes might include not only tidying up accumulated errors, but also repair of possible damage to the semi-autonomous cytoplasmic organelles, such as mitochondria and plastids, and even possibly other cell structures. There is some evidence of significant events concerning these organelles during critical stages of sexual reproduction. There is no suggestion of anything to counteract progressive deterioration in asexually reproducing clones.

Any, or all of these possibilities might be enough to account for the short-term advantages of sexual reproduction and thus its survival long enough to be significant in evolution.

## 22.4 Subsexual reproduction

*Principles*

1. In some life cycles, fertilization or normal meiosis is omitted in at least one of the sexes.
2. In animals and in flowering plants, unfertilized eggs may develop parthenogenetically to form a new individual with the same genetic complement as the parent.
3. In ferns, cells of the gametophyte develop by mitotic divisions to form a new sporophyte.

*Question and answer*

a ***What proportion of the genes of* Apis mellifera *are sex-linked?***

Males are haploid and the whole of their haploid chromosome set is transmitted intact to each sperm cell. Thus all the genes are sex-linked.

## 22.5 Variation in life cycles

*Principles*

1. There are three basic types of life cycle: haplontic, diplontic, and haplo–diplontic.
2. The types of life cycle differ in the distribution of mitoses in relation to fertilization and meiosis.

### STUDY ITEM

### 22.51 Life cycles

*Principles*

1. All sexual cycles have fertilization and a meiotic division in common, but they differ in the location in the cycle of those two events, relative to the mitotic divisions necessary for the development of multicellular individuals, or the asexual reproduction of unicellular organisms.
2. Mitoses may be limited to the haplophase, which forms gametes, the diplophase, which develops from the zygote and ultimately undergoes meiosis, or both.
3. The occurrence of mitoses in both haplophase and diplophase results in an alternation of multicellular generations which are often of very different structure and appearance.
4. The plant and fungal kingdoms include a wide range of life cycle types, and their distribution through the green plant groups suggests an evolutionary progression from the haplontic, through haplo-diplontic cycles with progressively more dissimilar generations and thus reduced haplophase, to the diplontic cycle (see Study item 22.53). Nearly all

plants have haplo-diplontic cycles alternating between the gamete-forming haplophase (gametophyte) and the meiotic spore-forming diplophase (sporophyte). Most animals have the apparently more advanced diplontic cycle.

5 Statements such as 'gametogenesis involves meiosis', 'gametes are formed by meiotic division', or 'the products of meiosis are male and female gametes' are not universally true, although they are frequently presented as being so by those ignorant of the majority of plant life cycles.

*Questions and answers*

a ***When do mitoses occur?***

They are restricted to the haplophase; the zygote divides meiotically.

b ***By what type of cell division are gametes formed?***

Mitotic cell division.

c **Chlamydomonas, *an alga, exhibits this type of life cycle. Which other organisms fall into the group of haplontic organisms? You will need to consult texts to find out.***

This cycle is only found among the simpler eukaryotic algae and fungi, including: *Mucor*, *Rhizopus*, *Volvox*, *Ulothrix*, *Oedogonium*, *Spirogyra*, *Zygnema*, and the Desmids.

d ***When do mitoses occur in the life cycle?***

In the diplophase only: the zygote divides mitotically.

e ***What type of cell division gives rise to the gametes?***

Meiotic cell division.

f ***What organisms exhibit this type of life cycle? Consult texts if necessary.***

Diatoms and most animals, including humans and all other vertebrates.

g ***When do mitoses occur?***

In both diplophase and haplophase: the zygote divides mitotically.

h ***By what type of cell division are gametes formed?***

Mitotic cell division.

i ***This type of life cycle is said to show an alternation of generations. Why?***

Because two distinct multicellular organisms, distinguishable by their different reproductive structures and often by their morphology, alternate regularly through successive sexual cycles.

j ***In some algae, such as* Cladophora *(figure [S] 151), the two phases of the cycle appear exactly alike. In such a case, how would you set about deciding to which phase a given organism belonged?***

If the cycle is a normal sexual one, the chromosome number and nuclear volume of corresponding cells will be twice as large in the diplophase as in the haplophase. The only other distinguishing feature, restricted to mature individuals, is the appearance of the reproductive structures. The diplophase produces meiotic spores capable of germinating to form the haplophase. This, in turn, produces gametes which fuse to form a zygote before germinating. In many cases, the spores and gametes are distinguishable, even when the spores are motile.

k ***Which groups possess this type of life cycle?***

All green land plant groups and some of the more complex algae, including *Cladophora*, *Ulva*, *Fucus*, and *Laminaria*, have haplo-diplontic life cycles. Of these, only *Cladophora* and *Ulva* are of the isomorphic type.

l ***What type of life cycle do the following organisms possess? Consult suitable texts.***

*Mucor* and *Spirogyra* are haplontic. *Blatta*, *Homo*, *Lumbricus*, *Pinnularia*, and *Rana* are diplontic. *Dryopteris*, *Marsilea*, *Pelargonium*, *Pinus*, *Polytrichum*, and *Ulva* are haplo-diplontic □

**Practical investigation. *Practical guide 6*, investigation 22C, 'The life cycle of *Dryopteris filix-mas*, the male fern'.**

## STUDY ITEM
## 22.52 Alternation of generations

*Principles*

1 The alternation of generations represents a major switch in development at the spore stage, from sporophyte to gametophyte, and at the gamete/zygote stage, from gametophyte to sporophyte. This is particularly obvious in heteromorphic cycles, regardless of which generation is dominant.

2 The development switches occur, perhaps significantly, at the two unicellular stages of the life cycle, and they are usually accompanied by several conspicuous cellular changes. Of these, the change in chromosome number, characteristic of sexual cycles, is known not to be essential for the phase change, because generations alternate in apomicts where the chromosome number remains unchanged.

*Questions and answers*

a ***Which is the dominant phase in each of the following groups? Consult suitable texts to check your answer.***

***1 Angiosperms***
***2 Bryophytes***
***3 Gymnosperms***
***4 Phaeophytes***
***5 Pteridophytes***

The sporophyte is the dominant phase in **1**, **3**, **4**, and **5**, and the gametophyte in **2**.

b ***On the basis of what you already know about reproduction, what evidence is there to indicate whether the change in chromosome number, which usually accompanies the alternation of generations, is a causal factor, or even an essential feature, of the phase change?***

The fact that in some apomicts, such as *Dryopteris affinis*, there is an alternation of the morphologically distinct gametophyte and sporophyte generations with no difference in chromosome number between them reveals that the halving of the chromosome complement at meiosis and the doubling at fertilization does not cause, or even have to accompany, the morphological alternation.

c ***How might these features be involved in the phase change?***

Presumably each phase change involves the initiation of a new, genetically controlled developmental sequence. It seems likely that the influence of the preceding generation needs to be reduced to a minimum. It would seem necessary therefore to isolate the first cell(s) of the new generation from any surrounding tissues of the previous phase, and at the same time to erase any of the cytoplasmic legacies of earlier development within the cells. It is therefore plausible to suggest that feature **1** may be associated with isolating each generation from the previous one, while **2** and **3** reflect the elimination of cytoplasmic components inherited from the previous generation and their subsequent replacement by equivalent structures produced during, or after, the switch. There is, however, no direct evidence as yet that this is the correct interpretation of these events, or indeed that these events are connected with the phase change in any way at all.

d ***Are the three features listed above essential to the alternation of generations, or are they merely some of the several unique events involved in meiosis and gametogenesis? Use the information given above to plan investigations which would help to answer this question.***

Apogamy and apospory represent generation switches in the absence of gametogenesis and fertilization, in the former, and of meiosis in the latter. Both processes can be induced experimentally in bracken. If the cells which eventually give rise to the apogamous and aposporous outgrowths can be located and identified, their fine structure could be examined under the electron microscope. Comparison of these cells with corresponding cells of normal tissues with no induced apogamy or apospory, and with meiotic cells and gametes effecting the same generation switches within a normal sexual cycle, should be made. This might indicate whether the wall and cytoplasmic changes are likely to be directly associated with the processes of meiosis and gametogenesis, or with the generation switches, whenever they occur, or with neither. The crucial step in the interpretation would be the location of aposporous and apogamous cells at a sufficiently early stage in the transition. Growing the material under controlled and constant conditions will be important in this respect so that as far as possible the events occur at the same time and place in each individual. The use of genetically identical clonal material might help here, but it will still not be easy. □

**Practical investigations. *Practical guide 6*, investigation 22B, 'Reproduction in *Marsilea vestita*, the shamrock fern', and investigation 22E, 'Events leading to fertilization in angiosperms'.**

**STUDY ITEM**

22.53 The evolution of life cycles

*Principles*

1 Sexual life cycles have evolved, probably from an asexual origin, and from haplontic, through haplo-diplontic, to diplontic cycles.

2 Stages in the evolutionary sequence are represented by existing groups of plants, some of which have haplontic cycles and most of which have haplo-diplontic cycles, while a few have diplontic cycles.

3 Among haplo-diplontic cycles, progressive reduction of the gametophyte and its retention, first within the spore and then within the sporangium, has been combined with the loss of mobility of the sperm. These changes have reduced the vulnerability of the gametophyte and its dependence on surface water for fertilization. At the same time, they have created the need for alternative means of dispersal and gamete transfer.

4 These trends have implications for the breeding system and for ecology. Notably, they allow the invasion of drier habitats and the greater control of the proportions of self- and cross-fertilization.

*Questions and answers*

a ***Place the groups in sequence from primitive to advanced.***

Green algae, Pteridophytes, Gymnosperms, Angiosperms.

b ***What were your criteria for deciding the order?***

The main criteria are the age according to the fossil history and their structural complexity and diversity, which are assumed to increase with evolutionary development. Some fundamental similarities, such as photosynthetic pigments, indicate common ancestry for these groups.

**c** ***What evolutionary trends are revealed by this sequence in relation to:***
***1 the development of the gametophyte relative to that of the sporophyte?***
***2 the retention of the female gamete, the zygote/embryo, the gametophyte, and the spore within the structures in which they were formed?***
***3 the requirement for water for the male gamete to reach the egg?***

**1** The gametophyte becomes progressively reduced relative to the sporophyte (compare *Cladophora, Dryopteris, Marsilea, Pinus, Pelargonium*).
**2** The trend is towards greater reduction of these structures. The female gamete is retained in oogamous algae. The zygote is retained in all the higher groups with protected embryos. The surrounding gametophyte is retained within the released megaspore in the heterosporous pteridophytes. The megaspore is retained in the parental sporangium in the seed plants.
**3** This requirement is lost. Most of the algae, all of the bryophytes and pteridophytes, and a few of the gymnosperms have motile male gametes. These depend on water for their ability to swim to an egg to fertilize it. The remainder of the seed plants have non-motile male gametes, which are transferred to a female cone or flower by a pollinating agent, and then to an egg cell by a pollen tube.

**d** ***What are the genetical and other implications of these trends?***

**1** The reduction of the gametophyte means that (a) most mutations will occur in the diplophase where recessive mutations can be stored, protected by dominant alleles, and (b) the gametophyte can be retained within the spore and sporangium.
**2** The retention of the megagametophyte means that (a) the limited ecological tolerance of the gametophyte no longer restricts the distribution of the species, (b) spores are no longer able to function as the dispersal unit, a role taken over by a new and larger structure, the seed, and (c) spore retention is only possible in heterosporous plants, where the microspore can still disperse genetic information even though the megaspore is not released. Replacement of the spore by the seed as the dispersal agent allows a single propagule to introduce genetic heterogeneity through heterozygosity into a new site whereas a single haplophase spore of the lower plants can only produce homozygous sporophytes and identical progeny.
**3** Motile male gametes swimming over short distances in a water film are no longer effective as the sole means of bringing about cross-fertilization once the megagametophyte and associated egg cells are retained within the parental sporophyte. The replacement of motile male gametes by the transfer of microspores to ovules by external agents and then of male gametes to the egg by pollen tube growth eliminates the dependence of the plant on surface water.

e ***What evidence is there that diplontic cycles have evolved from haplo-diplontic ones and haplo-diplontic cycles from haplontic ones?***

The simplest forms of the oldest groups of plants have haplontic cycles. Most other plants have haplo-diplontic cycles. There is a progressive reduction of the gametophyte through the pteridophytes and the gymnosperms to the angiosperms. The last group have a near diplontic cycle with the gametophyte reduced to a few cells. Those protists with diplontic, or near diplontic cycles, are among the most specialized. The few groups which do not have diplontic cycles are ancient and relatively primitive. These include the Foraminifera, which have a haplo-diplontic life cycle, and some flagellates and sporozoans which have a haplontic cycle.

The simplest interpretation of these facts is that the first sexual cycles were haplontic, and that diplontic cycles have evolved by the progressive shift in emphasis from gametophyte to sporophyte. □

## 22.6 Gametogenesis and fertilization

**Practical investigations. *Practical guide 3*, investigation 10A, 'The relationship of the urinary system of a mammal to other organs of the body' (includes dissection of male and female reproductive systems), and *Practical guide 6*, investigation 22F, 'The structure and function of the mammalian gonads', and investigation 22E, 'Events leading to fertilization in angiosperms'.**

*Principles*

1. In the sexual life cycle gametogenesis always immediately precedes fertilization.
2. Meiosis does not always immediately precede gametogenesis.
3. In diplontic organisms the products of meiosis develop directly into gametes.
4. In haplontic and haplo-diplontic organisms the products of meiosis are spores which develop into multicellular gametophytes before some cells differentiate into gametes.

### STUDY ITEM

### 22.61 The life cycle of angiosperms

*Principles*

1. Angiosperms show the greatest reduction of the gametophyte generation of any terrestrial plant group. Both male and female gametophytes are reduced to a few cells.
2. The megagametophyte is retained within the tissues of the sporophyte.
3. The microgametophyte develops two male gametes and a pollen tube, which transfers the gametes to the embryo sac.
4. The polyploid endosperm is a tissue which develops to nourish the developing embryo.
5. The carpel is a structure, unique to angiosperms, which has allowed the elaboration of pollination mechanisms, monitoring of the pollen, and the development of a fruit.

**6** The angiosperms are the most advanced plant group because of the reduction of the gametophyte stage, the replacement of the haploid spore by the diploid seed as the dispersal unit, the loss of the motile male gamete and the development of pollination mechanisms, the protection of the ovule by the carpel, and the diversification of different seed dispersal mechanisms.

*Questions and answers*

**a** ***Comment on each of the numbered stages in* figure [S]157. *What stages must occur between 10 and 0?***

The numbered stages are:
**1A:** Microspore mother cell.
**2A:** First meiotic division of spore mother cell.
**3A:** Second meiotic division.
**4A:** Four haploid microspores (pollen grains).
**5A:** Microspore nucleus divides to form a vegetative, or tube nucleus, and a generative cell. This structure forms the microgametophyte.
**6A:** Generative nucleus may divide at this stage to form two male gametes.
**7A:** Pollen grain has germinated to form a pollen tube containing all three nuclei. The tube grows through the style towards the micropyle of an ovule.
**1B:** An enlarged cell in the nucellus undergoes a meiotic cell division.
**2B:** Three of the four products of the division play no further part and degenerate. The megaspore forms from the remaining cell.
**3B:** The megaspore nucleus divides mitotically to form two nuclei.
**4B:** And again forming four nuclei.
**5B:** A final mitotic division produces eight nuclei. This structure forms the megagametophyte or embryo sac.
**6B:** The nuclei arrange themselves, three at either pole and two centrally, in the enlarged embryo sac.
**7B:** Of the three nuclei at the micropylar pole, one forms the female gamete and two form synergid ('help') cells. The two centrally arranged nuclei are called the polar nuclei. The remaining three nuclei at the other pole form antipodal cells.
**8:** Prior to fertilization, the pollen tube reaches the embryo sac via the micropyle.
**9:** Fertilization has occurred. One male gamete has fused with the female gamete; the resulting zygote will develop to form the embryo. The other male gamete has fused with the polar nuclei to form a triploid ($3n$) nucleus. This divides and forms a tissue, the endosperm, which provides nutrient to the developing embryo.
**10:** An embryo has formed with two seed leaves (cotyledons), an embryonic root (radicle) and a shoot apex. It is surrounded by the tissues of the endosperm. On the outside of the endosperm are the seed coats and the ovary wall.

The stages which must occur between 10 and 0 are:
**i:** The seed must mature.
**ii:** It may be dispersed from the parent plant and it may go through a dormant stage.

**iii**: Germination under suitable conditions.
**iv**: Growth and development to become a mature plant.
**v**: Under suitable environmental conditions the development and maturation of flowers.

**b** ***In which phase of the life cycle does vegetative reproduction also sometimes occur? Is this the same in pteridophytes and bryophytes?***

It is the sporophyte of angiosperms which can reproduce asexually. In bryophytes, gametophytes frequently propagate vegetatively and the sporophytes rarely, if ever. In pteridophytes, the gametophytes of some species and the sporophytes of most species reproduce in this way.

**c** ***Why is it incorrect to refer to a pollen grain as a male gamete?***

Because a male gamete is a single specialized cell which fuses with an egg. A pollen grain is either (initially) the single-celled meiotic spore, or (later) the two- or three-celled microgametophyte, which is so reduced that it can be contained within the spore wall. Fully developed pollen grains sometimes contain male gametes along with the vegetative cell.

**d** ***What happens to each of the components of the embryo sac after fertilization? You will need to consult suitable texts.***

The fertilized egg becomes the embryo. The fertilized polar nuclei become the endosperm. The synergid cells degenerate, one of them usually before fertilization. The antipodal cells usually degenerate, but occasionally divide and enlarge for a while and may even penetrate ovule tissue like parasitic haustoria.

**e** ***In what ways is the endosperm unlike other tissues? Can you suggest any possible reasons for its unusual features?***

The endosperm is unlike the zygote in having two female chromosome complements and one male, instead of one of each. It is unlike the parent sporophyte tissue in having one male complement in addition to the two female ones. It is unlike all other tissues in being polyploid when first initiated.
It has been suggested that the hybridity likely to be brought in by the male contribution and the polyploidy makes the endosperm a more vigorous tissue, better able to compete with other tissues for parental resources which can then be passed on to the embryo. Another suggestion is that the 2:1 ratio of maternal and paternal genes makes it an effective intermediary facilitating interchange of materials between the purely maternal (ratio 2:0) ovule and the embryo with its evenly balanced (ratio 1:1) male and female contributions. Endosperm is also unusual in that, at least in the early stages of development, the nuclear division is not accompanied by cell division, so that a coenocytic mass of multinucleate cytoplasm is formed. This might reflect the fact that, in a time of uniform function with no internal differentiation, separation into cytoplasmic compartments is unnecessary. Thus resources need not be diverted into wall formation during the enlargement of the endosperm.

f ***In evolutionary terms what features of the angiosperm's life cycle reflects its position as the most advanced plant group?***

The reduction of the microgametophyte to three cells and the megagametophyte to about eight. The reduction of the latter within the sporangium throughout its development and its replacement by the seed as the dispersal unit. The loss of the motile male gamete and the development of elaborate pollination mechanisms, including complex flower structures to achieve dispersal of the microspore, and of the pollen tube to transport the gametes within the flower. The protection of the ovule by the unique carpel which allows elaboration of the pollination mechanism, monitoring of the pollen, and the development of a fruit. This, in turn, leads to the diversification of seed dispersal mechanisms. The formation of endosperm following double fertilization, is found in no other plant group. □

### Sexual cycles in mammals

*Questions and answers*

a ***What are the advantages of having a breeding season?***

Both sexes are available for mating at the same time. The probability of individual females becoming pregnant is higher. The young will be born at a time of the year suitable for their survival.

b ***What are the advantages of the unusual reproductive cycles of the bat and badger?***

In both cases the birth is delayed through the adverse winter period. However, there is also an advantage in the young being born as early in the year as possible, so that they have a long period of favourable conditions before the next winter.

## STUDY ITEM

### 22.62 The effect of males upon oestrus in female mice

*Principle*

1 The presence of male mice synchronizes and increases the frequency of oestrus in female mice.

*Questions and answers*

a ***What do you think was the value of the last period of the experiment when the males were removed and conditions were the same as at the beginning? Calculate the percentage of time for which the mice were in oestrus.***

It acted as a control. Percentages are 10 %, 19.6 %, 10 %.

b ***What do these figures suggest is the effect of having:***
***1 a number of females together without a male?***
***2 males in the immediate vicinity of females?***

In the first period of 12 days there were on average 1.2 days when a female mouse was found to be in oestrus. After the males were introduced, the incidence of oestrus per mouse rose to an average of 2.35 days, that is almost double.

c ***When males are present what do you notice in the majority of females about:***
***1 the timing of oestrus?***
***2 the frequency of oestrus?***

**1** There is a certain amount of synchronization of oestrus.
**2** Oestrus occurs more frequently.

d ***What hypothesis can you suggest to explain the effect of the males on the female oestrus?***

Sight, smell, and sound of the males might affect the female.

e ***What modifications in the experimental procedure would you make to test your hypothesis?***

This question is intended for a group discussion.

f ***What practical benefit could the social effect on the incidence of oestrus have in managing a stock of mice for studies of development?***

Females can be induced to be in a receptive state at the same time, hence breeding programmes might be synchronized more easily and generations produced more frequently. □

**STUDY ITEM**
22.63 Control of the menstrual cycle

*Principles*
**1** The menstrual cycle involves a complex system of hormone interaction.
**2** The cycle is controlled by systems of negative feedback.

*Questions and answers*
a ***What may prevent the maturation of further eggs during***
***1 an individual menstrual cycle?***
***2 pregnancy?***

**1** Once ovulation takes place during an individual cycle, the level of FSH begins to drop, preventing the maturation of any further follicles.
**2** Chorionic gonadotrophin, produced by the embryo, inhibits the production of FSH and consequently the maturation of follicles.

b ***If pregnancy does not occur, what may explain the drop in progesterone level at day 25?***

Progesterone inhibits the production of LH, so as it builds up after day 18 there is a drop in LH production which, in turn, causes the corpus luteum to cease production of progesterone – so it turns itself off, so to speak.

c ***What are the effects of the drop in progesterone level?***

It removes any inhibition of the production of FSH and LH.

d ***What information about the menstrual cycle could a daily record of body temperatures provide?***

On the day of ovulation there is a slight drop in body temperature. It then rises, and, until menstruation, the body temperature is maintained at about 0.3 °C above 'normal'. Thus by taking daily body temperatures it would be possible to determine the day on which ovulation occurred.

e ***The hormone FSH has been used as a 'fertility drug'. How might it act?***

FSH stimulates the maturation of follicles, so if infertility was due to the ovary not producing ripe ova, taking FSH might effect a cure. In practice it has been found that the dosage is critical. If too much FSH is taken, multiple births may result, as FSH not only affects the rate of follicle development, but also the number of follicles that are stimulated to develop.

f ***Hormone preparations, taken orally, are used as contraceptive pills. What hormones might you use in developing a suitable pill? Why would they possess a contraceptive effect?***

The combined contraceptive pill contains hormone preparations similar to oestrogen and progesterone. These affect the pituitary, by negative feedback, so that there are no mid-cycle surges in the production of FSH or LH, the surge by the latter hormone being essential for ovulation. This pill also prevents the release of any FSH, so that follicles in the ovary are not stimulated to mature.
Another type of pill contains only a progesterone preparation. This operates by affecting the mucus at the cervix (the entrance to the uterus) and so producing a barrier to sperm. It also affects the uterine wall, making it less suitable for implantation.

g ***Since such hormone preparations can be taken orally, what does this fact tell you about the chemical structure of the hormones?***

They cannot be composed of compounds, such as carbohydrates or proteins, which would be broken down by the gut enzymes. They are, in fact, steroids.

h ***At menopause, the ovaries cease the production of oestrogen, although they continue to produce a hormone, testosterone, which in the male affects the development of secondary sexual characteristics. What effects do these changes have on the female? How could such effects be reduced?***

Testosterone may encourage the growth of facial hair, a secondary sexual characteristic in most males. The effect might be reduced by taking a hormone preparation containing oestrogen.

i ***Explain the menstrual cycle in terms of a feedback system.***

During the early part of the oestrous cycle in women, around days five to eleven, the increase in oestrogens affects, via a negative feedback loop, the production of FSH by the pituitary. When the oestrogen levels reach a maximum, about three days before ovulation, they cause a positive feedback and as a result the production of both FSH and LH is increased.
After ovulation the corpus luteum produces progesterone. This, via another negative feedback loop, inhibits the production of LH and FSH. If pregnancy does not occur the reduction of LH levels causes degeneration of the corpus luteum and a consequent decrease of progesterone levels. □

## Events leading to fertilization in humans

**Practical investigations. *Practical guide 3*, investigation 10A, 'The relationship of the urinary system of a mammal to other organs of the body' (includes dissection of male and female reproductive systems), and *Practical guide 6*, investigation 22B, 'Reproduction in *Marsilea vestita*, the shamrock fern', and investigation 22E, 'Events leading to fertilization in angiosperms'.**

*Questions and answers*

c ***Why do you think this is so?*[*This question refers to the statement that if there is a reduced concentration of sperm in the ejaculate infertility is likely.*]**

If the normal count is 100 million sperm per $cm^3$, only 200 to 300 sperm reach the vicinity of the egg. Thus, if the sperm count is one-fifth, only 40 to 60 sperm will reach the egg. On average, 10 per cent of sperm are abnormal, and this might be even higher where the sperm count is abnormally low. It is possible that the presence of a number of sperm around the egg assists in breaking down the barriers near the egg, so that one sperm can enter. If this is correct, then in circumstances of low sperm counts fertilization is virtually impossible.

d ***Write an essay describing, from a sperm's point of view, the process from its formation in the testis until it achieves union with an egg in the oviduct.***

This essay is intended to provide an opportunity both for summarizing the processes involved and for a piece of creative writing.

## Events leading to fertilization in flowering plants

*Question and answer*

e *How would you set up an investigation to find evidence for a chemical attractant in the stigma?*

Set up a series of pollen grain preparations, as in *Practical guide 6*, investigation 22E. Place a small piece of crushed, mature stigma at the side of the coverslip in contact with the pollen germination medium. Record the directions in which the pollen tubes grow. Analyse the results statistically to see if there is any relationship between the direction of the pollen tube growth in relation to the position of the stigma. A suitable control is needed such as slides with germinating pollen but no stigma pieces.

## STUDY ITEM

### 22.64 Inheritance of plastids in *Pelargonium*

*Principles*

1 In animals, all the cytoplasm in the zygote is derived from the female gamete. The male gamete only contributes a nucleus. There is maternal inheritance of all cytoplasmic genetic factors, such as mitochondrial DNA.

2 It is usually assumed that higher plants also show this maternal inheritance of cytoplasm, which would affect not only mitochondria, but also plastids, which are similarly self-replicating and contain their own DNA.

3 Observations, described in this Study item, suggest that this is not always the case, and that, in some cases at least, paternal inheritance of cytoplasm occurs.

*Questions and answers*

a *Why do albino plants not survive beyond the seedling stage?*

Albino plants are unable to produce chlorophyll, and therefore cannot carry out photosynthesis. Once their food supply has been exhausted they will die.

b *What can you conclude from the data in table [S]34 about the plastids and how they are inherited?*

Selfing the crosses G × G and W × W produces progeny similar to the parents; the chloroplast types are stable. Selfing W × W reproduces very few seeds capable of germination. The progeny of the reciprocal G × W and W × G crosses include green and white seedlings, but they differ in the relative frequencies of the different seedling types: thus the plastid mutation is in the cytoplasm and not in the nucleus. The occurrence of both normal and mutant plastids within one cell indicates that the cytoplasmic mutation is in the plastid DNA.
In both cultivars, the results of G × W and W × G crosses show that plastids from the male (pollen) parent enter and replicate in at least some of the progeny. It may be that those seedlings which resemble the

female parent receive no plastids from the male parent. Alternatively, all the zygotes may have initially contained both normal and mutant plastids, but some have already 'sorted out' by the seedling stage. This would easily account for those seedlings which apparently have only a parental type of plastid. Those which have not sorted out by the seedling stage are still variegated.

The proportion of variegated seedlings varies with the cultivar, especially after G × W crosses. It is low in 'Flower of Spring' and high in 'Paul Crampel'.

The rate of sorting out is likely to vary with, for example, the absolute and relative numbers of plastids contributed by the two gametes and the absolute and relative rates of replication of the normal and mutant plastids. Cultivars differ in these respects, and there is evidence that a maternal nuclear gene, controlling plastid replication after fertilization, is the dominant controlling factor.

The two varieties have roughly similar proportions of seedlings inheriting the maternal plastid type but seedlings with white plastids are always fewer than those with green plastids. This is misleading. Further experiments showed that the level of maternal transmission of plastids in both G × W and W × G crosses is determined by the nuclear genotype of the female parent and is thus constant for one cultivar, regardless of the male parent variety, but differs in other varieties. The level of maternal transmission of green plastids is always greater than that for white plastids for each variety. Maternal transmission of green plastids is usually between 50 and 100 per cent, depending on the cultivar, and for white plastids, between 0 and 65 per cent. This means that (a) in W × G crosses involving a cultivar with low female transmission as white female parent, plastid inheritance is predominantly *paternal*, and (b) in W × G crosses involving a cultivar with high female transmission as white female parent, white plastids are inherited by a majority of seedlings.

c ***What can you deduce from these results regarding the inheritance of cytoplasm in* Pelargonium*?***

Clearly, plastids are transmitted to at least some of the zygotes from the pollen as well as from the egg. There is biparental inheritance of plastids. As plastids act as a cytoplasmic 'marker', this must mean there is biparental inheritance of cytoplasm. Thus, contrary to the generally held view about plants, at least some of the male gamete cytoplasm is capable of entering the zygote, with the male nucleus. As would be expected, plastids, albeit in an undifferentiated state, can be identified in electronmicrograph sections of male cells.

Several other plants also have plastids in the male gametes and they too have biparental plastid inheritance, revealing the entry of male cytoplasm at fertilization. These plants include *Oenothera, Hypericum, Rhododendron*, and *Secale*. In gymnosperms, plastid inheritance seems to be almost entirely paternal. However, the majority of flowering plants show only maternal transmission of plastids. Unfortunately, it is not possible to draw conclusions from this about the fate of the male cytoplasm at fertilization, because none of the plants examined have recognizable plastids in the male cell cytoplasm. It is therefore not possible to distinguish by breeding experiments between the exclusion of male cytoplasm from the zygote and the entry of male cytoplasm lacking plastids. Electronmicrograph studies have also, as yet, been unable to provide the answer. □

### The transfer of male gametes to the female

**Practical investigation. *Practical guide 6*, investigation 22D, 'Pollination mechanisms in angiosperms'.**

*Questions and answers*

**f** ***Why are these species with such specific mechanisms vulnerable? What modifications would reduce this vulnerability?***

The pollination mechanism in species such as *Yucca* is very vulnerable in that flowers cannot be naturally pollinated in the absence of the one pollinating agent. Thus, for example, *Yucca* is a relatively common garden plant in the UK and flowers easily, but no seed is set because of the absence of the moth. Modifications to reduce this vulnerability would be self-pollination or the development of a method of asexual propagation.

### STUDY ITEM

### 22.65 Pollination syndromes in *Trichostema* species

*Principles*

1. Flower characteristics are mainly concerned with mechanisms for transferring pollen from anthers to stigmas of the same or different flowers.
2. During their evolution, angiosperm species have become specialized to varying degrees to exploit different vectors. Floral features reflect these adaptations, resulting in a more or less characteristic pollination syndrome.
3. There are major differences between most wind-pollinated flowers and animal-pollinated flowers. Among animal-pollinated species, the plants and the pollinator have evolved together and differences in flower size, colour, pattern, shape, scent, pollen and nectar production, relative position of anthers and stigmas, and timing of their maturity, reflect differences in the size, strength, agility, hairiness, colour and scent perception, food preferences and requirements, and foraging behaviour of the animals which pollinate them.
4. While some plant species with unspecialized flowers are pollinated by a variety of animals with non-specialized feeding requirements, in many cases the floral syndrome indicates at least one of the major groups used as pollen vectors.
5. Wastage of resources through loss of pollen and failure of ovules to be fertilized is common in sexual reproduction and varies in extent according to the pollination method and vector. Cross-pollination generally requires more resources, because of greater wastage, compared with self-pollination.

*Questions and answers*

a ***How is each of the species* T. lanatum, T. micranthum, T. laxum, T. austromontanum, *and* T. ovatum *pollinated? Give reasons for your conclusions.***

*T. laxum* and *T. ovatum*: obligate cross-pollination by Hymenoptera because no seed set unless visited for the available nectar.
*T. micranthum* and *T. austromontanum*; automatic self-pollination because they are the two species of the remaining three with similar characteristics, including high seed set, even without flower visitors, and because absence of nectar combined with small flowers are characteristic of self-fertilized species.
*T. lanatum*: cross-pollinated by humming birds. The large flower size and unusually large nectar volume and pollen: ovule ratio is consistent with this, despite the fact that it is clearly also self-fertile.

b ***How are the flower size, nectar volume, and pollen:ovule ratio related to***
***1 breeding system (self-fertilization or cross-fertilization), and***
***2 pollinating agent?***

Flower size, nectar volume, and pollen: ovule ratio are typically small where automatic self-pollination eliminates the need for animal visitors and ensures efficient pollen transfer from anthers to stigma with minimal loss. Larger flowers, nectar volumes, and pollen: ovule ratios are necessary in insect-pollinated obligate out-breeders in order to attract pollinators and to compensate for greater losses in transit. Larger flowers and pollen: ovule ratios, and unusually large quantities of nectar typically characterize plants pollinated by humming birds, which are less efficient pollinators than bees, and require much more energy.

c ***What do your conclusions imply about the commitment of resources to pollination mechanisms?***

Selection has favoured economy of resources to the lowest level which allows good seed-set. The reduced energy expenditure necessary to ensure seed set in self-pollinating species releases resources for other purposes. □

## STUDY ITEM
## 22.66 Speciation in campions

*Principles*

1 Red and white campions will interbreed, producing pink-flowered plants with other intermediate characteristics.
2 Under natural conditions the habitats of the two types are sufficiently far apart that they probably do not normally interbreed. It is only under the conditions of a man-made habitat, the hedgerow, that conditions may occur when the two forms overlap and interbreed.

*Questions and answers*

a ***What possible explanations are there to account for the populations with pink-flowered individuals?***

There are three main possibilities:

**1** There is one variable species in which flower colour ranges from red through pink to white, even within a population.

**2** There are two distinct species, in at least one of which flower colour can vary under the influences of genetic or environmental factors.

**3** There are two distinct species, one red- and one white-flowered, which in certain circumstances hybridize to give forms intermediate in colour.

b ***How could you distinguish between your suggested explanations?***

**1** At least some other major characters would vary independently of colour, so that red, pink, and white forms would show no clear cut morphological differences and would freely interbreed.

**2** On the basis of a number of morphological characters, the plants would fall into two distinct and probably intersterile categories, at least one of which would include a range of flower colours.

**3** Red-flowered and white-flowered plants would be clearly distinguishable on the basis of several other morphological characters, while pink-flowered forms would have different combinations of, or intermediate, characters and could be produced by crossing the interfertile red- and white-flowered plants.

c ***What conclusions about the relationship between red and white campions can you draw from these data?***

Red and white campions differ in characters other than just flower colour. Pink-flowered campions are intermediate in many respects, not only colour. Most populations of red campions are ecologically isolated from white campions, and this isolation is reinforced by differences in pollination timing and vector, and slightly reduced pollen performance. Despite this, hybridization between red and white campions is possible. The ecological distributions overlap in man-made habitats of wayside hedgebanks. Stable, undisturbed hedgebank communities favour the red form, while disturbance, such as road widening, favours the white. Where there is a mosaic of periodic, very localized, disturbance, as caused by small landslips, the two types are able to grow in close proximity. It is also in this situation, where the sites of less recent landslips provide habitats intermediate between the open, newly disturbed areas and the stable, fully established hedgebank community, that the intermediate pink-flowered forms become established.

d ***Do you consider that red and white campions are two separate species, or two extreme forms of one species? Give your reasons.***

Although at least one author, applying the biological definition of species, has treated the red and white forms as subspecies of the same species (Clapham *et al.* (1981), *Flora of the British Isles*), several other taxonomists treat them as two distinct species, *Silene alba* (white campion) and *Silene dioica* (red campion), which can form hybrids naturally under certain conditions. It seems likely that the two species evolved independently over a long period, isolated mainly by different ecological preferences. Relatively recently, human activities have increased the number of disturbed sites, allowing *S. alba* to spread more widely, and created a new habitat, the hedgerow, where the two species are brought together again. As the genetic isolation which developed during their separate evolution is not complete, hybrids are sometimes produced when the species grow within pollination range of each other. Presumably there are some insects which visit both species, if only occasionally. The fact that both species are dioecious and thus always cross-pollinated will increase the chances of hybridization. Once formed, the hybrids can establish successfully in habitats intermediate between the parental preferences, such as the later successional stages in the re-establishment of hedgerow communities after disturbance, provided these occur within the limited seed-dispersal range of the two parents. Hedgebanks disturbed by small landslips provide all the conditions necessary for the establishment of hybrids and, because the hybrids are fertile, for subsequent interbreeding and backcrossing to form a 'hybrid swarm' of intermediates ranging from one parental type to the other. However, despite the free gene exchange between the parent types taking place under these specialized conditions, during much of their history, and even now in most of the populations, the integrity of the two distinct forms is maintained. It is only human influence which has brought about the breakdown of the naturally effective barriers. It is this that has influenced the decision of most taxonomists. □

## Notes

1. This exercise would benefit from having plant material available for study during the discussion. It could equally well be presented as a practical investigation.
2. There are other, similar, examples in the British flora of the breakdown of isolation. In eastern Europe, water avens (*Geum rivale*) occurs in shaded, marshy habitats in upland regions while wood avens (*Geum urbanum*) is a lowland plant colonizing disturbed woodland, and they rarely have the opportunity to hybridize. In Britain, they sometimes grow together in damp hedgerows and woods (mostly disturbed in Britain) and then they hybridize to produce hybrid swarms (see Briggs and Walters, 1984).
3. This project provides a good starting point for a closer study of hedgerows as a habitat – not only as a refuge in regions of intensive agriculture or suburban development where other semi-natural habitats have largely disappeared, but also as a unique meeting point for species which otherwise have quite distinct distributions, in, for example, woodland, grassland, disturbed ground, or streamsides.

## 22.7 Breeding systems

*Principles*

1. Conservation of much of the genotype is necessary for the success of an organism.
2. The chromosomes provide a means of preserving many advantageous combinations of alleles.
3. Asexual reproduction provides a means of restricting variation.
4. In sexual reproduction, variation is reduced when there are a small number of chromosomes and few chiasmata formed during meiosis.
5. Inbreeding reduces the level of heterozygosity.
6. An increase in variation requires cross-fertilization between gametes of distantly related individuals.
7. Higher plants are generally hermaphrodite so a balance between self- and cross-fertilization will allow generation of new genotypes and conservation of the most successful ones.

*Questions and answers*

a ***What can you say about the genetic make-up of the zygote produced by such inbreeding?***

The zygote will be homozygous at all genetic loci.

b ***What will be the characteristic of the genetic make-up of the gametes of the next generation?***

The gametes will be genetically identical.

c ***In what way will the products of such inbreeding be different in their genetic make-up from clonal individuals?***

Asexually produced clones still retain heterozygosity at various loci. However, the products of this inbreeding will be homozygous at all loci and will have lost any benefit that heterozygosity may confer.

### STUDY ITEM

### 22.71 Reproductive structures in flowering plants

*Principle*

1. The nature of the reproductive structures found in a flower affect the proportions of self- and cross-pollination.

*Questions and answers*

a ***Find out the meaning of these terms by consulting suitable texts.***

Dioecious: having male and female flowers on different plants.
Monoecious: having male and female flowers on the same plant.
Hermaphrodite: having male and female organs in the same flower.

b ***What is the effect on the frequency of selfing of:***
***1 being dioecious?***
***2 being monoecious?***
***3 being hermaphrodite?***

**1** As there are separate male and female plants, selfing is not possible.
**2** The possession of unisexual flowers on the same plant reduces, but does not necessarily prevent, selfing.
**3** Plants with anthers and stigma in the same flower would appear to be capable of the highest degree of selfing. However, the reproductive organs may mature at different times. In addition there are several more subtle factors operating, such as flower arrangement, positions of parts within a flower, structural differences in pollen and stigmas, and physiological interactions between pollen and stigma. All these factors may either prevent selfing, or influence the breeding system by affecting the relative proportions of self- and cross-pollination.

c ***What characters or mechanisms found in hermaphrodite plants would reduce the likelihood of self-pollination?***

□ See the answer to **b** above.

## STUDY ITEM
## 22.72 Self-incompatibility in plants

*Principles*

1. Many plants have physiological mechanisms which inhibit the growth of pollen in stigmas having the same phenotype for the self-incompatibility loci, thus preventing most or all self-fertilization, even after self-pollination. It may also prevent fertilization after cross-pollination, if the other plant shares the same incompatibility allele(s).
2. In some species, the incompatibility phenotype is determined in the parent sporophyte and all the pollen has the phenotype of the active allele(s) at the incompatibility locus (sporophytic incompatibility). One allele is usually dominant over the other and dominance relationships are usually the same in the pollen and the stigma. Thus self-fertilization is prevented. In other species, the self-incompatibility phenotype is determined individually by the allele in each pollen grain after meiosis, and then both alleles sometimes operate in the stigma too (gametophytic incompatibility).
3. These mechanisms reduce or prevent self-fertilization and homozygosity and can be exploited in the production of heterozygous seed in economic crops, but they create problems when inbred lines of uniform homozygous seed are required. For this, selfing has to be carried out in buds which have not yet developed the physiological incompatibility mechanism.
4. Self-incompatibility is thought to have arisen early in angiosperm evolution. It has subsequently been lost from those species which are now naturally self-fertile.

*Questions and answers*

a ***Why does the bud-selfing result in a good seed-set?***

The physiological incompatibility reaction in the stigma only develops as the flower opens.

| | | *As pollen parent* | | |
|---|---|---|---|---|
| | | A | B | C |
| *As seed parent* | A | SI | I | C |
| | B | I | SI | C |
| | C | C | C | SI |

**Table 17**
Summary of results.
C = compatible
I = incompatible
SI = self-incompatible

**b** ***How might bud-selfing be useful?***

In producing otherwise unobtainable homozygous plants during *Brassica* breeding.

**c** ***Which of the genotypes included in table [S]36 are self-incompatible?***

All of them.

**d** ***In the crosses involving genotypes A, B, and C, which are the incompatible combinations?***

A × B and B × A.

**e** ***What can you say about seed-set in self-incompatible (SI), incompatible (I), and compatible (C) combinations?***

Within C, I, and SI combinations, seed-set is not uniform, some genotypes yielding more seeds than others. This indicates that other genotypic factors have a modifying effect on seed-set. SI and I combinations usually yield a few seeds, indicating that the incompatibility response is not always completely effective. Environmental factors may influence this.

**f** ***If only three S alleles are involved in total in genotypes A, B, and C, what are the three genotypes with respect to the S locus?***

□ $S_1S_2$, $S_1S_3$, and $S_2S_3$.

## 22.8 Embryo protection and nutrition

*Principles*

1. Plant embryos are protected by parental tissue.
2. In angiosperms the seeds can provide a means of both protection and dispersal.
3. During the development of the embryo, nutrients are provided by the parent.
4. In egg-laying animals the eggs often possess some form of protection.
5. After laying, the eggs of many animals are cared for by one or both parents.

6 Eutherian mammals show the highest degree of parental protection by the development of a special structure, the placenta. This provides a means of exchanging materials between parent and embryo.

*Questions and answers*

a ***The fact that fewer seeds are produced per flowering plant than spores per fern of similar size suggests that the seed possesses significant advantages over the spore. What might be the advantages?***

The multilayered seed coat offers more protection than the spore wall. The internal tissues can accommodate much larger food reserves, thus allowing the young sporophyte to grow larger during establishment before it is self-supporting. When the seed begins to germinate, the embryo already consists of many cells and is therefore larger and more robust than the single-celled spore. The seed disperses the diplophase (embryo) and the spore disperses the haplophase (germinates to form a gametophyte). This has repercussions for propagules entering new sites. The seed can 'carry' heterozygosity and can alone give rise, by self-fertilization and seed production, to a population which, at least initially, has some variability. A single colonizing spore can give rise to a population, but only by intragametophytic selfing which will produce homozygous sporophytes and a uniform population in the next generation.

b ***What are the advantages of such behaviour?***

First, the embryo is more protected from mechanical damage and from predators. Second, as the animal is motile, the embryo is maintained in a favourable external environment. In true viviparous animals an additional factor is that the nutritional requirements of the embryo can be provided for more efficiently, as compared with a zygote isolated from the parent.

c ***What materials would need to be exchanged?***

The embryo requires a supply of nutrients and oxygen, and needs to dispose of waste substances, such as carbon dioxide and nitrogenous wastes. The embryo also produces hormones which must pass into the maternal blood circulation, such as chorionic gonadotrophin which maintains the corpus luteum in the ovary.

d ***How might such information be passed?***

If fertilization does not occur then the level of progesterone circulating will drop and the lining of the uterus will no longer be maintained. In humans, assuming that fertilization takes place around day 16 of the menstrual cycle and menstruation will commence on day 28, there are 12 days in which the embryo's presence must be signalled to the mother. The signal is made by a hormone. In this case the chorion secretes chorionic gonadotrophin. This hormone is similar in structure to luteinizing hormone and maintains the production of progesterone by the corpus luteum.

e ***In the human placenta what barriers remain between the embryonic and maternal blood circulations?***

Four layers: two layers of chorionic epithelium, derived from the trophoblast; embryonic connective tissue; embryonic endothelium of the embryonic blood vessel. When the placenta has fully developed only two or three layers, all embryonic remain. These are: chorionic epithelium, which has reduced down to one layer; connective tissue (very much reduced); and endothelium of the embryo's blood vessels.

f ***Why would a direct connection between the two circulations be disadvantageous?***

The mother and embryo are not genetically identical, so there could be immunological reactions and the mother could develop antibodies to the embryo. Because the barrier is not a complete one, this does happen as in the case of the rhesus factor (see *Study guide I*, Chapter 4, page 118). The barrier also prevents the passage of micro-organisms that might be circulating in the maternal blood. However, as noted, the barrier is not perfect: a number of harmful materials can pass through including teratogens, such as thalidomide, the rubella virus, and drugs such as heroin (a baby can be born a heroin addict).

# PART II The *Practical guide*

The investigations related to this chapter appear in *Practical guide 6, Development, control, and integration.*

## INVESTIGATION 22A Reproduction in fungi

### ITEMS NEEDED

*Mucor hiemalis* on potato dextrose agar plate, plus and minus strains 1/group
*Penicillium expansum* on malt agar plate 1/group
*Saccharomyces cerevisiae*, culture 1/class
Microscope slides, prepared, of the three fungi (optional)
Forceps 1/group
Incubator
Microscope, binocular stereo 1/group
Microscope, monocular (× 100, × 400) 1/1
Microscope slides and coverslips
Mounted needle 1/group
Pipettes, dropping 1/group

(*Study guide* 22.2 'Asexual reproduction' and 22.6 'Gametogenesis and fertilization'.)

***Saccharomyces cerevisiae* culture.** Make up two hours in advance by adding 2 g of fresh baker's yeast to 100 $cm^3$ of 5 % sucrose solution, stirring well and leaving at 20–25 °C.
***Penicillium expansum*.** Inoculate onto malt agar plates about 4 days in advance and incubate at 25 °C.
***Mucor hiemalis* plus and minus strains.** Inoculate 4 cm apart on potato dextrose agar 2–3 days in advance and incubate at about 20 °C. Before incubation, turn the dish over and mark the points of inoculation. In addition, set up a few plates inoculated as above, but only using the plus strain, and others only using the minus strain.

*Assumptions*

1 Competence in the use of microscopes.
2 A knowledge of sexual and asexual processes involving mitotic and meiotic cell divisions, including the process of crossing-over.

3 Familiarity with safety precautions for handling micro-organisms and with sterile techniques.

*Principles*

1 *Saccharomyces cerevisiae* exhibits the basic asexual process of cell division.
2 Moulds show an elaboration of the asexual process where divisions in aerial hyphae gives rise to spores.
3 Two mating strains, with no male or female characteristics, are needed for sexual reproduction in *Mucor hiemalis*.
4 *Mucor hiemalis* possesses a haplontic life cycle where mitoses only occur between meiosis and fertilization and the body of the fungus is haploid. Meiosis occurs immediately after fertilization, so restoring the pre-fertilization genome.
5 Sexual and asexual structures in *Mucor hiemalis* have different survival values for the organism, depending upon the environmental conditions.
6 Sexual reproduction and the increase in numbers are seen as distinct processes in *Mucor hiemalis*.
7 Recombination of genetic material during meiosis has disadvantages as well as advantages for the offspring.

This practical asks students to make observations without much background knowledge. They may find this difficult. However, the observations they should be able to make will form a useful basis for the elaboration of the principles of sexual and asexual processes.

The practical could be introduced by a brief examination of mouldy bread and reference should be made to the classification of the fungi. Students must be familiar both with the safety precautions for handling micro-organisms and with sterile techniques. Appropriate precautions should be taken to avoid the unnecessary dispersal of spores.

Prepared microscope slides, showing the reproductive structures of these fungi, will be a useful supplement. The conidiophores of *Penicillium* are extremely delicate, so students will have to prepare the slide with care and look intently at it in order to locate a reasonable example. Further, air bubbles around the conidiophores are often a problem in examining such a preparation.

An alternative technique for making a preparation of *Penicillium* is as follows:

**1** Cut a strip of transparent adhesive tape about 4 cm long.
**2** Holding it gently by the two ends, press the middle 2–3 cm gently onto the surface of a *Penicillium* culture.
**3** Remove the tape and press it, sticky side down, onto the surface of a clean microscope slide.

By using this technique students may find it easier to obtain preparations of whole conidiophores.

*Questions and answers*

**a** ***The simplest form of asexual reproduction is called budding. This process is found in* Saccharomyces cerevisiae. *What nuclear and cytoplasmic processes must occur before a new organism is formed by this process?***

The nucleus must undergo mitosis and the cell must undergo cytokinesis to provide the new organism with its genetic complement (*figure 16*).

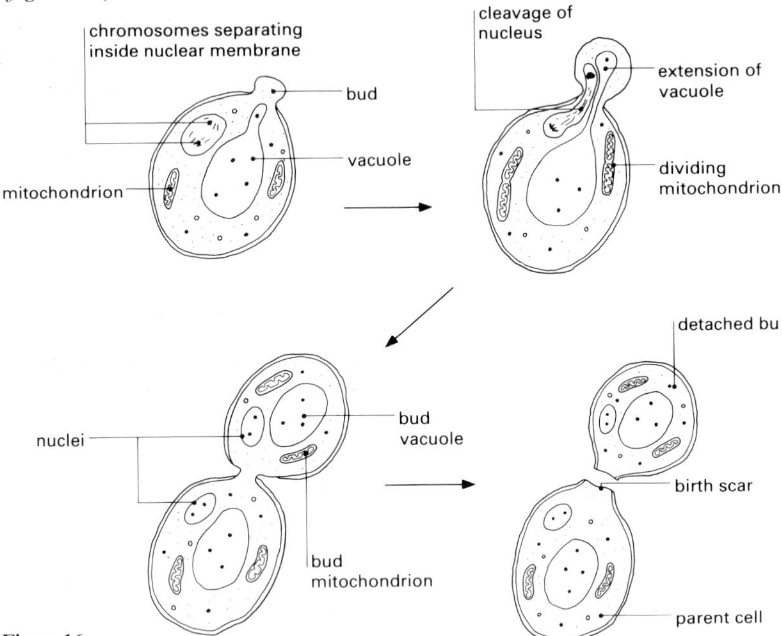


**Figure 16**
Budding in *Saccharomyces cerevisiae.*
*After Williams, J. I. and Shaw, M.,* Micro-organisms, *2nd revised edn, Bell & Hyman, 1982.*

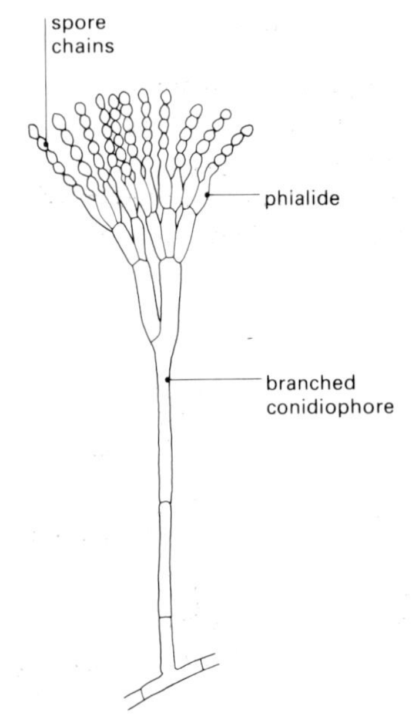


**Figure 17**
A conidiophore of *Penicillium.*
*After Williams, J. I. and Shaw, M.,* Micro-organisms, *2nd revised edn, Bell & Hyman, 1982.*

**b** ***Asexual reproduction in* Penicillium expansum *follows the same basic process, but with structural elaborations. What modifications did you observe?***

'Budding' in *Penicillium expansum* occurs in reproductive units of aerial hyphae called conidiophores (*figure 17*). Special cells at the tips of these hyphae cut off chains of spores, called conidiospores. These spores are probably better protected than the simply budded cells of yeast.

**c** ***What are the advantages of the two types of asexual reproduction?***

Budding is an effective method of reproduction in a liquid medium. For aerial spore dispersal the mechanism in *Penicillium* and *Mucor* is particularly effective. A large number of light resistant spores can be produced in a short time.

**d** ***What differences were you able to detect between the two* Mucor *strains?***

There is no physical difference detectable between the two strains in this species of mould.

e ***What evidence did you observe of sexual reproduction in* Mucor*? Form a hypothesis concerning the occurrence of the sexual process in this mould. Design an experiment that would test it.***

Special bodies, zygospores, are formed along the line of contact between the two *Mucor* strains. A hypothesis to explain this could be that sexual reproduction occurs at points of contact between the hyphae of two different mating strains of *Mucor hiemalis*. Experiments to test this could involve different arrangements of mating strains and situations where the contact of different mycelia was restricted. At this point examination of the plates which were inoculated with the same strain will show that under such circumstances no zygospores are produced.

f ***Why do you think that the two types of* Mucor *are named + and − strains, and not male and female?***

*Mucor* is a homogamous species; there is no difference between the two strains in the vegetative structure, or in the size, shape, or behaviour, of the gametangia produced. Thus the differences between the strains are on a biochemical rather than on a structural or behavioural level.

g ***Sexual and asexual structures are quite distinct in* Mucor*. How may their functions differ? How might this be related to the dispersal of the mould and to factors that adversely affect its survival?***

Sexual processes produce thick-walled zygospores that remain dormant for some time before germinating, resistant to adverse conditions and containing the offspring with potential genetic variability (*figure 18*). Asexual processes produce numerous spores, of identical genetic make-up, for dispersal and rapid colonization of suitable environments (*figure 19*).

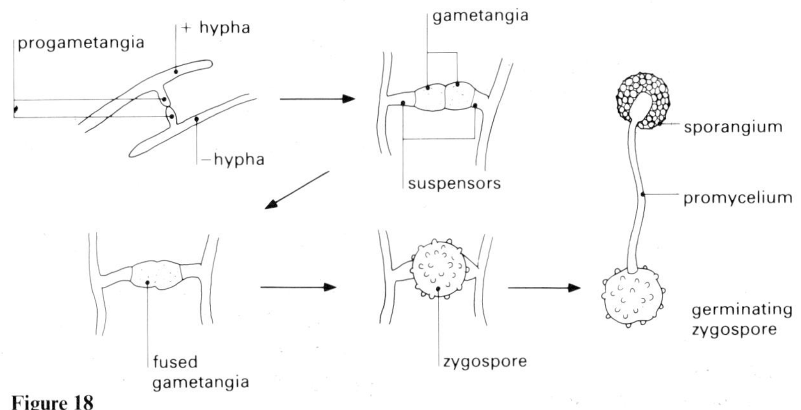


**Figure 18**
Formation of zygospores in *Mucor*.
*After Williams, J. I. and Shaw, M.,* Micro-organisms, *2nd revised edn, Bell & Hyman, 1982.*

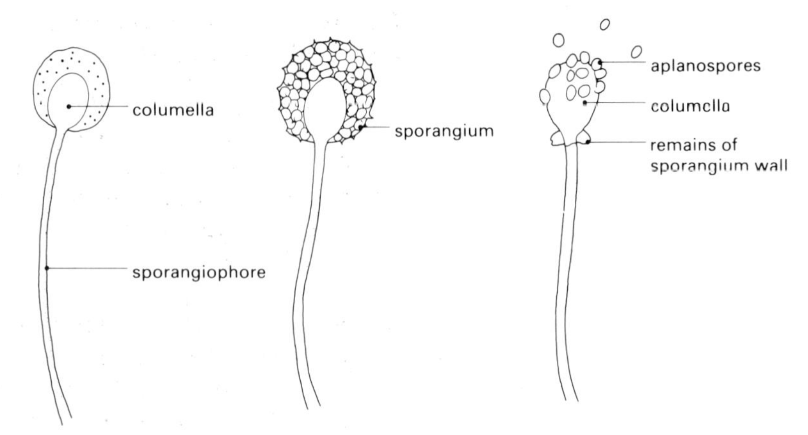


**Figure 19**
Asexual reproduction in *Mucor*.
*After Williams, J. I. and Shaw, M.,* Micro-organisms, *2nd revised edn, Bell & Hyman, 1982.*

**h** ***Draw a diagram showing the life cycle of* Mucor hiemalis, *including both sexual and asexual phases (you may need to consult suitable texts). Why do you think that scientists are still not certain of exactly when meiosis takes place in the life cycle?***

*Mucor hiemalis* exhibits a haplontic life cycle; the zygote is the only diploid phase (see *figure 20*). The nuclei of *Mucor* are very small and difficult to observe when the mycelia join. The zygospores develop a thick black coat, making it impossible to observe nuclear events taking place inside. The students have not been given the information that hyphae are haploid. You may wish to do this, or alternatively to let them arrive at a number of possible solutions. When a *Mucor* zygospore germinates it produces a promycelium which develops one sporangium. The spores germinating from this sporangium give rise to a normal mycelium.

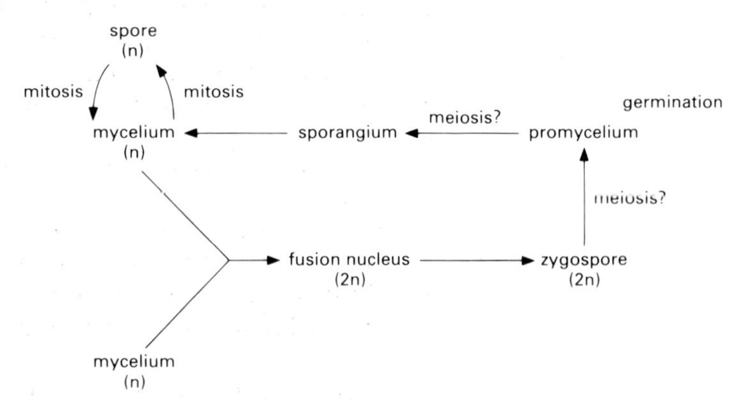


**Figure 20**
The life cycle of *Mucor hiemalis.*

i ***Why must meiosis occur during the life cycle of a sexually reproducing organism?***

Meiosis firstly generates and regulates the recombination of the genes and, secondly reduces the genome content to compensate for the doubling that occurs at fertilization. In *Mucor*, meiosis occurs after the fusion of nuclei from the + and − gametangia. It is uncertain exactly when it takes place, but it may be on the germination of the zygospore (see Ingold, 1978).

At this point you might like to raise two issues:

**1** Variation is not the only important factor in a breeding system (see *Study guide II*, section 22.1). In a stable environment a stable genotype shows a high survival value compared with variable genotypes.

**2** The vegetative body of a mould such as *Mucor* is haploid. As a result it may be more vulnerable to the effects of chance gene changes, as in mutation.

## INVESTIGATION
## 22B Reproduction in *Marsilea vestita*, the shamrock fern

### ITEMS NEEDED

*Marsilea vestita* sporocarps 1–2/group

Water, distilled

Microscope, binocular stereo 1/group
Microscope, monocular (× 100, × 400) 1/1
Microscope slides and coverslips
Petri dish 1–2/group
Pipette, dropping 1/group
Scalpel 1/group
Stopclock or watch 1/group

(*Study guide* 22.5 'Variations in life cycles' and 22.6 'Gametogenesis and fertilization'.)

Sporocarps of *Marsilea vestita* can be obtained from biological suppliers. They have a shelf life of many years.

*Assumptions*

**1** That students have a basic understanding of gamete structure and function.

*Principles*

**1** Many lower plants and most aquatic invertebrate animals liberate gametes and avoid the need for physical contact between mating types.

**2** Liberation of gametes has disadvantages: for success the process requires environmental water and mechanisms to ensure that the different types of gamete meet.

**3** Motile male gametes are not confined to the animal kingdom.

**4** The developing embryo depends upon the gametophyte for raw materials needed for development prior to the production of photosynthetic tissues.

**5** In all land plants, gametes are eventually produced from spores by mitosis in the gametophyte generation.

**6** *Marsilea vestita*, like seed plants, is heterosporous, producing unisexual gametophytes from microspores and megaspores.

*Marsilea* reproduces asexually by means of rhizomes and sexually by the liberation of microspores and megaspores from the sporocarps (*figure 21*). Once the sporocarp starts to imbibe water, the gelatinous ring inside swells, so pushing the attached sori out of the sporocarp (*figure 22*). This initial spore release takes place 45 to 60 minutes after immersion of the sporocarp. During the next 16 hours, the sori break

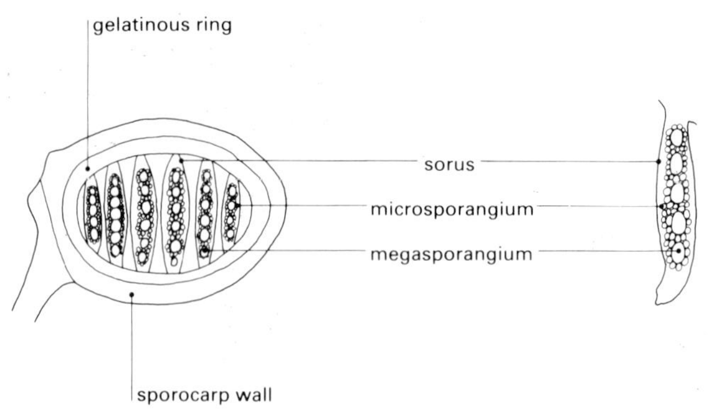


**Figure 21**
A vertical longitudinal section through a sporocarp of *Marsilea.*

down and release microspores and megaspores (*figure 23*). The microspores form small gametophytes which release male gametes, antherozoids, and then degenerate. The megaspore develops a single apical archegonium that secretes a gelatinous matrix. This matrix appears to attract the antherozoids. Release of motile antherozoids by the microspores and their active phase lasts about 15 minutes, but the time of release varies from between 16 and 20 hours after immersion of the sporocarp. The antherozoid is a complex structure as shown in *figure 24.*

This practical can serve as an introduction to the role of gametes, in general, and to the haploid/diploid phases in the life cycles of pteridophytes. It is relatively easy to grow plants from the embryonic sporophytes and it is worth encouraging students to do this. However, so far, the right conditions for the development of sporocarps have not been found. Try to obtain specimens of the adult plant from horticultural suppliers for the students to examine.

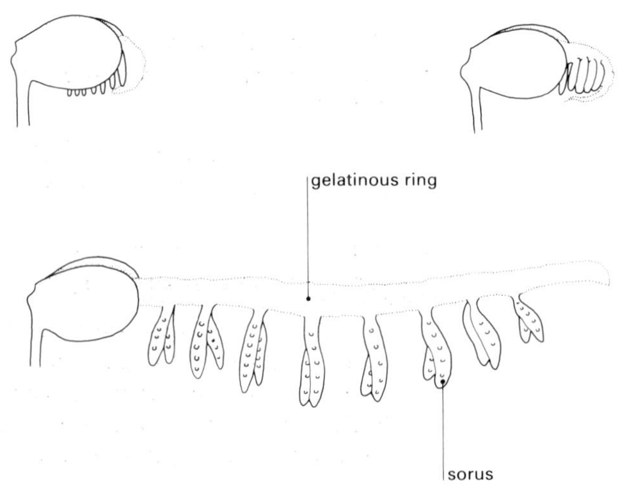


**Figure 22**
Stages in the extrusion of sori in *Marsilea.*

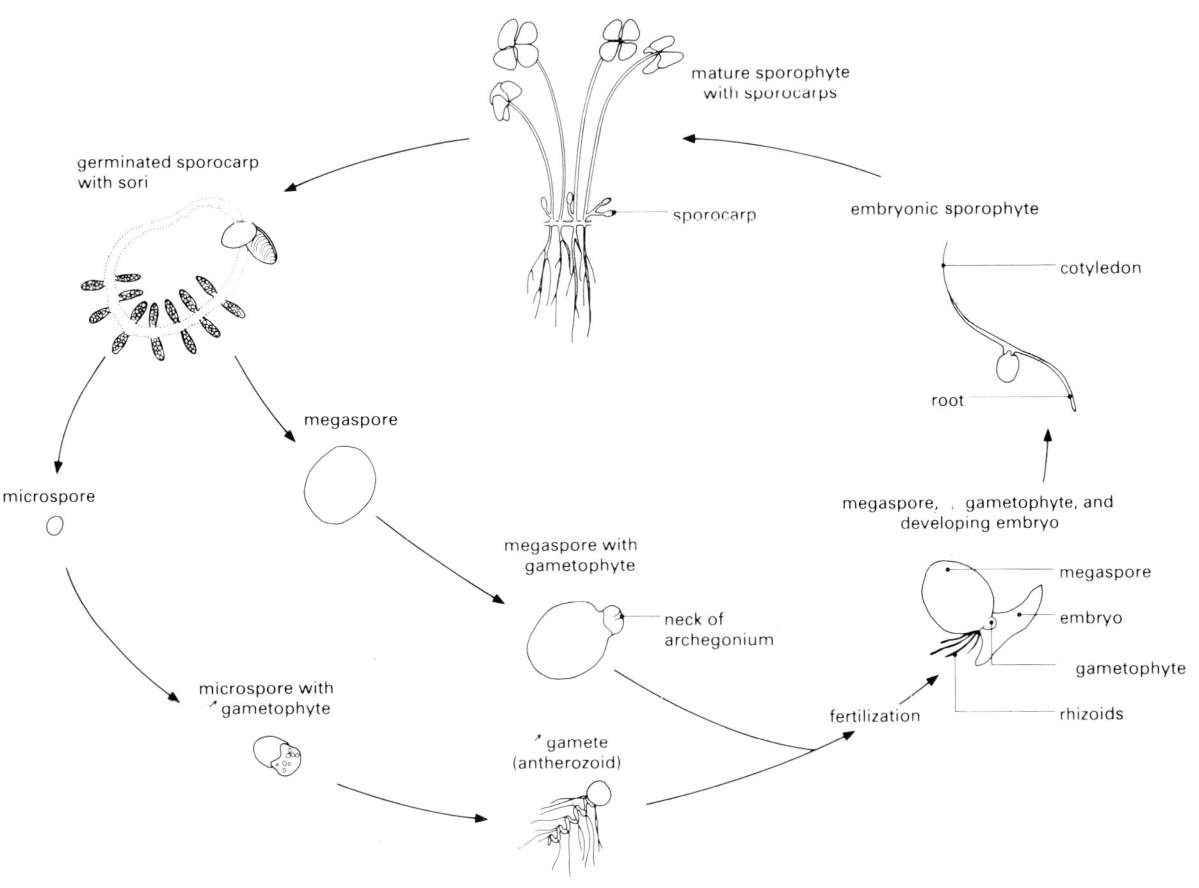


**Figure 23**
The life cycle of *Marsilea vestita*.
*After Philip Harris Biological, Marsilea vestita* shamrock fern sporocarps, instructions and teaching notes. *Philip Harris Biological Ltd, 1975.*

*Questions and answers*

**a** ***How many types of spore does* Marsilea vestita *produce? How do they differ?***

Two. A few large immobile female spores and numerous small immobile male spores that release motile antherozoids.

**b** ***What is the mechanism of spore release? Suggest the environmental conditions that are necessary for successful release and fertilization.***

The sporocarp imbibes water and bursts releasing 'pod-like' sori, which are attached to a gelatinous ring or sorophore. The sori breakdown and release microspores and megaspores. Spore release is likely to occur when water levels are high e.g. after periodic drought conditions and temperature is favourable for development.

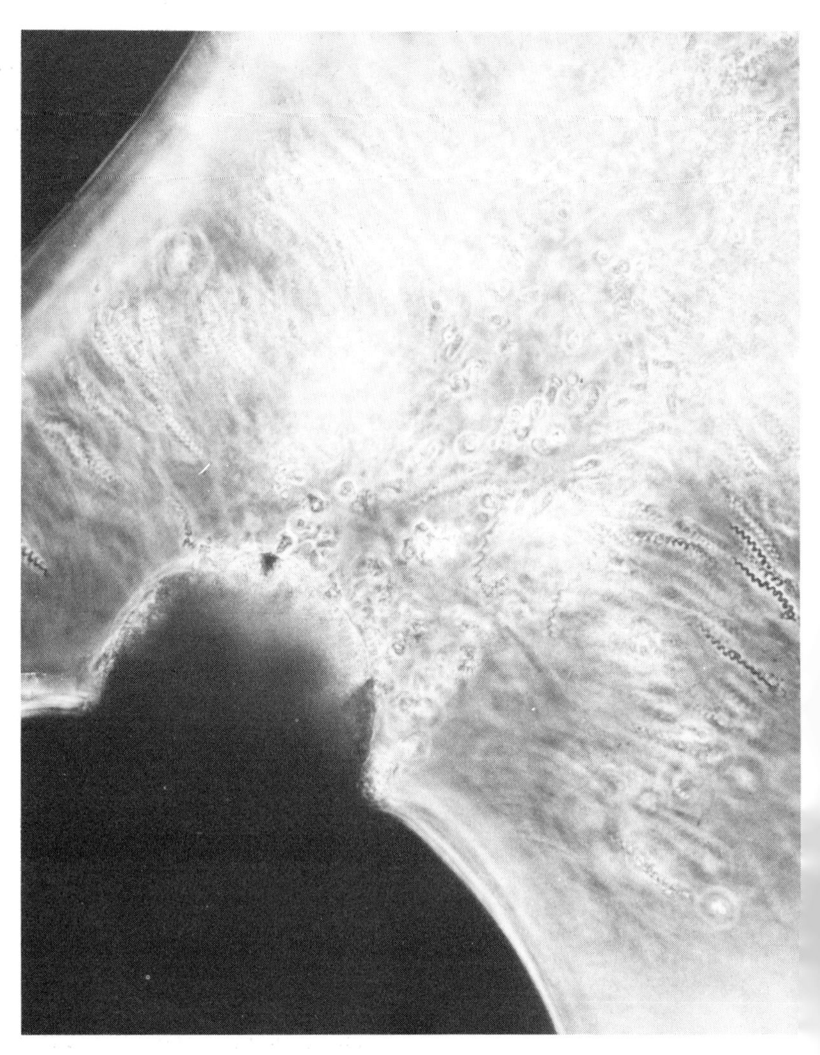

**Figure 24**
*Marsilea*; megaspore with protruding gametophyte with archegonium. The neck canal is clearly visible with a coiled male gamete about to enter. Other male gametes are seen trapped in the surrounding mucilage. ( × 56)
*Photograph, Philip Harris Biological Ltd.*

c ***What events take place immediately before fertilization? What mechanisms may help to increase the chances of gametes meeting?***

The megaspore develops a swelling at one end. This represents the female gametophyte with a single apical archegonium. This secretes mucilage that appears to attract the antherozoids. The motile antherozoids are released from the degenerating microspores. Antherozoids move through the mucilage towards the archegonium. Chances of fertilization are increased if the motile gamete is attracted towards the non-motile one.

d ***What are the advantages and disadvantages of organisms releasing their gametes directly into their surroundings?***

Advantages are that parent organisms do not have to come into physical contact, and it is a simple operation. Disadvantages are that the motile gametes are confined to a watery medium, it is a hazardous process, and special mechanisms are needed to increase the chances of gametes meeting.

e ***Based upon your observations of early development, how does the embryo obtain the nutrients needed for growth and development?***

The embryo obtains nutrients for initial development from the female gametophyte (megaspore), but early production, in seven to nine days, of photosynthetic tissue allows the embryo to become self-sufficient.

## INVESTIGATION
## 22C The life cycle of *Dryopteris filix-mas*, the male fern

### ITEMS NEEDED

*Dryopteris* sporophyte, mature fronds 1/group
Gametophytes, mature 1/group
Microscope slides, prepared sections through gametophyte sex organs

Fern medium

Forceps 1/group
Lamp, microscope 1/group
Microscope, binocular stereo 1/group
Microscope, monocular 1/1
Microscope slides and coverslips
Mounted needle 1/group
Petri dish 1/group
Pipette, dropping 1/group

(*Study guide* 22.5 'Variations in life cycles'.)

**Growing gametophytes.** Ferns spores, such as those from *Dryopteris filix-mas*, can be collected in the autumn. If the spores are kept dry in a specimen tube they will remain viable for several years. Alternatively they may be purchased from biological suppliers and also from some horticultural suppliers. Gametophytes can easily be grown from spores. It is best to grow the gametophytes on a liquid medium, or a semi-solid one as this makes them much easier to examine than if grown on soil.

In practice it has been found that *Pteridium*, bracken, may be a better choice for growing gametophytes. It produces antheridia and archegonia in culture more rapidly and more reliably than *Dryopteris*.

Fern gametophytes are very useful material to use to investigate factors affecting plant development. An experimental fern development kit to do this is available from Philip Harris Biological Ltd, Oldmixon, Weston-super-Mare, Avon BS24 9BJ.

**Fern medium.** A semi-solid medium can be prepared by adding 0.5 % mass per volume of agar. A suitable medium is as follows:
Magnesium sulphate, $MgSO_4 7H_2O$, 0.51 g $dm^{-3}$
Potassium nitrate, $KNO_3$, 0.12 g $dm^{-3}$
Calcium nitrate, $Ca(NO_3)_2 . H_2O$, 1.44 g $dm^{-3}$
Dipotassium hydrogen phosphate $K_2HPO_4$, 0.25 g $dm^{-3}$
Iron(III)chloride, $FeCl_3 . 6H_2O$, 0.017 g $dm^{-3}$
The solutions, if stored separately, can be made up at 100 × the concentrations given. Petri dishes are suitable containers for growing the gametophytes, but larger containers will be required eventually if it is intended to grow the sporophytes. For further details of culture techniques and experimental work see Dyer, 1983 (listed in the bibliography).

The student is expected to develop a thorough understanding of the life cycle of the fern. This is really essential for the appreciation of the process of reproduction in angiosperms, so this section should be covered before dealing with the life cycle of angiosperms. It is very worth while to encourage the students to grow the fern gametophytes and observe the development of sporophytes from them.

*Assumptions*

1 Knowledge of the terms gametophyte and sporophyte.
2 An understanding of the process of reproduction in *Marsilea*.

*Principles*

1 Typical ferns exist in two distinct forms and exhibit alternation of generations.
2 The distribution of ferns largely depends upon the morphology and ecology of the gametophyte phase.
3 Spores, and hence gametophytes, are produced by meiotic division of diploid cells.
4 Gamete production in ferns and higher plants is by mitotic division of haploid cells.
5 *Dryopteris filix-mas* is homosporous and normally produces bisexual gametophytes, whereas *Marsilea vestita* is heterogamous and produces separate male and female gametophytes.

*Questions and answers*

**a** ***In what ways is the fern gametophyte adapted to lead an independent existence?***

The rhizoids and photosynthetic tissue allow the gametophyte to have an existence independent of the sporophyte.

**b** ***How are the antheridia and archegonia distributed on the gametophyte?***

The archegonia are grouped together, just below the apical notch. The antheridia are grouped further down the gametophyte, near the rhizoids.

**c** ***How might the structure of the gametophyte affect the distribution of the fern?***

The small delicate nature of the gametophyte and the separation of the male and female sex organs indicates that the prothallus requires a fairly wet environment for growth and sexual reproduction. So sporophyte establishment is limited to damp habitats, but some are able to spread by vegetative growth beyond these limits.

**d** ***Given that the gametophyte is haploid, by what cytological process are gametes produced?***

Gametes must be produced by cell division involving mitosis.

**e** ***By what cytological process does the sporophyte produce spores?***

By meiosis.

**f** ***What evidence must have been obtained by scientists in order for them to understand the nature of this type of life cycle? Why is it called alternation of generations?***

Botanists must have recognized that spores do not, on germination, give rise to the typical fern plant, that the intervening plant structure contains reproductive organs, and that motile gametes are produced. Further, a typical fern plant grows from the site of one type of sex organ. It is called alternation of generations because there are two types of plant body in the life cycle, which alternate in appearance with each other. Also, one plant, the sporophyte, is normally diploid, whilst the other, the gametophyte, is haploid.

## INVESTIGATION
## 22D Pollination mechanisms in angiosperms

(*Study guide* 22.6 'Gametogenesis and fertilization'.)

### ITEMS NEEDED

Plants in flower, several types
Forceps 1/1
Hand lens 1/1
Microscope, binocular stereo 1/group
Microscope, monocular (× 100, × 400) 1/1
Mounted needle 1/1
Scalpel 1/1

*Assumption*

1 That students are able to recognize the parts of flowers from a variety of types.

*Principles*

1 Some hermaphrodite flowers show structural mechanisms which promote cross-pollination.
2 Wind- and insect-pollinated flowers show basic morphological differences.
3 The structure of many non-wind-pollinated flowers is related to the structure and behaviour of specific animal groups.
4 Pollination between flowers of the same or genetically identical individuals does not lead to cross-fertilization.
5 Cross-fertilization has important evolutionary implications in the generation of variation in the offspring.
6 Self-pollination has survival value for the offspring.

It is usually easy to see the stage of development of the male parts of a flower since, with the aid of a hand lens or binocular microscope, pollen can be seen on the dehisced anthers. The anther shrivels up when the pollen has been shed. The female parts may be more difficult. If a style has several lobes, these are usually spread apart as it becomes receptive. The stigma becomes more prominent and papillae may be observed. A good flora will often state whether a species is protogynous, protandrous, or homogamous. The flowering shoots provided for the practical should show a range of entomophilous and anemophilous structures. Tables 18 and 19 provide an indication of the sort of results that students might be expected to obtain.

*Questions and answers*

a ***Consider the insect-pollinated plants that you have studied. What evidence can you find that a plant species is adapted to a particular type of insect pollinator?***

Examples of the mechanisms adapted for particular insect groups will depend upon the species used.

| *Plant* | *Self-pollination* | *Cross-pollination* |
| --- | --- | --- |
| White dead nettle | | anthers mature first (protoandrous) |
| *Pelargonium* | | anthers mature first (protoandrous) |
| Plantain | | stigma matures first (protogynous) |
| *Arabidopsis* | anthers located near stigma, small insignificant flowers homogamous | |
| Grasses | | anthers project out of flower |
| Yellow toadflax | | anthers project out of flower |

**Table 18**
Mechanisms favouring either cross- or self-pollination.

**b** ***Why may pollination between flowers not necessarily ensure cross-fertilization?***

Pollination between flowers does not necessarily result in cross-fertilization, as the pollen may come from a flower on the same plant or clone.

**c** ***How might the reproductive system of angiosperms have contributed to their success?***

Flower structure that favours self-pollination gives a high chance of fertilization. This results in numerous similar offspring able to survive in an environment similar to that of the parent. These species are successful at rapid colonization of open uniform habitats. It is generally considered that cross-pollination was established early in angiosperm evolution. Subsequently, inbreeding by selfing developed among some as an adaptive modification to reduce variability, leaving the out-breeders as the main agents of evolutionary change. In such species adaptation of the flower towards specific pollinating agents has improved the chances of cross-pollination. The resultant range of variation, shown by the offspring, has contributed to the success of the species in complex, established plant communities. There are, of

| *Anemophilous flowers* | *Entomophilous flowers* |
| --- | --- |
| Small inconspicuous flowers | Large brightly coloured flowers with scent and nectar attractants |
| Stamens on long filaments, stand out from the flower | Stamens situated in flower in a position where insect likely to make contact |
| Large stamens loosely attached to filament | Smaller stamens firmly attached to filament |
| Large quantities of smooth light pollen grains | Smaller quantities of pollen with surface features for attachment to insects |
| Feathery stigmas protruding from the flower | Fat, or lobed stigmas usually inside the flower |

**Table 19**
General characteristics of anemophilous and entomophilous flowers.

course, other factors that have contributed to the success of the angiosperms. The seed habit is introduced in the next practical. A general discussion of other factors could be left until later.

## INVESTIGATION
## 22E Events leading to fertilization in angiosperms

### ITEMS NEEDED

Plants, *Pelargonium* sp. and *Impatiens* sp. in flower
Microscope slides, prepared, showing pollen tubes in a style
Pollen germination medium/culture solution

Sucrose
Water, distilled
Balance 1/class
Forceps 1/group
Graph paper
Graticule, eyepiece 1/1
Microscope, binocular stereo 1/group
Microscope, monocular (× 100, × 400) 1/1
Microscope slides
Mounted needle 1/group
Paper, filter, Whatman No. 1
Pencils, glass writing
Petri dishes 4/group
Pipettes, dropping 1/group
Scissors
Spatula 1/class
Stage micrometer 1/class
Stopclock or watch 1/1
Syringes, 1 $cm^3$, 10 $cm^3$, or measuring cylinders

(*Study guide* 22.5 'Variations in life cycles' and 22.6 'Gametogenesis and fertilization'.)

**Pollen grain germination medium**
*Stock solution*
Boric acid, $H_3BO_3$, 0.1 g
Calcium nitrate, $Ca(NO_3)_2 4H_2O$, 0.3 g
Magnesium sulphate, $MgSO_4 7H_2O$, 0.2 g
Potassium nitrate, $KNO_3$, 0.1 g
Water, distilled, 100 $cm^3$
*Culture solution*
Stock solution, 1.0 $cm^3$
Water, distilled, 9.0 $cm^3$
Sucrose, 0.5 to 1.5 g*
(*Optimum varies with the species. It may be necessary to use trial and error, although the pollen from many species will grow on a 10 % sucrose solution.)
The stock solution can be kept for some months in a refrigerator. However, the culture solution must be made up fresh immediately before use. In some cases pollen germination has been speeded up by the addition of a little crushed mature stigma.

A medium can also be obtained from Philip Harris Biological Ltd, Oldmixon, Weston-super-Mare, Avon BS24 9BJ. A pollen germination kit is also available with more detailed teaching notes suggesting further work and ideas for student projects.

*Principles*
1 Environmental conditions affect pollen tube growth.
2 Pollination leads to double internal fertilization and seed formation.
3 Retention of the ovule and embryo inside the parent plant possesses both advantages and disadvantages.

Sucrose concentration is only one environmental factor that could be varied. A discussion before the practical might raise other suggestions that the students could investigate and the procedure be modified accordingly.

There are alternative methods of pollen tube culture (see bibliography). Further samples of growing pollen tubes could be stained with, say lactopropionic orcein, to show up the nuclei. If time permits students should observe prepared microslides showing pollen tube growth down a style.

If the students are familiar with the use of stage micrometers to calibrate graduated eyepieces, then the units of growth could be converted into μm. Graphs should be drawn of length (μm or eyepiece units) against time.

*Questions and answers*

a ***What factors may have affected the rates of growth? Try to explain any differences in the growth rates shown in your graphs.***

Temperature, age of pollen, composition and concentration of growth medium, and so on, could all affect the growth rate. Graphs will vary according to the conditions.

b ***What differences did you find in pollen tube growth between* Impatiens *and* Pelargonium*? Propose a hypothesis to explain your answer, and outline a means to test it.***

Pollen tubes of *Impatiens* tend to grow straight with only the occasional change in direction. In contrast, those of *Pelargonium*, grow extremely contorted under these conditions. The 'deformed' pollen tubes of *Pelargonium* may be as a result of unsuitable osmotic conditions in the medium, insufficient sucrose, or lack of an 'organizer' substance produced by the carpel. Details of the outlined experiment will depend upon the wording of the hypothesis.

c ***Why is it wrong to call a pollen grain a gamete?***

The pollen grain and tube represents the male gametophyte. The pollen grain nucleus divides mitotically to form first the vegetative and generative nuclei and then the latter divides again to form two male nuclei (gametes); thus the pollen grain is much more than just a gamete.

d ***What happens between the germination of the pollen grain and the formation of the seed? You will need to consult suitable texts.***

Points that should be mentioned include: growth of the pollen tube down the style, release of two gametic nuclei, double internal fertilization, formation of the embryo, endosperm, and related structures, and their relationship to the finally produced seed.

e ***What are the advantages and the disadvantages of retaining the ovule and embryo inside the tissues of the adult?***

Advantages include:
**1** Decreased dependence upon environmental water.
**2** The provision of nutrients for early development.
**3** Dispersal of offspring with a food reserve increasing the probability of successful germination.
Disadvantages include:
**1** Problems of gametes meeting; the need for pollen tube formation for gamete movement, and a mechanism to 'direct' the growth of the tube.
**2** The complex nature of the seed, and increased use of energy for the food store – an energy expensive process compared with other means of reproduction.

## Angiosperm life cycle

The nature of the angiosperm life cycle should be discussed at this point, and the reduction of the male and female gametophyte to the pollen grain and the embryo sac considered. The similarities and differences between the fern and angiosperm life cycles should be stressed.
The nature of gametogenesis in angiosperms and mammals can now be compared, and the differences in the life cycles in relation to where meiosis takes place explored. Students should remind themselves of the reproductive system in mammals with reference to the provision of the fluid medium for gametes (see *Practical guide 3*, investigation 10A).

**f** ***Draw the life cycles of a mammal and an angiosperm, comparing the occurrence of meiotic and mitotic cell divisions.***

See *figure 25.*

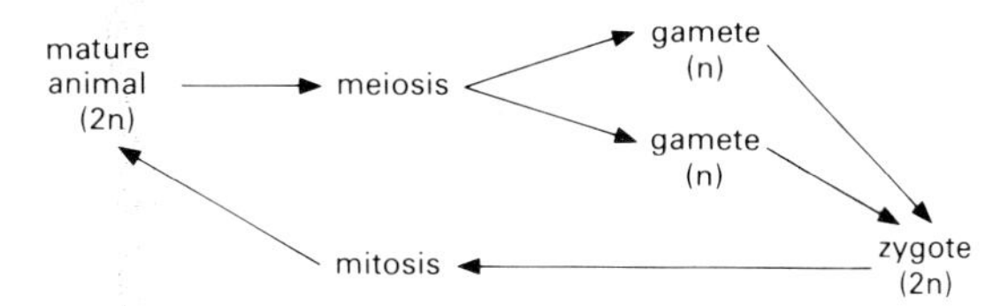


**Mammal**

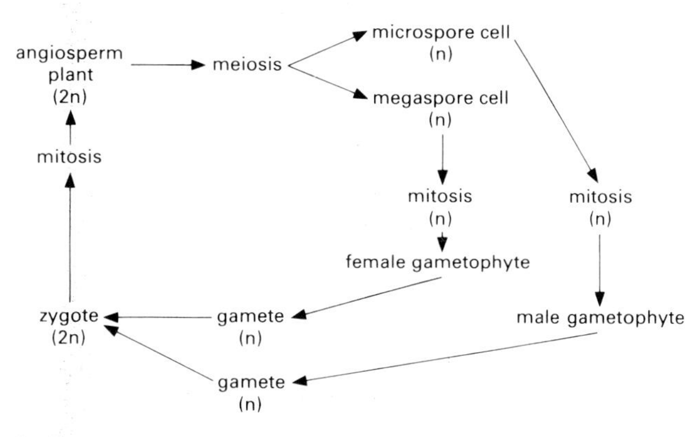


**Angiosperm**

**Figure 25**
Life cycles of a mammal and an angiosperm to indicate phases of cell division.

**g** ***Consider the life cycles of a fern, an angiosperm, and a mammal. What comparisons can you draw between the haploid and diploid structures?***

See *figure 26.*

**Fern**

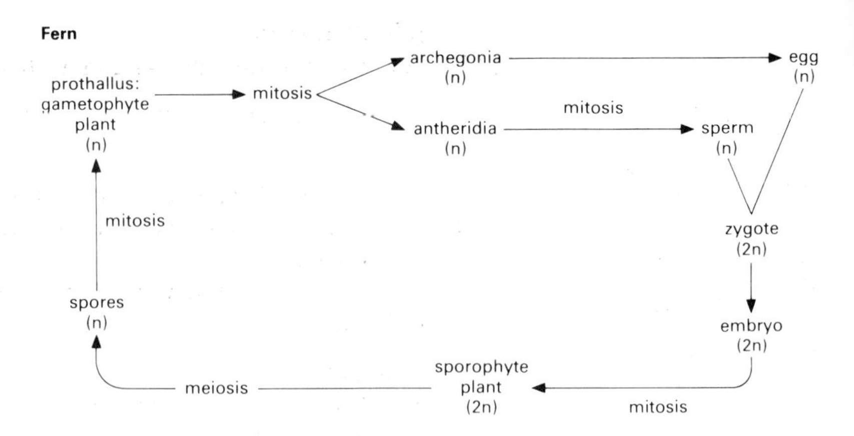


**Angiosperm**

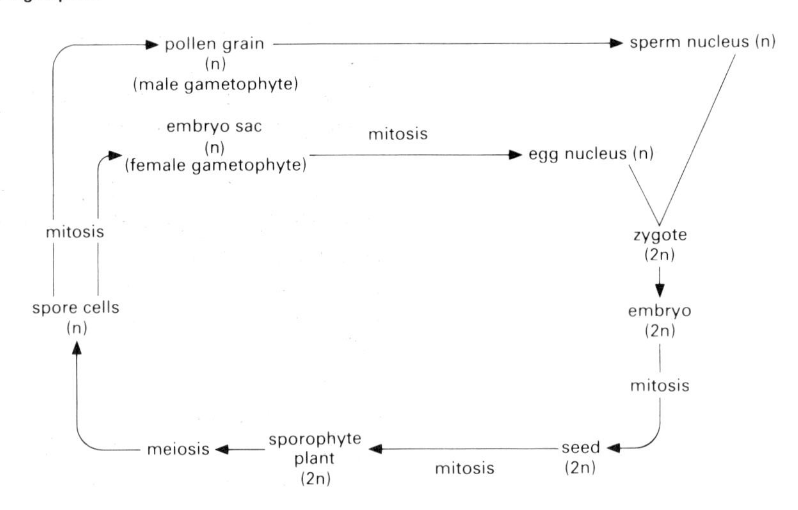


**Mammal**

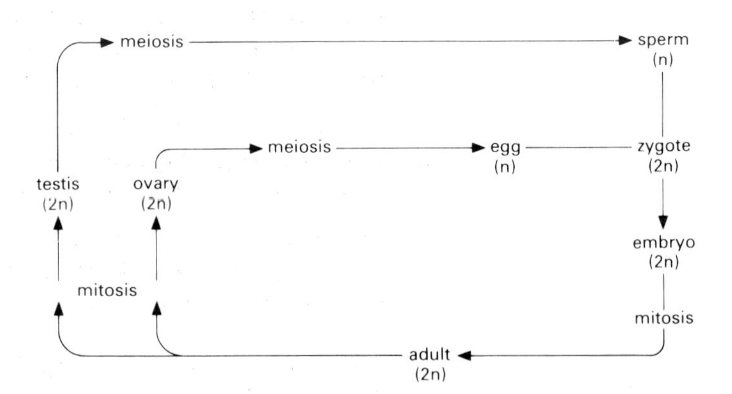


**Figure 26**
Life cycles of fern, angiosperm, and mammal.

## INVESTIGATION
## 22F The structure and function of the mammalian gonads

### ITEMS NEEDED

Microscope slides, prepared:
ovary, mammalian, TS
sperm, human
testis, mammal, TS
Graticule, eye-piece 1/1
Microscope, monocular
(× 60, × 100, × 400) 1/1
Ruler 1/1
Stage micrometer 1/class

(*Study guide* 22.6 'Gametogenesis and fertilization'.)

*Assumptions*

1 Knowledge of the process of gametogenesis in mammals.
2 Knowledge that hormones are involved in gametogenesis, sexual cycles, and maturity.
3 Ability to set up a microscope correctly and to use it as a measuring instrument.

*Principles*

1 The events of gametogenesis can be related to the structures seen within the gonads.
2 The differences in size and number between egg and sperm can be related to mobility and the chances of survival of each type of gamete.
3 Gonads secrete hormones involved in the regulation of their own correct functioning.
4 There are similarities and differences between gametogenesis in angiosperms and mammals.

Some students may have difficulty in identifying all the structures shown in the diagrams. The use of 35 mm colour slides would be a helpful addition to organize the students' observations. Few prepared slides are labelled with the type of stain used. This information can often be found in the suppliers' catalogues, or from the manufacturers themselves. Most prepared slides are, however, stained with haematoxylin and eosin. *Peacock's elementary microtechnique* (Peacock, 1973) gives details concerning staining schedules and their effects.

Sections of human gonads are not suitable for this practical as gametogenesis and gametes are not usually visible. As the testis and ovary sections are likely to be from different species of mammal, direct comparison of sizes is not possible. The practical offers an opportunity to review gametogenesis in animals and plants. It may be useful to compare the histology of the organs of gametogenesis in angiosperms with those of mammals at this point.

*Questions and answers (Part 1, Ovary)*

**a** ***Examine the development of the follicles in the preparation. At what stage of the oestrous cycle was the ovary? Explain your answer.***

The presence of a corpus luteum would indicate post-ovulation or pregnancy. The size of the most mature follicles will give an indication of the approximate stage of the cycle (*see figure 27*).

**b** ***How much does the size of 1 a follicle and 2 an oocyte increase during development?***

The results will depend upon the species used – which may not be known. Students can estimate this by measuring the size of an immature follicle and the most mature one they can find. The size increase of the follicle is greater than that of the oocycte.

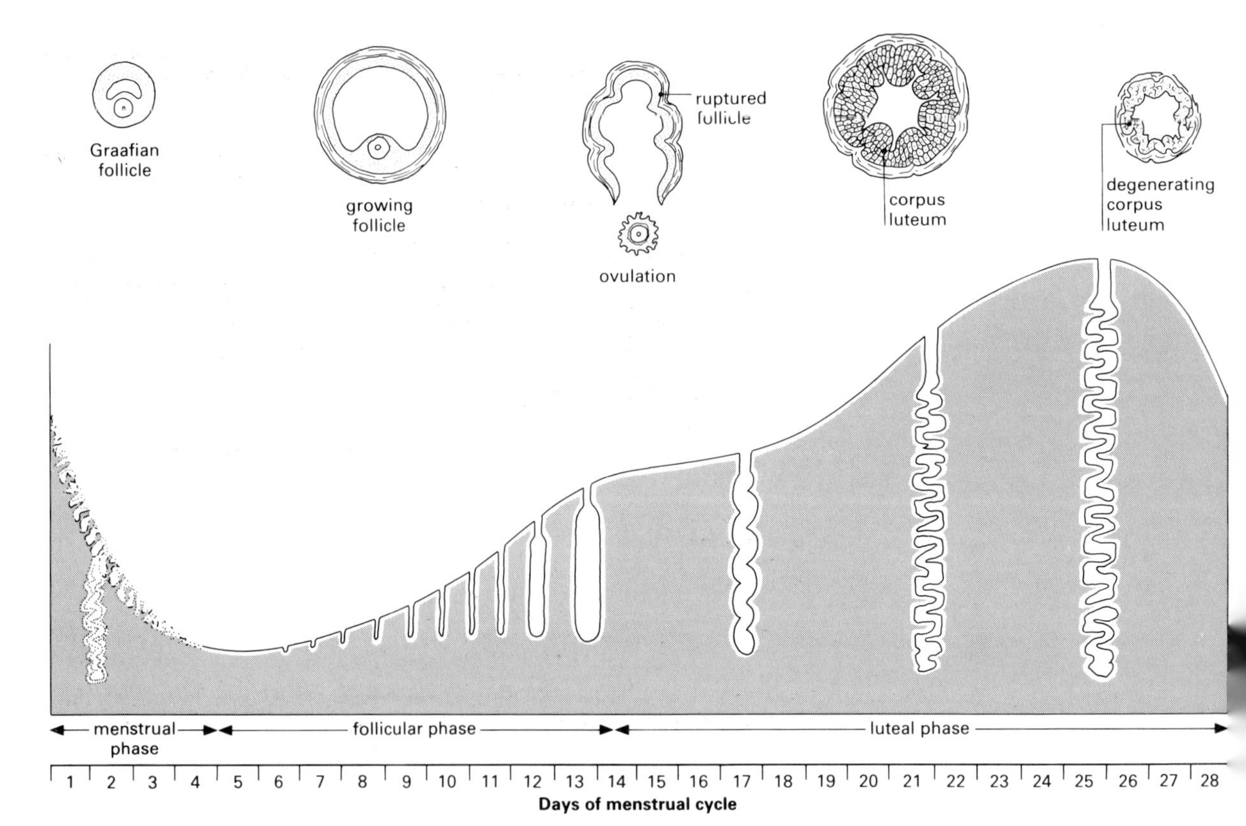


**Figure 27**
The condition of follicles at different stages of the menstrual cycle.

c ***How does the size of a mature human primary oocyte compare with that of the head of a human sperm?***

From the measurement and magnification of the oocyte shown in *figure (P) 7*, the oocyte is about 110 µm in diameter, as compared with the head of the human sperm, which, measured from the slide, is likely to be about 10 µm in diameter.

d ***What structures in the ovary secrete hormones? Consult texts to find out. Add annotations about this to the diagram you produced in 4 above.***

See *figure 28*.

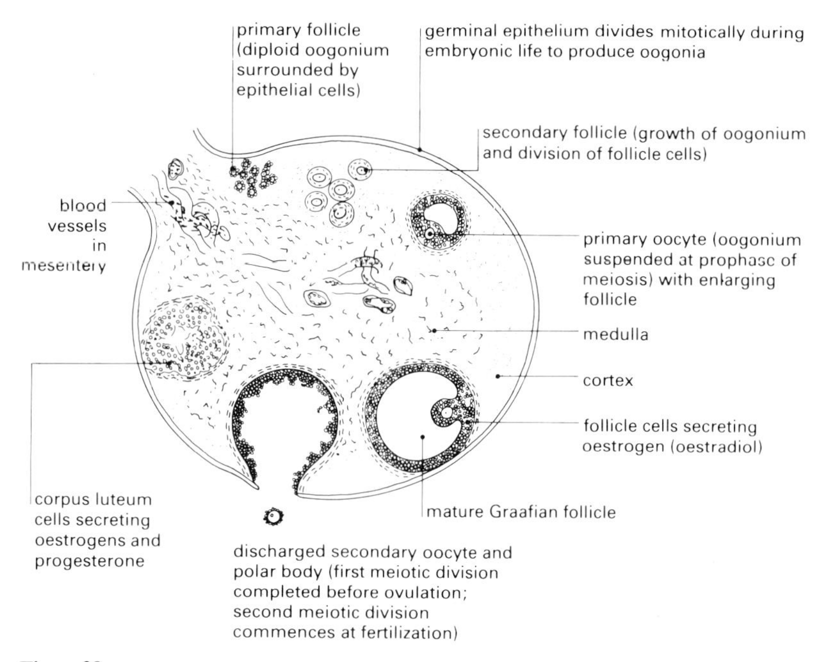


**Figure 28**
An annotated diagram of an ovary showing the stages in oogenesis and hormone secretion.
*From Clegg, P. C., and Clegg, A. C.,* Biology of the Mammal. *Heinemann, 1975.*

*Questions and answers (Part 2, Testis)*

**a** ***What is the size of the head of a mature sperm in your testis preparation? How many times longer is the tail compared with the head? Compare this result with the size of a human sperm.***

Typical results could be:

| | *Total length* | *Tail* | *Head size* | *Relative tail length* |
|---|---|---|---|---|
| Human sperm | 55 µm | 45 µm | 10 µm | 4.5 |
| Rabbit | 56 µm | 36 µm | 20 µm | 1.8 |
| Rat | 182 µm | 164 µm | 18 µm | 9.1 |

**b** ***What stains were used in making the preparation? Consult texts to find out what structures they would stain.***

If the preparation was stained with haematoxylin and counterstained with eosin, for example, then the haematoxylin stains nucleic acids black or blue, and the eosin colours cytoplasm red.

**c** ***Which cells were most darkly stained? Suggest why these cells should take up more stain than others.***

Cells in the germinal epithelium and the sperm heads themselves would be darkest. The cells of the germinal epithelium are actively dividing and so will have a high proportion of DNA and RNA relative to the amount of cytoplasm. The sperm head is mainly nucleus and the nucleus is more condensed compared with other cells.

**d** ***What parts of the testis produce hormones? Consult texts to find out. Add annotations about these to the diagram you produced in 5 above.***

See *figure 29.*

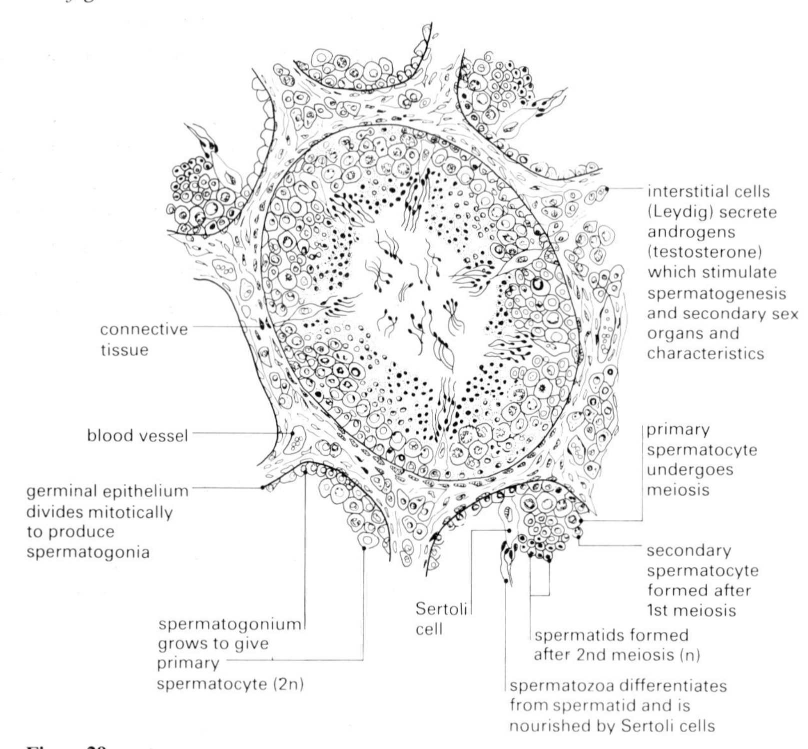


**Figure 29**
An annotated diagram of a testis showing the stages in spermatogenesis and hormone secretion.
*From Clegg, P. C., and Clegg, A. C.,* Biology of the Mammal. *Heinemann, 1975.*

**e** ***What was your estimate of the number of sperm present in the testis? Explain how you arrived at the number.***

This question is intended as extension work for the faster working student. It concerns skill in estimating rather than in observing microstructures. It will require the student to estimate:

**1** the number of sperm seen in one tubule
**2** the number of tubules in the section
**3** the size of the testis, and, assuming it to be a sphere sectioned through its centre, its volume (can be calculated using $4\pi r^3/3$). Then, having been told that the section is 5 µm thick a rough estimate of the number of sperm in the testis can be made.

**f** ***Why should there be more male gametes than female?***

The larger non-motile female gamete remains in the protected environment of the female's tissues. The more hazardous transfer of male gametes requires large numbers to enhance the chance of successful contact.

**g** ***What are the similarities and differences between gametogenesis in angiosperms and mammals?***

Gametogenesis is similar in the two groups in that differentiation of the cells containing the male nuclei occurs in both groups. Only one nucleus of the four produced in the female meiosis will give rise to an ovum. Many more small male gametes are produced than the larger female gametes.
Angiosperms and mammals differ in that gametogenesis of the female mammal occurs only during embryonic life, while in angiosperms, new embryonic mother cells are produced in each flowering season.
Another major difference is in the occurrence of meiotic and mitotic cell divisions. In the mammal, meiosis gives rise directly to the gametes. In angiosperms, gametes are formed from the gametophyte generation and are produced by mitotic cell division.

# PART III Bibliography

AUSTIN, C. R. and SHORT, R. V. *Reproduction in mammals 7: Mechanisms of hormone action.* Cambridge University Press, 1979. (Hormone action in reproductive cycles.)
AUSTIN, C. R. and SHORT, R. V. *Reproduction in mammals 2: Embryonic and fetal development.* 2nd edn. Cambridge University Press, 1982. (Blastocyst formation, placentae.)
BEACONSFIELD, P., BIRDWOOD, G., and BEACONSFIELD, R. 'The placenta'. *Scientific American,* **243**(2), 1980, pp. 80–89. (Development of the placenta and its use in research.)
BELL, P. R. and WOODCOCK, C. L. F. *The diversity of green plants.* 3rd edn. Edward Arnold, 1983. (Useful general introduction.)
BERRY, D. R. Studies in Biology No. 140. *The biology of yeast.* Edward Arnold, 1982. (Reproduction and genetics.)
BINGHAM, C. D. Schools Council EULO Project, *Plants.* Hodder & Stoughton, 1977. (Useful background information on plant maintenance and use.)
BOLD, H. C., ALEXOPOULOS, C. J., and DELEVORYAS, T. *Morphology of plants and fungi.* 4th edn. Harper & Row, 1980. (Useful in this general area.)
BRIGGS, D. and WALTERS, S. M. *Plant variation and evolution.* Weidenfeld & Nicholson, 1984 2nd edn. (*Geum* hybrids.)
CAMMELL, M. E. 'The black bean aphid, *Aphis fabae*'. *Biologist,* **28**(5), 1981, pp. 247–258. (Details of life cycle and control measures.)
CLAPHAM, A. R., TUTIN, T. G., and WARBURG, E. F. *Excursion flora of the British Isles.* 3rd edn. Cambridge University Press, 1980. (Details of the campion species.)
CLARK, J. T. 'Demonstrating the early stages of ferns and mosses'. *School Science Review,* **60**(211), 1978, pp. 283–4. (Germinating spores on agar plates.)

COHEN, J. *Reproduction.* Butterworths, 1977. (Takes a broad look at sexual reproduction.)
DALE, B. Studies in Biology No. 157, *Fertilization in animals.* Edward Arnold, 1983. (Broad coverage of the process.)
DIXON, A. F. G. Studies in Biology No. 44, *Biology of aphids.* Edward Arnold, 1973. (Details of life cycle.)
DYER, A. F. 'Fern gametophytes in culture – a simple system for studying plant development and reproduction'. *Journal of Biological Education,* **17**(1), 1983, pp. 23–39. (Methods of growing prothalli and an extensive account of possible investigations and their development.)
FREELAND, P. W. 'Some proposals for practical work with dried yeasts'. *School Science Review,* **60**(210), 1978, pp. 46–58. (Additional techniques such as investigating spore formation.)
FUCHS, F. 'Genetic amniocentesis, *Scientific American,* **242**(6), 1980, pp. 37–43. (Use and value of this technique.)
HARRISON, D. *Patterns in biology.* Edward Arnold, 1975. (General in this area.)
HAWKINS, P. W. 'The germination and growth of pollen'. *School Science Review,* **55**(192), 1974, pp. 511–513. (Pollen tube preparations, staining of stigma to show pollen tubes.)
HOGARTH, P. Studies in Biology No. 75, *Viviparity.* Edward Arnold, 1976. (Broad survey of its occurrence through the animal kingdom.)
INGOLD, C. T. Studies in Biology No. 88. *The biology of* Mucor *and its allies.* Edward Arnold, 1978.
JANZEN, D. H. 'Seed and pollen dispersal by animals: Convergence in the ecology of contamination and sloppy harvest'. *Biological Journal of the Linnean Society,* **20**(1), 1983, pp. 103–113. (Interrelationships of plants and animals with respect to pollen and seed dispersal, female humans as fruit mimics.)
LEWIS, D. Studies in Biology No. 110, *Sexual incompatibility in plants.* Edward Arnold, 1979. (Further details of this aspect of plant breeding systems.)
LLEWELLYN, M. 'The biology of aphids'. *Journal of Biological Education,* **18**(2), 1984, pp. 119–131. (An overview of aphid biology together with some suggestions for projects.)
LOWSON, J. M. *Textbook of botany,* 15th edn, revised by SIMON, E. W., DORMER, K. J., and HARTSHORNE, J. N. University Tutorial Press, 1981. (Plant life cycles, pollination, and embryo development.)
PEACOCK, P. (revised by BRADBURY, S.) *Peacock's elementary microtechnique.* 4th edn. Edward Arnold, 1973. (Alternative method for pollen tube growth.)
PHILIP HARRIS BIOLOGICAL. 11B 1680 '*Marsilea vestita* shamrock fern sporocarps, instructions and teaching notes'. Philip Harris Biological, 1975. (Details of the method of using *Marsilea* sporocarps.)
PROCTOR, M. and YEO, P. *The pollination of flowers.* William Collins, 1973. (Detailed account of pollination mechanisms and vectors.)
RAY, P., STEEVES, S., and FLUTZ, S. *Botany.* Holt, Rinehart & Winston, 1983. (Useful in this general area.)
SCHULTZ, A. H. *The life of primates.* Weidenfeld & Nicholson, 1969. (Sexual cycles in primates.)

SHEFFIELD, E. 'A simple artificial medium for the culture of cryptogamic plants'. *School Science Review*, **64**(226), 1982, p. 226. (Use of 'Instant Bio' soluble plant food.)
SHEFFIELD, E. and BASTIN, J. H. 'Simple methods for fern prothalli'. *School Science Review*, **60**(211), 1978, pp. 286–289. (Detailed accounts of methods, culture media needed to successfully grow fern prothalli.)
SLATER, P. J. B. 'Baby knows best: the physiology of pregnancy and lactation'. *Journal of Biological Education*, **15**(2), 1981, pp. 8–10. (Hormonal control, from the baby's point of view.)
SMITH, G. M. *Cryptogamic botany, Volume 2*. 2nd edn. McGraw-Hill, 1955. (Details of *Marsilea* life cycle.)
SPIRA, T. P. 'Floral parameters, breeding system and pollination type in *Trichostema* (Labiatae). *American Journal of Botany*, **67**, 1980, pp. 278–284. (Pollination mechanisms related to the pollination vector.)
STODDART, D. M. Studies in Biology No. 73. *Mammalian odours and pheromones*. Edward Arnold, 1976. (Odour signalling in reproduction.)
S202 COURSE TEAM. *Animal Physiology 1, Unit 18. Control mechanisms in reproduction*. Open University, 1981. (Mammalian reproductive cycles.)
TILNEY-BASSETT, R. A. E. 'Genetics and plastid physiology in *Pelargonium*'. *Heredity*, **18**, 1963, pp. 485–504. (Plastid inheritance in *Pelargonium*.)
TOOLE, A. G. and TOOLE, S. M. *A-level biology course companion*. Charles Letts Books, 1982. (Angiosperm life cycle.)
TRIBE, M. A. and ERAUT, M. R. Basic Biology Course, Unit 4, Book 11, *Hormones*. Cambridge University Press, 1979. (Programmed text approach to hormones involved in mammalian reproduction.)
WEIER, T. E., STOCKING, C. R., BARBOUR, M. G., and ROST, T. L. *Botany: an introduction to plant biology*. 6th edn. John Wiley, 1982. (Useful in this general area.)
WHITTEN, W. K. 'Modifications of the oestrous cycle of the mouse by external stimuli associated with the male'. *Journal of Endocrinology*, **17**, 1958. pp. 307–313. (Effects on the oestrous cycle of the introduction of males nearby.)
WILSON, C. M. and OLDHAM, J. H. 'Teaching the species concept using hybrid plants and habitats'. *Journal of Biological Education*, **18**(1), 1984, pp. 45–56. (The campion investigation.)

# CHAPTER 23 THE NATURE OF DEVELOPMENT

*A review of the chapter's aims and contents*

1. Three main processes appear to be involved in the development of a fertilized egg – morphogenesis, differentiation, and pattern formation.
2. Genes control the cell activities that are involved in the processes of development.
3. Although many of the answers are not available, there has been considerable progress in analysing development.
4. The basic steps in the development of a multicellular animal – early development, laying down of the body plan, organogenesis, and growth – are outlined.
5. The various cell activities that are used repeatedly throughout development are introduced. During early development, cells become different (differentiate); this is related to gene activity. Tissue rearrangements involve *cell motility* and lead to the laying down of the body plan. This also involves *cell–cell interactions* – induction – which specify the early nervous system. Cell–cell interactions also play an important role in organogenesis. Finally, during growth, *cell proliferation* is a predominant activity of cells.
6. The concept of determination of cell differentiation is discussed. The distinction between the fate of a cell and its determination is made.
7. A second major process, morphogenesis, changes in form, involves *changes in cell shape*, illustrated by the formation of the neural tube, and *cell movement*, illustrated by the development of the neural crest.
8. The third major process that takes place during development is pattern formation, which generates spatially organized arrays of differentiated cells, involves *cell–cell interactions*, and is illustrated by the development of the limb.
9. The developmental processes in plants and animals are compared. It is not the intention to consider the processes in plant development in any great detail, but to highlight the significant feature, the presence of meristems in plants.
10. A comparison between plants and animals raises highly speculative questions which relate back to the very early evolution of organized cellular life.

*Assumptions*

1. A knowledge of the basic processes of reproduction.
2. Some knowledge of the life cycles of organisms.
3. An understanding of mitotic and meiotic cell division.
4. A knowledge of introductory genetics.

# PART I The *Study guide*

## 23.1 The nature of development

*Principles*

1. Most animals start development from a single cell.
2. Development can be studied in terms of the activities of cells during the process.
3. The stages of development start with fertilization.
4. In higher animals these stages consist of: cleavage, gastrulation, organogenesis, and growth.
5. During development three main processes operate: morphogenesis, differentiation, and pattern formation.
6. Development provides a link between genetics and morphology. It is a process whereby genetic information is converted via cell activities into body form.

*Questions and answers*

**a** ***What animals are exceptions to this rule?***

A number of examples ranging from *Hydra* to honey bees are given in Chapter 22.

**b** ***What is likely to be the major internal mechanism controlling the egg's development?***

The action of the genes.

**c** ***What will be the advantage to us of understanding the molecular basis of cell motility during development?***

First, we need to find out which molecules are involved in movement, both structural and control molecules, that is the molecules that control for, say, the assembly of structural molecules. Then we need to find out which genes code for these molecules. Then one could investigate the patterns of activity of these genes, particularly those coding for the control molecules, during development and how these patterns are controlled. Thus one could study the link between cell activities, molecules and genes. But this is a tall order!

**d** ***Why might this fact be of interest to evolutionary biologists?***

During early development the body plan of all vertebrates is very similar. The fact that during embryonic development structures apparently homologous with embryonic gill slits, as found in the lower vertebrates, are found in mammals suggests a common ancestor.

**e** ***How might a biologist set about finding experimental evidence that the genetic information is the same in all the cells of an embryo?***

The experimental evidence to suggest that this is so is introduced as Study item 23.21. However, this question could be used for an introductory class discussion.

f ***What other examples of congenital defects have you come across? What do you know about their causes?***

Spina bifida is itself often associated with hydrocephalus, although the latter can exist on its own. In both cases the underlying causes of the malformations are unknown. They may be due, in part, to environmental factors and/or, in part, to genetic factors. There are several clear examples of genetically inherited congenital defects of which Down syndrome and sickle cell anaemia are probably the best known (see *Study guide II*, sections 15.4, 'The "one gene–one polypeptide" hypothesis', 16.3, 'Polyploidy and aneuploidy', and 20.5, 'Polymorphism').

## 23.2 Early development

*Principles*

1 During cleavage there is no overall growth of the embryo.
2 The type of cleavage depends upon the amount of yolk present in the egg.
3 At early stages of cleavage the individual cells have the potential to develop into separate complete organisms.
4 This ability is lost in later stages of cleavage.

**Practical investigations. *Practical guide 6*, investigation 23A, 'Tracing the early development of a nematode worm, *Rhabditis*', and investigation 23B, 'Embryonic development in shepherd's purse, *Capsella bursa-pastoris*'.**

*Questions and answers*

a ***How could you show that there was no overall growth of the embryo during cleavage?***

By determining whether the mass remains constant.

b ***Most mammalian eggs contain little yolk, compared with the large quantities of yolk in birds' eggs. What explanation can you give for this?***

Birds' eggs must contain all the nutrient required for the embryo to complete its development. Thus much food reserve in the form of yolk is required. In mammals, apart from the small group of egg-laying forms, all embryos develop an attachment with maternal tissues and receive supplies of nutrient as they develop.

c ***How could you detect whether human twins were the result of an embryo that had later split, or of the fertilization of two separate eggs?***

This is, of course, normally no problem, but the question has been included because of its value for discussion, particularly in terms of the notion that, scientifically, it is almost impossible to *prove* that twins are identical. If the twins have developed from the same zygote, then they must be of the same sex. As they are genotypically identical any character determined by genes alone could be used to provide evidence. If the character being measured differs in the twins, then they are non-identical. However, it is not really possible, however many tests you make, to *prove* that the twins are identical. This is because two eggs separately fertilized could be identical in respect of any particular gene. However, a large number of tests all showing the two to be similar would support (but not prove) the view that they are identical.

The technical problems, mentioned in the *Study guide*, concerned with testing the ability of daughter cells from late cleavage divisions to give rise to new individuals in mice, are as follows. Isolated blastomeres of the 8-cell stage can give rise to blastocyst-like structures *in vitro*. However, single blastomeres from this stage do not form embryos when implanted *in utero*. Thus to find out what each blastomere can give rise to, they are combined with 'carrier' blastomeres of a different genotype. This aggregated mixture of blastomeres is then implanted into a foster mother to continue development.

## STUDY ITEM

## 23.21 Genes and cells

*Principles*

1. The information that is encoded in the genes of a fertilized egg is used during development to make a new organism.
2. Cells become different very early in development, long before they form the specialized cells of the tissues of the body.
3. Nuclei from cells of specialized adult tissues contain all the information to direct development.

In table (S)37, there are columns headed '1st transfers' and '1st and serial transfers'. The former refers to the results of experiments of the type illustrated in figure (S)188. The latter – serial transfers – refers to experiments in which a nucleus is transferred to an enucleated egg, the development of this egg is allowed to proceed to the blastula stage, and then the nuclei from these blastula cells are transferred to enucleated eggs.

These serial transfers are more successful than 1st transfers. This may be because one is selecting nuclei or because the nuclei become conditioned in some way.

*Questions and answers*

a ***How do the nuclei from the different sources differ in their ability to support the development of the eggs?***

Nuclei from cells grown from adult skin only very occasionally support development of the egg to the tadpole stage; nuclei from epithelial cells of the intestine of tadpoles are more successful in supporting development while nuclei from blastula or gastrula endoderm cells are by far the most successful in supporting development.

b *What do the results indicate about whether the genes of specialized cells are irreversibly changed?*

Although only a tiny percentage of nuclei from cells of the highly specialized tissues tested, adult frog skin and intestinal epithelium, can support the development of the egg up to the tadpole stage, nevertheless this positive result shows that these nuclei contain all the information for the stages of development up to the tadpole stage. In view of the technical difficulties involved, it is amazing that any positive results are obtained. One can conclude that cell specialization does not involve irreversible changes in the genes. (Further evidence is the fact that when nuclei from brain cells, which normally never divide, are injected into enucleated eggs, they can synthesize DNA.)

c *What might be the advantages of this type of cloning in animals? What ethical problems could it pose?*

Even for amphibians, no adult frogs have ever been obtained by injecting nuclei from adult tissues into enucleated eggs. Recent experiments in which success for cloning mice has been claimed, are now suspect. In principle, if adult animals such as mammals could be produced, this might provide a way of improving livestock herds. For example, cows with exceptional milk yields could be cloned. However, at present, this seems a remote possibility and even if possible, it would be very expensive. If the technology were developed for mammals such as cows it would most probably be applicable to human beings as well. Thus it would be theoretically possible to produce a clone of individuals from any human being. They would, of course, be identical as to sex and all other characteristics determined by the genes. Such an application was the subject of a book *The Boys from Brazil* by Ira Levin (1978) (later made into a film), in which clones were made using nuclei from the cells of Adolf Hitler. □

**STUDY ITEM**

**23.22 How do cells become different?**

*Principles*

1 The future of a cell in the 16-cell mouse embryo depends on its position.

2 Teratoma cells injected into embryos can participate in forming normal tissues.

*Questions and answers*

a *What result would you expect if the position of a cell does determine whether it participates in forming the embryo proper or the placenta and membranes?*

The cells on the outside of the chimera, although some were originally from the inside of an embryo, will form the placenta and protective membranes. The embryo proper will develop from the cells on the inside of the chimaera irrespective of their origin.

b ***What coat colour would the baby mouse possess?***

Since the cells on the inside of the chimaera are those that will participate in forming the embryo and were taken from brown mice, the baby mouse will have a brown coat.

c ***What conclusions can be drawn about the behaviour of teratoma cells from this experiment?***

Teratoma cells can respond to cues in a normal embryonic environment and thus participate in forming normal structures. Such experiments provide insights into the problem of cancer. Thus, the development of cancers may involve failures of differentiation. In addition, these experiments can provide information about developmental cues in embryos. □

## 23.3 Development of the body plan

**Practical investigation. *Practical guide 6*, investigation 23C, 'Morphogenesis in amphibians'.**

*Principles*

1 Vital staining techniques are used to construct fate maps of developing embryos.
2 In a blastula, all the main layers of the body – ectoderm, mesoderm, and endoderm – are on the outside of the embryo.
3 Gastrulation brings the layers into a proper relationship with each other.
4 Gastrulation involves changes in cell shape, adhesiveness, and cell migration.

*Question and answer*

a ***Why do you think such stains are referred to as 'vital stains'?***

The stain is taken up by the cells, but apparently does not affect the vital activities of the cell in any way.

## 23.4 Tissue interactions and organogenesis

Investigation 23C is a simulation of the process of gastrulation in amphibians. It requires the students to show that they can interpret two-dimensional drawings in terms of three-dimensional structures.

*Principles*

1 During development interactions occur between different tissues.
2 The mesoderm induces the overlying ectoderm to form the neural plate.
3 At later stages in development the fate of the ectoderm depends upon the nature of the underlying mesoderm.

## 23.5 Growth

*Principles*

1. Growth is largely brought about by cell multiplication.
2. The growth plate is the main structure involved in the growth of the long bones.
3. The growth of the long bone and its associated muscles are co-ordinated.

**STUDY ITEM**

### 23.51 Measuring the growth of humans

*Principles*

An examination of the growth curve readily shows several features:

1. The curve is steepest at the beginning showing that we grow most rapidly when young.
2. The curve has a strange shape around the time of puberty when there is a transitory increase in steepness. This reflects the adolescent growth spurt.
3. The growth spurt occurs earlier in girls and their final height is less than that of boys (see question **a**).
4. After puberty growth slows down and eventually ceases.

The first two points are best illustrated by constructing a growth velocity curve (*figure 30*). This shows the rate of growth at different ages. The curve is constructed by drawing up a table that gives the growth increase at each age. This is most easily obtained by measuring the increase in height that occurred during the previous year. (Strictly, it is

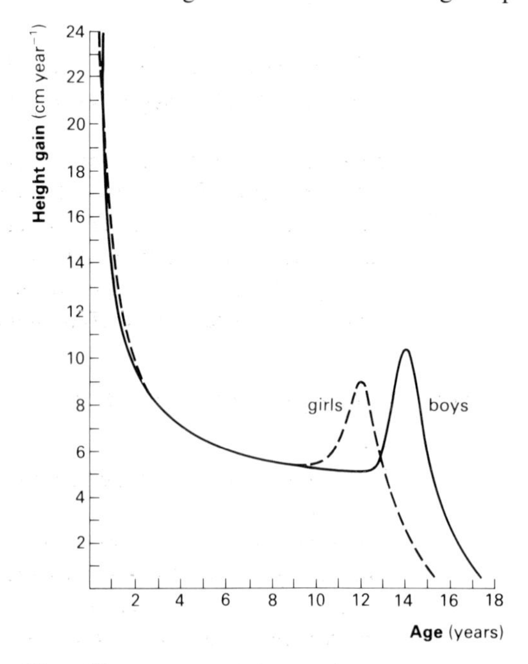


**Figure 30**
Typical growth velocity curves for English boys and girls.
*From Tanner, J. M., Whitehouse, R. H., and Takaishi, M.* Arch Dis. Childh., **41**, *1966, 454–471.*

the slope of the curve at a particular point and a more accurate value for, say, year 2 would be the increase from 1.5 to 2.5 years.) When the values have been measured, the increase for each year is plotted against age. It will be seen that the value falls from over 20 cm/year to about 6 cm/year until puberty when it increases to about 10 cm/year. After this the velocity decreases to zero.

*Questions and answers*

a ***What differences are there in growth rate changes between boys and girls?***

As mentioned previously, the adolescent growth spurt occurs earlier in girls compared with boys, but their final height is less.

b ***What do you think determines the different rates of growth?***

Most of the increase in height is due to growth in the cartilaginous growth plates of the skeleton. Their growth is under hormonal control. While this control is not fully understood, growth hormone, a protein made by the pituitary, causes the release of another protein hormone, somatomedin, from the liver. Somatomedin stimulates cartilage growth. The growth spurt is also associated with increased levels of testosterone.

c ***Why do we stop growing?***

There is no clear answer to this question – except perhaps, fortunately we do. Some animals, such as lobster and carp, continue to grow throughout their lives. □

**Practical investigation, *Practical guide 6*, investigation 23D, 'Growth and development in the fore-limb of mice'.**

Investigation 23D involves making measurements of the elements of the fore-limb in mice embryos at different ages, using a set of photographs. It uses a technique very similar to that used by developmental biologists studying limb development. It is probably more appropriate to relate the investigation to this section of the *Study guide* than to Section 23.8 which considers the mechanisms controlling limb development.

## 23.6 Fate maps and determination

*Principles*

1. Fate maps indicate what structures different parts of a blastula would normally form.
2. At early stages of development presumptive tissue has a greater potentiality for development than is indicated by the fate map. This is lost as development proceeds.
3. Embryos of vertebrates and sea urchins possess considerable ability to regulate their development.

## 23.7 Morphogenesis

### Neural tube morphogenesis

*Principle*

1 Changes in the shape of the cells of the neural plate lead to curving of the sheet and formation of the neural tube.

To make this easier to visualize a model can be made by sticking a row of elongated cubes of polyurethane foam together, side by side, onto a cardboard strip to represent the line of cells of the neural plate. Thread can be stitched along just under the upper surfaces of the 'cells' to represent the apical microfilament bundles. When this thread is tightened, the model will curve, illustrating the effect of contraction of the microfilament bundles.

Another possible way of visualizing the process would be to cut out cardboard shapes of the cells, both columnar as in the flat sheet, and tapered, as in the curved sheet. When fitted together these show that, whereas the columnar cells fit side by side to give a flat row, when cells are apically constricted, the sheet that they form must be curved.

### STUDY ITEM

### 23.71 The formation of the neural crest

*Principles*

1 The cells of the neural crest disperse and give rise to a number of different cell types in their appropriate positions.

2 When transplanted in the hind-brain region, forebrain neural crest cells form cranial ganglia, structures that they do not usually form.

*Questions and answers*

a ***Suggest two possible mechanisms that could account for the formation of tissues from the neural crest cells in their proper positions.***

One possibility is that neural crest cells are determined to form specific tissues before migration and the cells then migrate to take up their proper positions. The second possibility is that neural crest cells migrate into the body and differentiate into those tissues appropriate to the position in which they end up.

b ***What does this result suggest about the determination of neural crest cells?***

This result is consistent with the hypothesis that neural crest cells are not determined before migration but differentiate according to the local environment into which they migrate.

c ***What result would be expected from the reciprocal experiment in which quail hind-brain neural crest replaces chick forebrain neural crest cells?***

If the neural crest cells differentiate according to the local environment into which they migrate, the transplanted quail hind-brain neural crest will form the structures normally formed by neural crest cells migrating from the forebrain position. The cranial ganglia will be formed by chick cells migrating from the undisturbed hind-brain crest region of the host embryo. □

## 23.8 Pattern formation in the development of limbs

**Practical investigation. *Practical guide 6*, investigation 23D, 'Growth and development in the fore-limb of mice'.**

*Principles*

1 Experiments on the developing chick wing show that the outgrowth of the limb depends on the apical ectodermal ridge at its tip.

2 If the apical ridge is removed, truncated limbs result, the level of truncation depending on the stage in development at which the ridge is removed.

3 Since experiments show that the apical ridge of mammalian limb buds has the same property as that of chick limb buds, the principles can be applied to human limb development.

*Question and answer*

a ***Give a possible explanation for the basis of congenital malformations in which limbs are missing or truncated.***

Congenital malformations in which the limb is absent or truncated could be interpreted as being due to defects in the apical ridge. In general, congenital malformations may have a genetic basis, or they may be caused by environmental agents.

### STUDY ITEM

### 23.81 How do cells know where they are?

*Principles*

1 Positional information that informs a cell of its position along the proximo–distal axis of the developing limb is specified by a mechanism intrinsic to the tip of the limb, rather than by interactions with tissues already laid down.

2 The length of time spent in the progress zone (the undifferentiated region of a mesenchyme at the tip of the bud) informs a cell of its position along the proximo–distal axis. Cells that leave the progress zone early form proximal structures; cells that leave later form more distal ones.

*Questions and answers*

a ***What is the reciprocal experiment?***

The tip of a young bud is grafted in place of the undifferentiated tip of an old wing bud.

**b** ***Predict results of this experiment that would be consistent with the idea that the tip develops independently.***

The wing that develops would have a humerus and forearm followed by another forearm and hand. This sequence of structures can be deduced by working out the sets of structures that would develop independently both from the stump and the grafted tip. Thus, the stump of the old bud must give rise to a humerus and upper arm since its tip (as shown in the first example) gives rise to a hand. The grafted tip of the young bud must give rise to a forearm and a hand since its stump (as shown in first example) gives rise to a humerus.

**c** ***According to this hypothesis, what would happen if most of the cells in the undifferentiated zone at the tip of the limb bud were killed by a toxic drug or radiation?***

If the cell population at the tip of the limb bud is severely reduced, the surviving cells will have to undergo a number of divisions before sufficient cells are present to spill out of the zone. Thus surviving cells will spend longer in the progress zone than usual and when they spill out they will form distal structures rather than proximal ones. One might compare the progress zone to a bucket full of water into which water is dripping from a tap (representing addition of new cells). If one then throws away half of the water in the bucket (by analogy, kills half of the cells in the progress zone), the drips from the tap will first have to fill the bucket before water can overflow.

**d** ***If the surviving cells could continue to proliferate, what would the pattern of the limb be like?***

Proximal structures will be affected whereas distal ones will be relatively unaffected. In the most severe cases, the limb may consist of just a hand attached to the shoulder. If the effect is less severe, a humerus may be present followed by a hand, but the lower arm structures will be absent. Such results have been obtained when a chick wing bud was treated with high doses of X-irradiation at an early stage in development. Limbs in which both the upper arm and the lower arm are missing and the hand is attached to the shoulder are a characteristic limb malformation in thalidomide victims. Although the way in which thalidomide acts to bring about such malformations is controversial and not fully understood, the production of similar limb malformations experimentally by X-irradiation suggests that the killing of cells in the early bud could be a possible explanation. □

## STUDY ITEM

### 23.82 How does the polarizing region exert its effect?

*Principles*

1 Distance from the polarizing region informs a cell of its position acros the antero–posterior axis of the developing wing. Cells nearest the polarizing region participate in forming the most posterior digit, digit 4; cells a bit farther away from the polarizing region form the middle

digit, digit 3; and cells farther away still form the anterior digit, digit 2.

**2** The hypothesis is that the polarizing region produces a diffusible morphogen, which sets up a concentration gradient across the antero–posterior axis of the wing (*figure (S)206*).

*Questions and answers*

**a** ***How is the number of digits affected as the amount of tissue between host and grafted polarizing region is reduced?***

As an additional polarizing region is grafted to successively more posterior positions along the antero–posterior axis of the wing bud, the number of digits that develop between the grafted and host polarizing regions decreases because the amount of tissue available is progressively reduced. It should be noted that only the number of digits between grafted and host polarizing regions is considered. In buds to which the additional polarizing region is grafted to the apex of the bud (in the centre of the bud rim opposite the border between somites 17 and 18) digits can also be formed anterior to the grafts, the complete digit pattern being, for example, 234434.

**b** ***Study the pattern of digits formed as the grafted polarizing region is placed in successively more posterior positions. What comparisons can you make of the effects on the formation of digit 2 and digit 4?***

Digit 2 is the digit that forms farthest away from the polarizing region. As the additional polarizing region is grafted in successively posterior positions, the distance between the grafted and host polarizing regions progressively decreases. Thus cells between them are not sufficiently far away to form digit 2. Digit 2 does not form and patterns such as 4334 and 434 result. In contrast, digit 4, since it is the digit formed nearest to the polarizing regions, is always formed between the grafted and host polarizing regions until there is no space left between them, that is until the additional polarizing region is grafted in the position of the host polarizing region (opposite somite 19, see table (S)38).

**c** ***Using* figure [S] 206, *make sketches to show how this hypothesis could account for the data in* table [S] 38.**

The graft of an additional polarizing region introduces a second source of morphogen. When the graft is placed at the anterior edge of the bud, the gradient of morphogen diffusing from both the grafted and host polarizing regions will have a 'U' shaped form, thus specifying a second set of digits in mirror image symmetry with the normal set (*figure 31a*). When the graft is placed in the centre of the bud rim a shallower 'U' shaped gradient of morphogen will be formed between host and grafted polarizing regions (*figure 31b*). The concentration of morphogen between host and grafted polarizing regions will be too high to specify digit 2, and digit patterns 4334 and 434 will instead result. A gradient of morphogen will also be established across the part of the bud anterior to the graft and this can result in the formation of additional digits here too; the complete pattern of digits may be 234434. Note the same rules are obeyed: digit 4 is the digit formed

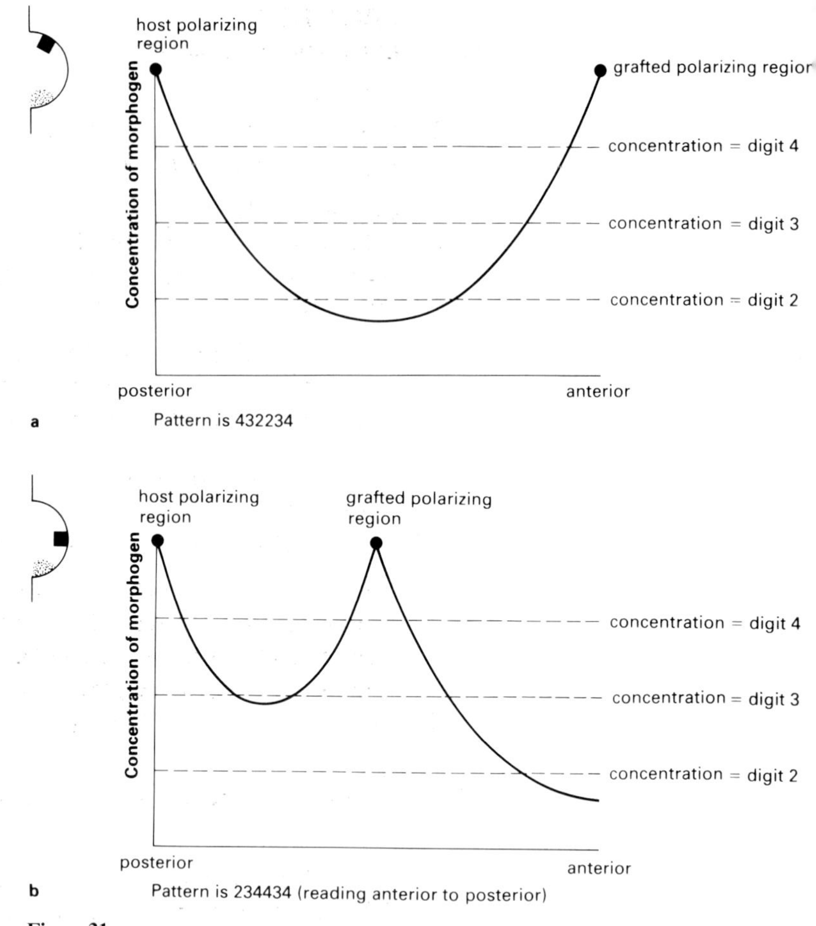


**Figure 31**
The effects of grafting an additional polarizing region to **a** the anterior edge of a limb bud, and **b** the centre of a limb bud rim.

nearest the grafted polarizing region both on its anterior and its posterior side. An illustration (*figure(S)207*) is provided of a wing that developed following a graft that contained a very small number of polarizing region cells that was placed at the anterior margin of the bud.

**d** ***Identify the digits that have formed.***

The digit pattern is 2234.

**e** ***According to the hypothesis that polarizing region tissue produces a diffusible morphogen, how could you account for this result?***

After the operation, only an additional digit 2 is formed, suggesting that cells in the anterior part of the bud are exposed to only a small increase in morphogen concentration. If the morphogen concentration depended on the number of polarizing region cells, this would account for the result. □

**STUDY ITEM**

### 23.83 Establishing connections

*Principle*

1 Developing motor axons make connections by growing towards appropriate target muscles.

*Questions and answers*

a ***What would be the predicted result if the normal pattern of innervation arises by nerves 'seeking' out their appropriate target muscles?***

Despite their displacement the nerves would still link up with the correct muscles in the developing leg. This result is usually obtained when short segments of the spinal cord are reversed.

b ***What other operations could be carried out to perturb the relationship of the nerves and their target muscles, which would provide evidence to distinguish between the two possibilities?***

One possibility is to shift the position of the target tissue rather than that of the nerves. This can be done by, for example, cutting off the early leg and grafting it in a more anterior position. For small shifts in the position of leg tissues, the nerves often still make the correct connections, taking aberrant pathways to do so. □

## 23.9 The pattern of plant and animal development compared

*Principles*

1 The evolution of the meristem is linked with the fundamental difference between plants and animals, that is their mode of nutrition.
2 A sedentary autotrophic plant evolved adaptations to reduce the effect of predator damage. The major adaptation was the meristem.
3 Plants show a plastic form of development compared with animals.

*Question and answer*

a ***There are examples in mammals of cells which differentiate throughout life. Name one example.***

Examples are cells of the Malpighian layer of the skin and germinative cells in the gut wall.

**STUDY ITEM**

### 23.91 Growth of a root

*Principle*

1 Cell division takes place in the tip of a root, and cell elongation and differentiation takes place successively behind the tip.

*Questions and answers*

a *What two alternative hypotheses could account for this result?*

The two major hypotheses are:

**1** The cells between the marks have elongated. There has been greater cell elongation in some parts of the root.

**2** Where the marks have separated most widely cell division has taken place.

b *For which of your hypotheses would you expect to find all the cells in the two sections the same size?*

The hypothesis that was based upon increased cell division.

c *For which would you expect to find the same number of cells in the two sections?*

The hypothesis based upon cell elongation.

d *Do your results agree with your predictions?*

The cells show a decreasing surface area, and therefore volume, in the photomicrographs **b** to **d** , supporting the hypothesis based upon cell elongation.

e *Can either of your hypotheses be rejected as false? If so, which?*

Yes. Since the prediction made in the hypothesis based upon cell division does not match up to the observations, it must be rejected as false.

f *When you have decided how you can best account for the elongation of the roots, examine the photomicrograph* **a** *in* **figure [S]210**, *showing a longitudinal section of a young root. Where is cell division most likely to be occurring in it?*

At the tip, where the cells are smallest and most numerous. Cell multiplication would be preceded by nuclear division. If this were not the case then some cells would be without nuclei, a point that can be checked in the photomicrographs **b**, **c**, and **d**. However, of course, there could be other reasons why cells in a photomicrograph do not possess nuclei, such as the nuclei not being in the plane of the section.

g *Can you see any sign of this process in photomicrographs* **b**, **c**, *or* **d**? *If so, which?*

Some mitotic division figures can be seen in the cells from the section cut 1 mm behind the tip (figure (S)210d).

h *What other processes will take place in the cells during development?*

Apart from elongation and nuclear division, the cytoplasm becomes vacuolated and changes occur in the cell wall. Some cells will become differentiated, for example, into xylem elements. □

**Practical investigation. *Practical guide* 6, investigation 23E, 'Growing new plants from old'.**

## STUDY ITEM

23.92 Differences between plants and animals

*Principle*

1 The nature of the earliest eukaryotic cells is uncertain, but two forms of life evolved: autotrophic and heterotrophic organisms.

*Questions and answers*

a ***Why is locomotion not essential for autotrophs?***

Apart from underwater, all parts of the earth's surface are bathed in sunlight; thus sessile plants can obtain light energy for photosynthesis. There is no selective advantage in mobility.

b ***What plants, or motile cells of plants, have you met that show locomotion?***

Male gametes of *Marsilea vestita*, antherozoids of ferns, etc. Unicellular green protists, such as *Euglena*, and colonial forms, such as *Volvox*.

c ***How can angiosperms survive without a motile gamete?***

By producing a male gametophyte that grows rather than swims towards the ovule.

d ***Why would this be an advantage?***

It eliminates the delicate free-living stage of a male gamete as found, for instance, in mammals.

e ***What are the disadvantages to an organism, such as an angiosperm, of being sessile?***

It is subject to predation and, unlike most animals, cannot attempt to escape by movement. If the environment becomes less suitable, for example through drought or lack of minerals, the plant, being unable to move, may survive only through reproductive propagules which are dispersed from the parent. □

*Questions and answers*

b ***How many different types of cells have you met in a mammal and in an angiosperm? Make a list of them.***

There are over a hundred easily recognizable different cell types in a mammal. In comparison, an angiosperm possesses around 20 to 30 cell types, such as collenchyma and palisade cells; these can usually only be recognized because they are in a particular place in a plant, or because they possess a different cell wall structure. Alberts *et al.*, *Molecular biology of the cell* (Chapter 16, appendix), has a list of the types of mammalian cell.

c ***What examples of regeneration can you find amongst animals?***

In Chapter 22 mention is made of the starfish *Linckia* which can regenerate a complete animal from one arm. Planaria can be cut into several pieces and regenerate. Earthworms show limited powers of regeneration. In the vertebrates, fish can regenerate fins and newts a new limb. In higher vertebrates regeneration is much more limited, being confined largely, as in humans, to replacement of cells in organs, such as the liver, and in wound healing.

d ***What parts of a plant are capable of regeneration? List specific examples.***

In artificial propagation almost all parts of a plant are used. However, for particular species certain parts are more suitable, such as the use of *Begonia* and *Saintpaulia* leaves. Roots can regenerate a complete plant, as can be shown using dandelion (*Taraxacum sp.*), and a wide range of plants are propagated by stem cuttings.

e ***Much of our knowledge of development has come from the study of a relatively small range of organisms, such as the sea urchin, frog, and mouse. How far do you think that it is reasonable to generalize from the conclusions from such work?***

This question is intended either for a group discussion or for an essay.

# PART II The *Practical guide*

The investigations related to this chapter appear in *Practical guide 6, Development, control, and integration.*

## INVESTIGATION
## 23A Tracing the early development of a nematode worm, *Rhabditis*

### ITEMS NEEDED

Earthworm 1–2/class
*Rhabditis* culture, live

Forceps 1/1
Graticule, eyepiece 1/1
Microscope, binocular stereo 1/group
Microscope, monocular (× 100, × 400) 1/1
Microscope slides and coverslips
Pipette, dropping, fine 1/group
Watch-glass, solid 1/1

(*Study guide* 23.2 'Early development'.)

**Rhabditis culture.** *Rhabditis* can be obtained by first cutting a live earthworm into sections about 20 mm long (contracted length). Teachers may prefer to anaesthetize the earthworm in MS 222 first (see *Teachers' guide I*, page 283). Each segment should then be cut open along the ventral surface. The earthworm pieces should be laid, epidermal surface uppermost on wet unsterilized soil in a Petri dish. The soil should be wet, rather than just damp, as damp soil encourages fungal growth. Place the lid on and leave the Petri dish at room temperature (18 ± 2 °C) for two to three days. The anterior and the clitellum regions of earthworms have been reported to contain the most nematodes.

*Rhabditis* is apparently found in many species of earthworm and certainly in *Lumbricus terrestris*. It is not practical to attempt to identify the species of nematode, indeed there is confusion in the literature regarding what species are found. As is general in nematodes, there are four moults during the life cycle. The nematode develops to the third stage in the earthworm, but the final moult, to produce the mature adult, only takes place on the death of the earthworm.

Cultures of *Rhabditis* are also available from biological suppliers.

The vinegar eel-worm, *Anguillula aceti*, is another nematode that has been used in studies of development. Similar embryonic stages will be found in the females. However, the process of development is much slower than in *Rhabditis*, taking about eight days. Cultures can be obtained from biological suppliers. They are easy to maintain, provided that suitable natural vinegar can be found that does not contain additives, such as preservatives.

*Assumptions*

1 Proficiency in the use of the microscope, including use of eyepiece graticules for measurement.
2 Knowledge of mitotic cell division.

*Principles*

1 Early development commences with mitotic cell division.
2 Cleavage in animals is not accompanied by cell growth.
3 Cell division restores the nuclear:cytoplasm volume ratio of normal somatic cells.
4 Products of cleavage have identical genetic complements.

Students should each be provided with a piece of earthworm. If, after washing it, the water in the watch-glass is left undisturbed, the soil particles and nematodes will settle to the bottom. A *gentle* squirt of water from the pipette will cause the nematodes to be disturbed with a little soil. They can then be sucked up and transferred to a slide.

Finding nematodes containing dividing eggs is largely a matter of luck, particularly as most eggs from mature females will be in later stages of development. A mechanical stage would be a great help, as the location of suitable eggs can be recorded. By this means, a search pattern for the eggs can be repeated over a period of time until one is found to be dividing. As soon as one is found attention should be concentrated upon it. The light should be switched off when observations are not being made to avoid over-heating and consequent drying out of the preparation. Should it begin to dry out, a drop of water can be placed at the edge of the coverslip, but this is likely to move the embryo being observed.

If good material is obtained, suggest that the time of cell divisions is recorded so that a graph can be drawn of cell number against time.

*Questions and answers*

**a** ***What pattern of cell division did you observe?***

The pattern of cell division is shown in *figure 32*.

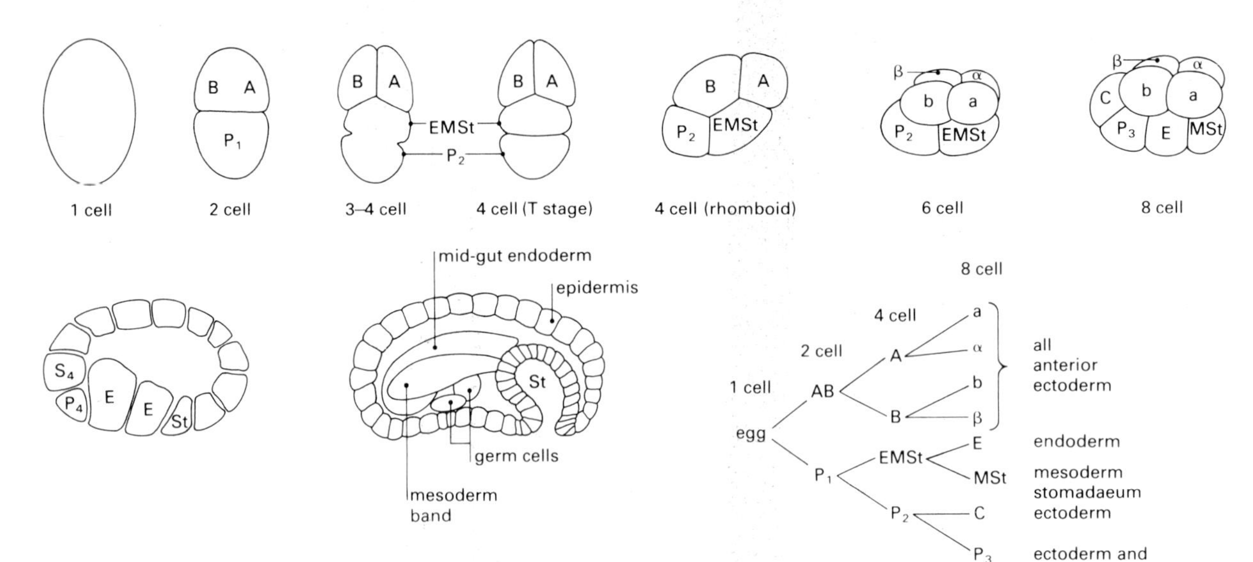


**Figure 32**
The pattern of early cleavage in *Rhabditis*.
*After Hinchcliffe, J. R., 'Observation of early cleavage in animal development: a simple technique for obtaining the eggs of Rhabditis (nematoda)'*, Journal of Biological Education, 7(6), *1973, pp. 33–37.*

**b** ***What is the rate of cell division? Compare your results with others in the group and see if any pattern emerges.***

Rates will depend on environmental conditions. Hinchliffe (1973) suggests:

**1** the time from one to a four-celled stage takes about 30 minutes;
**2** cell division from two to four-cell stage takes 5–7 minutes;
**3** after a period of 40 minutes the six-cell stage is produced and 10 minutes later the eight-cell stage is formed.

However Barrett (1983) indicates that a much more regular pattern of division occurs, and that, with suitable material, it should be possible to draw a graph of the results (*figure 33*).

**c** ***What happened to the egg size during cleavage? Explain the significance of your observations.***

The size of the egg remains constant during cleavage, irrespective of the number of cells that it contains. This means that no growth occurs during this phase, and cells formed by division get progressively smaller.

**d** ***In the light of your answer to (c), what happens to the ratio of nuclear to cytoplasmic volume during cleavage? Why may this ratio be important?***

The ratio of nucleus:cytoplasm is increasing, that is, there are more nuclei for a given volume of cytoplasm. Each nucleus may only be able to direct the activity in a certain volume of cytoplasm. Cell division without overall growth restores an optimum balance between nucleus and cytoplasm.

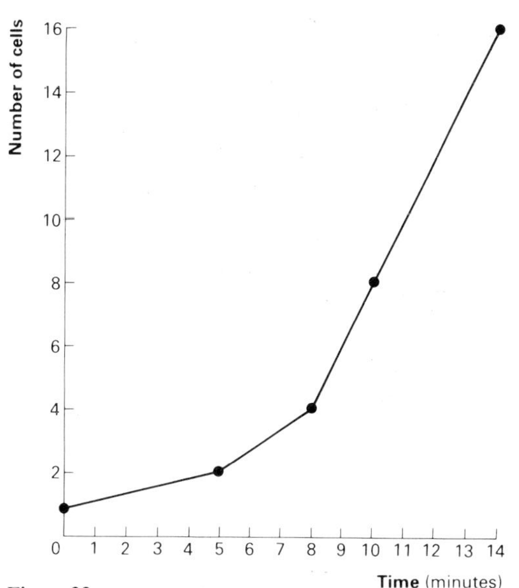


**Figure 33**
Graph illustrating the rate of cell division in *Rhabditis*.
*From Barrett, D. R. B., 'Observing cleavage in* Rhabditis *spp.'*, School Science Review, **64**(*228*), *1983, pp. 483–485.*

e ***Consider the genetic information that is contained within each cell of the developing embryo. Would you expect it to be the same? Give reasons for your answer. What developmental problems does this raise?***

Yes, because cleavage divisions are mitotic the genetic information within each cell will therefore be identical, providing that the division occurs normally and no somatic mutation occurs. However these same cells develop to form the varieties of cell types found in the various tissues and organs. Therefore there must be mechanisms to direct such changes and to control the action of the genome.

## INVESTIGATION
## 23B Embryonic development in shepherd's purse, *Capsella bursa-pastoris*

### ITEMS NEEDED

*Capsella bursa-pastoris* plants in flower
Microscope slides, prepared, of stages in *Capsella* embryo development

Glycerine, 5% solution
Hydroxide, potassium or sodium, 5% solution

Forceps
Microscope, binocular stereo 1/group
Microscope, monocular (× 100, × 400) 1/1 *(continued)*

(*Study guide* 23.2 'Early development'.)

*Capsella bursa-pastoris.* Extraction of the ovules is relatively easy. However it requires some skill and patience to dissect out the embryos, particularly in the early stages. It is best for students to start working on more mature ovules and work up the flowering shoot towards the apex and younger embryos (*figure 34*).

The crude squash preparations should be supplemented with prepared microslides of *Capsella*. Other common plants of the cruciferae family, such as thale cress (*Arabidopsis thaliana*) and hairy bittercress (*Cardamine hirsuta*), may well be worth investigating.

A practical pack and tape/slide materials, based on *Arabidopsis*, are available from Philip Harris Biological Ltd, Oldmixon, Weston-super-Mare, Avon, BS24 9BJ.

## ITEMS NEEDED *(continued)*

Microscope slides and coverslips
Mounted needles 2/1
Scissors
Watch-glass, solid 1/1

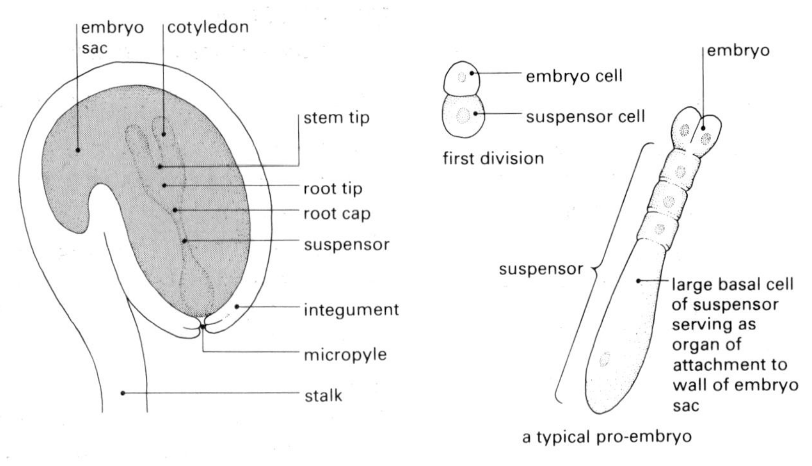


**Figure 34**
Tissue map of the ovule of *Capsella* and a typical pro-embryo.

*Assumptions*

1 Knowledge of the process of fertilization in angiosperms.
2 Knowledge of the process of seed formation in angiosperms.

*Principles*

1 Cleavage in plants is associated with growth.
2 There is an early differentiation of structures during plant development.
3 Major growth and development in angiosperms often occurs after a period of dormancy.
4 The seed habit has advantages over the spore habit.

*Questions and answers*

a ***In what ways is the process of cell division different from that found in* Rhabditis*?***

The pattern of cleavage is totally different. The first division produces an embryo cell and the basal cell of the suspensor. Subsequent divisions give rise to other cells of the suspensor and the pro-embryo. The suspensor pushes the pro-embryo into the endosperm.

b ***At what stage were any future organs first discernible? Which organs were you able to identify?***

Embryos from ovaries 3–4 mm long normally show well developed cotyledons, easily recognized as leaf-like structures.

c ***In the life cycle of many flowering plants the seed forms a dormant stage. What advantages does this have?***

This dormant phase can be advantageous for overcoming unfavourable environmental conditions, such as a cold season or drought. Many fruits (and seeds) possess specialized mechanisms to aid dispersal from the parent plant.

## INVESTIGATION
## 23C Morphogenesis in amphibians

### ITEMS NEEDED

Microscope slides, prepared, for additional information (optional)

Compasses, pair 1/group
Dividers, pair 1/group
Pens, felt colouring
Scalpel or other suitable sharp thin blades 1/group
Spheres, polystyrene about 50 mm diameter 2/group

(*Study guide* 23.3 'Development of the body plan'.)

**Pens.** Spirit-based overhead projector pens may be suitable though some give too dark a colour. Try a range of pens first, not all mark the polystyrene effectively.
**Polystyrene spheres.** The type sold for constructing atomic models is suitable. They can be cut using a polystyrene hot-wire cutter, although this must only be done in a fume cupboard with the fan on. Alternatively a very sharp thin blade, such as a Swan-Morton scalpel, can be used.

*Assumptions*

1. An understanding of the terminology used in describing sections.
2. An outline knowledge of amphibian egg structure and events from fertilization to blastula formation.

*Principles*

1. Vital stains can be used to map the developing embryo from the zygote stage. This enables a study to be made of the origin and development of different structures within the embryo.
2. A gastrula composed of three cell layers is formed by co-ordinated cell movements.
3. Adult structures are formed by differentiation and growth of these three embryonic cell layers.
4. The fate of cells depends upon their position relative to other cells in the embryo.

The changes in form during gastrulation are difficult to follow but, by completing these exercises, students should be able to understand the pattern. It is important that they are familiar with the relationship between two-dimensional sections and the three-dimensional structures they represent. Therefore, before starting this investigation, it is worth carrying out the following preliminary exercise.

**Relating 2-D sections to 3-D structures**
Present the group with an orange. Ask them to draw what they would see if it were cut into halves across the meridian. After they have made the drawing, ask them to cut the orange and compare the drawings. Repeat, asking for a drawing of the orange after a vertical cut, and finally after a cut made at 45° to the vertical.

A more difficult object to use for this is an apple, cuts being made in the same orientations. For most groups it is worth carrying out this exercise using both fruits. It has the advantage that the materials can be consumed afterwards – outside the laboratory of course.

The key for figures (P)3 and (P)4 in the *Practical guide*, contains terms with which the students may not be familiar. Complete the exercise first and then discuss the terms in relation to the results, at which time an outline of the whole process of gastrulation and subsequent development can be considered.

The orientation of the diagrams showing developing embryos often causes confusion. One reason is because most textbooks show the animal/vegetal axis as vertical in drawings of the blastula and almost horizontal in drawings of the gastrula. All drawings here show the pole axis horizontal.

The account of gastrulation in most current texts is based upon the original work of Vogt (1929) and shows the blastula to be composed of a single cell layer. Recent work by Keller (1975, 1976) however, has shown that the blastula is made up of a continuous layer of cells, which is more than one cell thick.

**Models and drawings**

In the exercise the students are asked to construct models from polystyrene spheres and to make drawings from these. Thus they are exercising their skills in constructing a three-dimensional object from a two-dimensional drawing and then using some of the information from the model to make a two-dimensional drawing.

The following models and drawings are made:

1 On a polystyrene sphere, the prospective fate map of the outer region. A drawing, based on the model of dorsal view of the blastula. See *figure 35*.

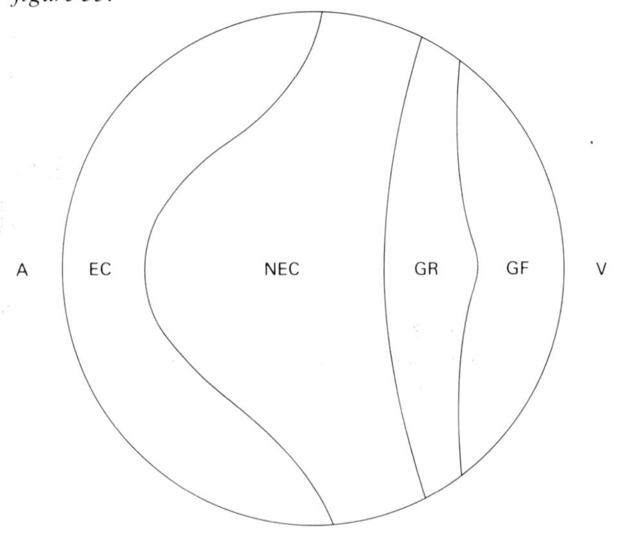


**Figure 35**
A dorsal surface view of the frog embryo blastula.

2 On the cut surface of a half sphere, a view of a vertical section through the middle of the blastula cut in the plane of the paper. A drawing of a vertical section of the blastula. See *figure 36*.

3 On the remaining hemisphere, a view of a vertical section cut at right angles to the plane of the paper, viewed from the posterior end. A drawing of the same view. See *figure 37*. On the surface of the hemisphere the fate areas. See *figure 35*.

4 Using figure (P)15, a drawing of a dorsal view of the gastrula, and a vertical section through the middle at right angles to the plane of the paper, viewed from the posterior end. See *figures 38* and *39*.

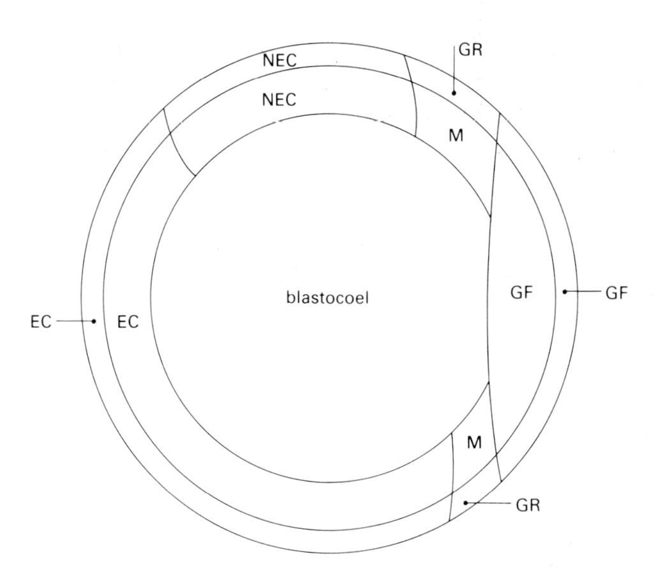


**Figure 36**
A vertical section of the frog embryo blastula from the lefthand side.

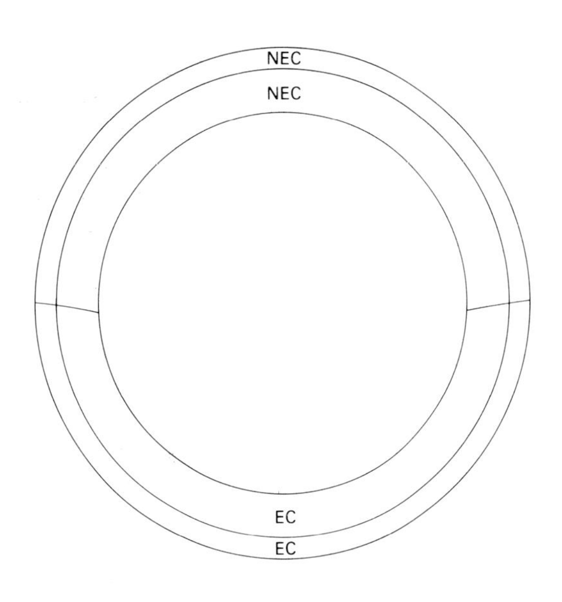


**Figure 37**
A vertical section of the frog embryo blastula from the posterior end.

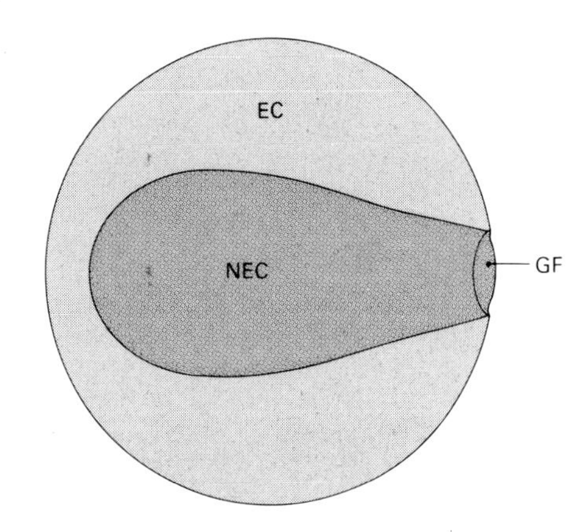


**Figure 38**
A dorsal view of the frog embryo gastrula.

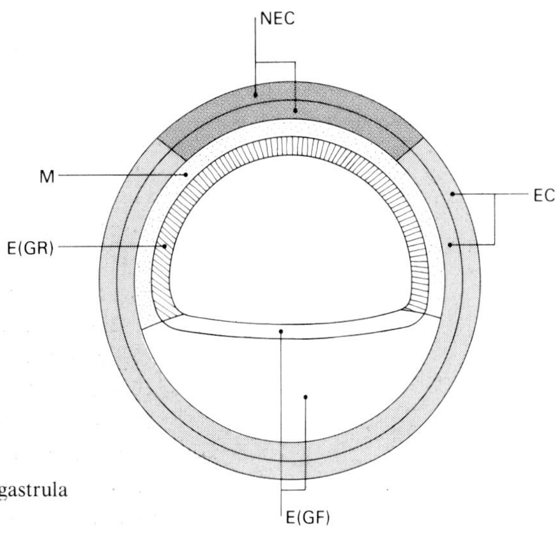


**Figure 39**
A vertical section of a frog embryo gastrula viewed from the posterior end.

*Questions and answers*

a ***What happens to the prospective mesoderm cells of the blastula during gastrulation?***

The prospective mesoderm moves into the blastocoel and forms a cell layer between the prospective ectoderm (EC and NEC) and the prospective endoderm (GR and GF).

b ***What are the developmental fates of the cell layers in the gastrula?***

The ectoderm will form the epidermis (skin) and the spinal cord and brain. The prospective mesoderm, with some ectoderm, will develop into many of the internal body organs and connective tissues. The endoderm forms the lining of the gut and is involved in the development of the associated organs (liver, pancreas, lungs).

c ***Compare the structure of the gastrula with that of the neurula* (figure [P]16). *What changes in form occur during neurulation?***

The neural ectoderm layers have formed the neural tube; the notochord arises from the mesoderm layer beneath. The mesoderm has extended downwards to completely enclose the endoderm. Dorsally the mesoderm splits to form the beginnings of a new body cavity, the coelom. The embryo itself becomes elongated along the antero–posterior and dorso–ventral axes.

## INVESTIGATION
## 23D Growth and development in the fore-limb of mice

ITEMS NEEDED

Graph paper
Vernier callipers 1/1 or group

(*Study guide* 23.5 'Growth' and 23.8 'Pattern formation in the development of limbs'.)

*Assumptions*

1 A knowledge of the structure of a pentadactyl limb.

*Principles*

1 Growth can be represented in a number of ways.
2 Percentage growth is a better measure than absolute growth as the amount of growth at a given time is dependent upon the amount of growth that has already occurred.
3 Differential growth rates affect the proportions of the limb components.
4 Percentage growth curves show a characteristic bell-shape.
5 Cartilage formation and growth occurs at both ends of the limb components.

Students need to realize the inaccuracies involved in the measurements of limb components, and therefore the need to use a consistent method of measurement. The main errors will be in measuring the humerus, wrist, and the components of the younger limb. With some groups it may be preferable to use clear plastic rulers, rather than callipers. To emphasize these points, consider collating the group measurements after procedure 1 and looking at the variation in the data derived from identical photographs. Depending upon the number of students in the group, calculations of standard deviation might prove useful at this point.

Individual and group averages for the limb components should be tabulated. Table 20 gives a specimen set of group averages. *Figures 40* and *41* show a growth curve and a growth rate curve.

| *Day* | *Humerus* | *Radius* | *Wrist* | *Middle-digit* | *Total* |
|---|---|---|---|---|---|
| 12 | 0.89 | 0.48 | 0.17 | 0.53 | 2.07 |
| 13 | 1.08 | 0.61 | 0.28 | 0.78 | 2.75 |
| 14 | 1.52 | 0.97 | 0.30 | 1.03 | 3.82 |
| 15 | 2.24 | 1.58 | 0.33 | 1.48 | 5.63 |
| 16 | 3.22 | 2.36 | 0.37 | 1.97 | 7.92 |

**Table 20**
Average measurements in mm of components of mouse fore-limb.

Finally, the students are asked to plot a percentage growth curve by plotting increase in length as a percentage of the previous day's length. The calculation is:

$$\text{Percentage growth} = \frac{\text{increase in length}}{\text{previous day's length}} \times 100$$

For example,

Humerus: day 12 = 0.89 mm
day 13 = 1.08 mm

$$\text{Percentage growth} = \frac{0.19}{0.89} \times 100 = 21\,\%.$$

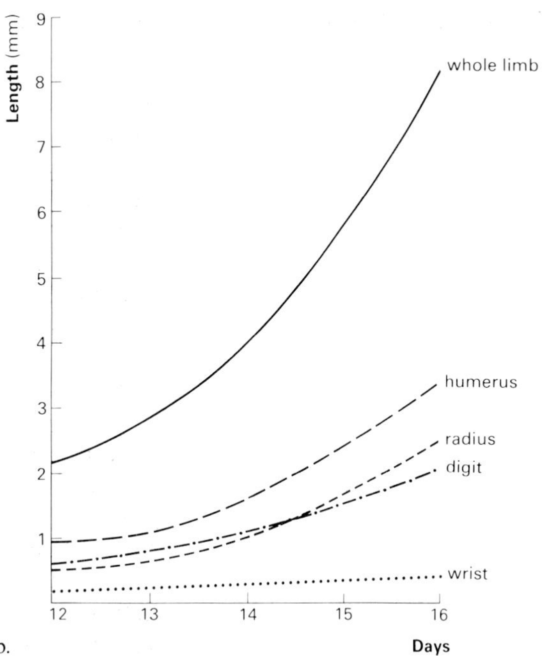


**Figure 40**
A growth curve for mouse fore-limb.

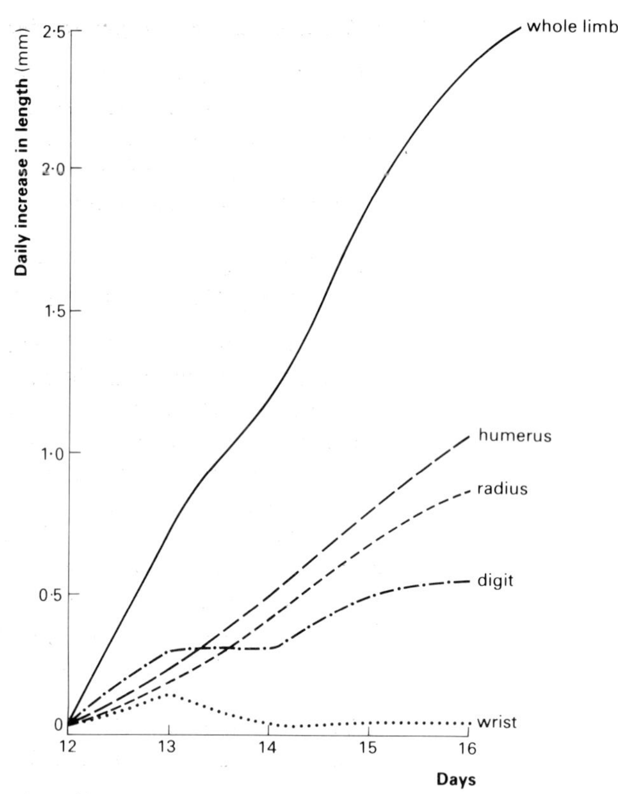


**Figure 41**
A growth rate curve for mouse fore-limb.

**Limb development research**
The majority of work has been carried out using chick embryos. Summerbell (1976) may be of interest in this respect for comparison.

*Questions and answers*

a ***What can you deduce from the three types of growth curve that you have drawn?***

**1** *Growth curve.* The whole limb, humerus, radius, and digits show exponential growth, while the wrist shows very slow linear growth.
**2** *Growth rate curve.* The rates of increase in length of the whole limb, humerus, and radius show a linear increase between days 12 and 15, with a slight drop in growth rate after day 15. The digit and wrist show similar growth rates to the other limb components during day 12, but a marked drop in rate during day 13 occurs. After day 14, digit growth resumes in a similar manner to the other limb components, but wrist growth shows only a very slight change.
**3** *Percentage growth.* Digit and wrist curves clearly show the drop in growth rate mentioned previously. Whole limb, humerus, and radius curves show that peak growth occurred at about day 15 and that the growth rate is decreasing after that day.

**b** ***What is the pattern of growth during mouse fore-limb development?***

Absolute growth shows an exponential increase during early development. The growth rate increases linearly for some days, before starting to level off leading to a zero value when full size has been reached. The growth of the digit and wrist show slightly different patterns with a period of slower growth.

**c** ***Examine the growth rate curve from procedure 7. What changes in proportion of the limb elements have occurred during development?***

The increase in the length of the humerus is slightly greater than that for the radius so although the two components remain in approximately the same proportion to one another, the humerus becomes slightly larger. The initially high rate of digit growth means that early in development the digits are longer than the humerus and radius. The following drop in digit growth rate reverses the proportion of main bones to digits. An early high rate of wrist growth makes a large contribution to the total limb length, but the subsequent very low growth rate means that its proportion of the limb decreases rapidly. The wrist is the only limb component to be set approximately in its correct size early in development.

**d** ***What two processes could explain the drop in growth rate of the wrist and middle digit during day 13?***

Decreased cell division or increased cell death.

**e** ***Using the percentage growth curves from procedure 8, estimate the age of the mouse at which the fore-limb would stop growing. Explain how you arrived at your answer and what assumptions, if any, you made.***

If the graph line for the whole limb is extended from day 16 it crosses the axis at day 19, around the time of birth. It assumes a constant decrease in growth from day 16 and that growth will not occur after birth. The latter assumption is, of course, untrue.

**f** ***Why might percentage growth curves be more useful to developmental biologists than absolute growth curves?***

Absolute growth curves represent the total cumulative growth. Developmental biologists are more interested in changes in the rate of growth and the amount of growth occuring relative to that which has already taken place.

**g** ***What is the nature of the light unstained regions that appear in the photographs of 15 and 16 day-old limbs?***

These regions are collars of bone, the cartilage having been broken down and replaced by bone. Bone does not take up the Alcian Green stain used in the preparation.

**h** ***What does the location of these lighter regions suggest about the growth and differentiation of cartilage during long bone development?***

The middle parts of the long bones must be the oldest regions as they ossify first. Limb clements have two sites of active growth and differentiation (growth plates). The addition of new cartilage tissue occurs at both ends of each limb element.

## INVESTIGATION
## 23E Growing new plants from old

ITEMS NEEDED

Cauliflower, fresh 1/class
Cauliflower, seeds (packet) 1/class
Bleach, *e.g.* Domestos
Disinfectant, *e.g.* Milton's sterilizing fluid
Meristem tissue, culture medium, bottles 2–3/group

Water, distilled
Water, distilled, sterile, in

Hand lens, × 8 or × 10 1/group
Inlculation chamber 1/class
Inoculating loop 1/group
McCartney bottles 3/group
Measuring cylinders, 10 $cm^3$, or burette 1/group
Microscope, binocular stereo 1/group
Paper towels
Scalpel 1/group
Tray, seed 1/group

(*Study guide* 23.9 'The pattern of plant and animal development compared'.)

**Culture media.** Cauliflower meristem culture medium in McCartney bottles is available from Philip Harris Biological Ltd, Oldmixon Weston-super-Mare, Avon BS24 9BJ, either separately, or as a tissue culture kit. Media for a range of different explants are also available from Flow Laboratories, Harefield Road, Woodstock Hill, Rickmansworth Herts WD3 1PQ and Gigco Ltd, PO Box 35, Trident House, Renfrew House, Renfrew Road, Paisley, PA3 4EF, Scotland. Table 21 shows the composition of a culture medium for cauliflower explants.

*Assumptions*

1 Proficiency in sterile techniques.
2 Knowledge that cauliflower curd is the term for the florets.

*Principles*

1 Cells contain the genetic information necessary to form any part of the whole organism.
2 Many cells have the ability to redifferentiate.
3 The genetic potential of a plant cell is normally restrained by the influence of neighbouring cells.

| *Constituent* | *Concentration* ($mg\ dm^{-3}$) | *Constituent* | *Concentration* ($mg\ dm^{-3}$) |
|---|---|---|---|
| **Macro-nutrients** | | **Iron source** | |
| Ammonium nitrate, $NH_4NO_3$ | 1650 | Disodium ethane-1, 2-diamine, tetra-ethanoate, $Na_2EDTA$ | 37.3 |
| Calcium chloride, $CaCl_2.2H_2O$ | 440 | Iron II sulphate, $FeSO_4.7H_2O$ | 28.8 |
| Magnesium sulphate, $MgSO_4.7H_2O$ | 370 | **Amino acid source** | |
| Potassium dihydrogen phosphate, $KH_2PO_4$ | 170 | Glycine | 2.0 |
| Potassium nitrate, $KNO_3$ | 1900 | **Vitamins** | |
| **Micro-nutrients** | | Meso-inositol | 100.0 |
| Boric acid, $H_3BO_3$ | 6.2 | Nicotinic acid | 0.5 |
| Cobalt chloride, $CoCl_2.6H_2O$ | 0.025 | Pyridoxine hydrochloride | 0.5 |
| Copper sulphate, $CuSO_4.4H_2O$ | 0.025 | Thiamine hydrochloride | 0.1 |
| Manganese sulphate, $MnSO_4.4H_2O$ | 22.3 | **Growth substances** | |
| Potassium iodide, KI | 0.83 | Indole ethanoic acid | 8.0 |
| Sodium molybdate, $Na_2MoO_4.2H_2O$ | 0.25 | Kinetin | 2.5 |
| Zinc sulphate, $ZnSO_4.4H_2O$ | 8.6 | **Carbohydrate source** | |
| | | Sucrose | 30 000.0 |
| | | pH 5.8 | |
| | | For callus cultures add 0.8 % w/v agar | |

**Table 21**
Modified Murashige and Skoog (111S) medium.

**4** Understanding of cellular totipotency can be applied to horticulture and agriculture.

The investigation demonstrates the principles of micro-propagation and describes their use in horticulture and agriculture. Propagation by seed and by tissue culture are compared.

To set up explants it is best to use an inoculation chamber as this will considerably reduce the likelihood of contamination.

To extend the investigation, the explant tubes could be kept in a range of environments to study the effect of conditions on explant growth and development. If interest can be maintained after about 10 weeks the plantlets can be transplanted into peat blocks or other appropriate substrate and grown on alongside the seedlings.

Other material which can be used for such investigations are carrots and tobacco. Like the cauliflower tissue culture, kits are available. You first have to produce callus, either direct from a piece of carrot, or by growing seedlings first as with both carrot and tobacco. In these cases it takes some weeks to produce the callus and then a further period of time to regenerate a plantlet from callus.

*Questions and answers*

**a** ***What would the cells of the curd normally have developed into? From your observations of the explants, what changes in structure and function did these cells undergo?***

The curd is composed of young flower buds. The explants eventually produce chlorophyllous tissue, roots, stems, and leaves. Thus the cells of the flower buds must have given rise to the range of cells found in such plant tissues and organs.

**b** ***What conclusions can you draw concerning the ability of the explant cells to develop their genetic potential?***

As flower bud cells and cells derived from them can produce any of the major plant structures, it would seem that plant cells can use any portion of their genetic component.

**c** ***Why does the curd of a cauliflower not show the sort of changes that you found in the explants?***

The cells of the intact flower buds must be restrained, by the presence or absence of plant hormones, from expressing genetic information that would lead to anything other than floral structure and function.

**d** ***How many new cauliflower plants could be grown by taking explants from curd? Explain how you calculated your answer.***

Students should explain how they estimated the surface area of the curd in $mm^2$ and divided by 9. Several hundred thousand explants could be obtained from a large cauliflower.

e ***Plants grown by such a culture technique are called clones. What is the meaning of the term? Why may the process be useful in horticulture and agriculture?***

A clone is a descendant from a single organism produced by asexual reproduction. The members of a clone have an identical genetic constitution to one another and to their parent. Clones are useful in any situation where the occurrence of variation due to a sexual process is undesirable. For instance, they enable the conservation of genetically pure stocks or the proliferation of a plant with the exact characteristics required for a given situation.

f ***Compare the two methods of propagation used in this investigation. What are the advantages and disadvantages of each?***

**1** *Cell culture.* Advantages: large numbers of offspring from one parent; exact characteristics of offspring known; more plantlets can be raised in a given space; may be produced at any time of the year. Disadvantages: specialized equipment and knowledge required; more capital outlay; much greater energy input required for mass production.
**2** *Seeds.* Advantages: simple to grow; variation in offspring useful for growth in an uncontrolled environment; little energy expenditure required. Disadvantages: large growing areas needed; producing seed requires skill and time; production is dependent on climate, pests, and so forth.

g ***Review the different methods of tissue culture. What are the potential applications of these techniques?***

Students will need to do some background reading; see the bibliography.

A 35 mm colour slide set, 'Basic techniques of plant tissue culture' by M. M. Yeoman is available from Philip Harris Biological Ltd, Oldmixon, Weston-super-Mare, Avon BS24 9BJ. This would be useful material for the students to view at this point.

# PART III Bibliography

ALBERTS, B., BRAY, D., LEWIS, J., RAFF, M., ROBERTS, K., and WATSON, J. D. *Molecular biology of the cell.* Garland Publishing, 1983. (Chapter 15, 'Cellular mechanisms of development', Chapter 16, 'Mammalian cell types'.)
BARRETT, D. R. B. 'Observing cleavage in *Rhabditis* spp.'. *School Science Review*, **64**(228), 1983, pp. 483–485. (Methods of culturing and observing *Rhabditis.*)
BILLETT, F. S. and WILD, A. E. *Practical studies of animal development.*

Chapman & Hall, 1975. (Information on *Rhabditis*, small mammals, and suggestions for work on fish eggs, amphibia, etc.)
BROWSE, P. MCM. *Plant propagation*. Mitchell Beazley, 1979. (Use of the different parts of a plant for propagation.)
CLARKE, C. A. Studies in Biology No. 20, *Human genetics and medicine*. 2nd edn. Edward Arnold, 1977. (Congenital defects.)
COX, F. E. G. Parasites of British earthworms. *Journal of Biological Education*, **2**(2), 1968, pp. 151–164. (Concise account of their occurrence.)
DOUGHERTY, E. C. *et al.* 'Axenic cultivation of *Caenorhabditis briggsae* (Nematode: Rhabditidae) with unsupplemented and supplemented chemically defined media.' *Annals of the New York Academy of Sciences*, **77**, 1959, pp. 176–217. (Cultivation of the nematode *in vitro*.)
GARDNER, E. J. *Human heredity*. John Wiley, 1983. (Congenital defects.)
GERLACH, D. 'How to clone plants'. *Carolina Tips*, **44**(9), 1981, pp. 37–40. (Methods of plant tissue culture.)
GURDON, J. B. *The control of gene expression in animal development*. Oxford University Press, 1974. (Account of frog nuclei transplantation experiments.)
HINCHLIFFE, J. R. 'Observation of early cleavage in animal development: a simple technique for obtaining the eggs of *Rhabditis* (nematoda)'. *Journal of Biological Education*, **7**(6), 1973, pp. 33–37. (Culturing and examination of the developing embryo.)
HOCKING-CURTIN, J. Plant tissue culture and the carrot. *Carolina Tips*, **45**(12), 1982, pp. 45–47. (Details of how to set up carrot tissue cultures.)
JOHNSON, G. E. 'On the nematodes of the common earthworm'. *Quarterly Journal of Microscopical Science*, **58**, 1912–1913, pp. 605–652. (Culture methods of nematodes, account of species found, their life cycles.)
KELLER, R. E. 'Vital dye mapping of the gastrula and neurula of *Xenopus laevis*, 1: Prospective area and morphogenetic movements of the superficial layer., *Developmental Biology*, **42**, 1975, pp. 222–241. (Structure of the blastula and process of gastrulation.)
KELLER, R. E. Vital dye mapping of the gastrula and neurula of *Xenopus laevis*, 2: Prospective areas and morphogenetic movements of the deep layer. *Developmental Biology*, **51**, 1976, pp. 118–137. (Structure of the blastula and process of gastrulation.)
LEVIN, I. *The boys from Brazil*. Michael Joseph, 1976. (Fictional account of human cloning.)
McLAREN, A. *Mammalian chimeras*. Cambridge University Press, 1976. (Mouse experiments.)
NORTHCOTE, D. H. Carolina Biology Readers No. 44, *Differentiation in higher plants*. 2nd edn. Carolina Biological Supply Company, distributed by Packard Publishing, 1980. (Meristem development.)
POTTS, G. D. 'Ernest Haeckel: a biographical sketch'. *American Biology Teacher*, **38**(9), 1976, pp. 544–545 and 556. (Concise account of his work and ideas.)
REINERT, J. and YEOMAN, M. M. 'Plant cell and tissue culture'.

Springer-Verlag, 1982. (Contains a wide range of investigations, some possible for advanced work in schools.)
SLACK, J. M. W. *From egg to embryo, determinative events in early development*. Cambridge University Press, 1983. (Early development in mouse, sea urchin, and amphibian.)
SMYTH, J. D. *Introduction to animal parasitology*, 2nd edn. Hodder & Stoughton, 1976. (Describes the life cycle of *Rhabditis*.)
STREET, H. E. (Ed.) *Botanical Monographs II, Plant tissue and cell culture*. 2nd edn. Blackwell Scientific, 1977. (Standard text on the subject.)
SUMMERBELL, D. 'A descriptive study of the rate of elongation and differentiation of the skeleton of the developing chick wing'. *Journal of Embryology and Experimental Biology*, **35**, 1976, pp. 241–260. (Detailed description of growth and differentiation.)
TICKLE, C., SUMMERBELL, D., and WOLPERT, L. 'Positional signalling and specification of digits in chick limb morphogenesis'. *Nature*, **254**, 1975, pp. 199–202. (Chick limb data.)
TRINKAUS, J. P. *Cells into organs: forces that shape the embryo*. 2nd edn. Prentice Hall, 1984. (Morphogenesis.)
VIDAL, G. 'The oldest eukaryotic cells'. *Scientific American*, **250**(2), 1984, pp. 32–41. (Evolution of the first autotrophic eukaryotes.)
VOGT, W. 'Gestaltungsanalyse am Amphibienkeim mit örtlicher Vitalfärbung. II Teil. Gastrulation und Mesodermbildung bei Urodelen und Anuran'. *Wilhelm Roux' Archiv für Entwicklungsmechanik der Organismen*, **120**, 1929, pp. 384–706. (Classical picture of early frog development.)
WINCHESTER, A. M. and MERTENS, T. R. *Human genetics*. Charles Merrill, 1983. (Congenital defects.)

# CHAPTER 24 CONTROL AND INTEGRATION THROUGH THE INTERNAL ENVIRONMENT

*A review of the chapter's aims and contents*

1. This chapter introduces some of the fundamental concepts of endocrinology and the common features of endocrine systems. Specific examples are used as illustrations.
2. A third level of control through the neurocrine system is introduced.
3. The structures of endocrine and neurosecretory cells are described and are related to their functions.
4. The mechanisms of hormone action are discussed at the subcellular level.
5. Feedback control (negative and positive) is introduced as an important component of many endocrine systems.
6. Hormonal control in the development of animals is examined, in particular the physiology of moulting in insects and of metamorphosis in amphibians.
7. The evidence for the existence of hormones as messengers in plants is considered.
8. An example of a plant hormone or auxin (IAA) and the mechanism of its action is given.
9. The role of different plant growth substances as regulators in development is described.
10. The immune response of vertebrates is introduced as a system whereby foreign material is recognised and eliminated from the body.
11. The mechanisms involved in the different forms of immunity are examined.
12. The role of lymphocytes in the different kinds of immunity are given particular consideration.
13. The possibility of a mechanism which gives a similar protection in plants is mentioned briefly.

# PART I The *Study guide*

## 24.1 Endocrine communication and control in animals

*Assumption*

1. An understanding of the role of the nervous system in co-ordination.

*Principles*

1. In contrast to the nervous system, the endocrine system often functions intermittently to bring about more diffuse and usually more prolonged actions.
2. Neurosecretory cells provide a third level of control.

3 The combined response of a target organ to two hormones acting together is often different from the individual responses to those hormones.

4 The structure of endocrine cells is related to the type of hormone being produced.

### What is endocrinology?

Endocrinology began as a branch of medicine and was at first essentially a study of the 'ductless glands' and their 'internal secretions'. Recently, it has grown in quantal leaps as new endocrine tissues have been discovered and studied in a wide variety of animals. Textbooks often reflect this origin, dealing with the subject organ by organ, chapter by chapter. This approach can over-emphasize to the student differences in detail between endocrine systems, while obscuring the more important common features. There are certain fundamental concepts in endocrinology, however:
endocrine involvement in co-ordination and control systems:
the structure and functioning of endocrine cells;
endocrine activity – what affects it and how it can be assessed;
the mechanisms of hormone action at the subcellular level.
These are discussed as general concepts and illustrated by a few specific examples.

### Nerves, hormones, and neurohormones

*Question and answer*

a ***What features of neurosecretory cells make them more suitable than conventional neurones for controlling endocrine glands?***

One clue may lie in the differences between the natures of nervous and endocrine (or neurocrine) control. It is essential that all cells of a similar function within an endocrine gland act simultaneously to effect a sustained increased (or reduced) discharge of hormones. Nervous control, by virtue of its short-lived and localized nature, is unsuitable. The neurosecretory cells, however, are part of the nervous system; they receive and conduct electrical signals, and yet they can broadcast chemical messages about the body to every cell that is programmed to respond to the messages. They are thus ideally suited to be the essential link between the nervous system, which monitors the environmental conditions, and the endocrine system, which brings into play physiological mechanisms which react to these conditions.

### The structure of endocrine cells

**STUDY ITEM**

#### 24.11 The ultrastructure of endocrine cells

*Questions and answers*

a ***What differences are apparent in the structure of these two types of cell and in the storage of the hormone?***

The major characteristics of a protein- (or peptide-) producing endocrine cell are the rough endoplasmic reticulum, a prominent Golgi complex, and large numbers of membrane-bound storage/secretory vesicles. The hormone is synthesized in the rough endoplasmic reticulum and packed into secretory granules in the Golgi complex. Steroid-synthesizing cells are quite different; they have extensive smooth-surfaced endoplasmic reticulum and a prominent Golgi complex. Steroid hormones, unlike peptides, are not stored prior to release; numerous lipid droplets in the cytosol contain, not the finished hormone, but its precursor materials. Characteristically, the mitochondria possess cristae which have a tubular or vesicular appearance reminiscent of smooth endoplasmic reticulum and quite distinct from the typical lamellar arrangement of 'normal' mitochondria.

**b** ***In what ways do they [neurosecretory cells] resemble other protein-synthesizing cells?***

Like other protein-synthesizing cells, neurosecretory cells have a well-defined Golgi complex, abundant rough endoplasmic reticulum, and numerous membrane-bound vesicles which are elaborated in the Golgi complex and contain the neurohormones.

**c** ***In what features does a neurosecretory cell differ from a conventional neurone?***

The axons running from the cell body are often swollen at intervals and vesicles are found stored in abundance in their dilatations. Release of the neurohormones into the blood occurs at the terminal dilatations of the axons (usually many axon terminals are found together in the neurohaemal organs). □

It is of interest to note that the material stored in the non-terminal dilatations of neurosecretory axons is not immediately lost to the pool of releasable hormone; freshly synthesized hormone in the main axon and terminal dilatations appears to be released preferentially. This has been termed the 'last in, first out' effect. The non-terminal dilatations probably represent sites of hormone storage for use when a massive release of hormone is required, but they are also sites of vesicle degradation. The mechanism by which stored neurosecretory vesicles can be recognized as being 'too old' and fit only for destruction is unknown.

In many animals, peptides which function as 'true' hormones in the body have been found to be present also in the brain. One important group of these peptides in mammalian brain is the endorphins. The discovery of specific receptors for opiates like morphine suggested that natural substances must exist within the brain which react (bind) with such receptors. Indeed, the endorphins (named from *endo*genous and m*orphine*) are such endogenous substances. Their physiological role is still uncertain but perhaps they control feelings of pleasure or pain, or states of arousal or sleep. These and many of the other peptides found within the brain probably act as neurotransmitters or modulators of neuronal function. Nevertheless, the fact that many neurones in the nervous system synthesize peptides suggests that neurosecretory cells are not particularly unusual nerve cells in using peptides as chemical messengers. Indeed, this may be a general or even a primitive property of neurones.

## 24.2 Hormone synthesis and release

*Principles*

1. The method of synthesis of hormones varies with the type of hormone being produced, as does their release.
2. The control of hormone release can involve nervous, endocrine, and metabolic interactions.
3. The release of hormone and the action of the target tissue operate as closed loop systems with a feedback of information which may be negative or positive.
4. The release of hormone is initiated as a result of a stimulus which causes depolarization of the plasma membrane, thus altering its permeability to inorganic ions (especially $Ca^{2+}$).
5. cAMP may also be involved in hormone release.

The synthesis of hormones of different types is dealt with briefly in the *Study guide*. Further information is given here about catecholamine synthesis and the role of prostaglandins.

### Catecholamine synthesis

In the adrenal gland, an enzyme converts noradrenaline to adrenaline The activity of this enzyme is controlled by corticosterone hypophysectomy causes a marked decrease in activity whereas treatment with ACTH or with large doses of corticosteroid restores activity. The enzyme is absent from adrenergic nerve endings, which explains why it is only in adrenal tissue that appreciable quantities of adrenaline are produced. Catecholamines and a wide range of other bioamines are found in invertebrates too, but little is known of their synthesis, although it is generally assumed to follow pathways similar to those operating in vertebrates.

### The role of prostaglandins

Prostaglandins are involved in reproduction (ovulation, luteolysis gamete transport, menstruation, abortion, parturition) and in the control of body temperature by the hypothalamus. The effects of several renally active hormones are mediated by prostaglandins produced within the kidney. Prostaglandins also play a role in blood clot formation, and in the inflammatory response. The true importance of these apparently ubiquitous molecules is only now beginning to be appreciated.

**Practical investigation.** ***Practical guide 6*****, investigation 24A, 'The microscopic structure of endocrine glands'.**

## Control of hormone release

**The mammalian hypothalamus.** It is mentioned in the *Study guide* that the higher centre in the vertebrate CNS which acts in closed loop systems is usually the hypothalamus. Further information about the hypothalamus is given here.

As vertebrates evolved, the hypothalamus came to occupy a central role as an integration centre, capable of processing numerous and diverse information inputs; appropriate hormonal responses are directed via the anterior and posterior pituitary.

It is worthwhile to examine in detail the hormonal feedback which influences hypothalamic output of releasing factors during reproduction in mammals. The effect of gonadal steroids on hypothalamic release of gonadotropin releasing hormone (Gn-RH) is usually negative in character. For example, injection of testosterone lowers circulating luteinizing hormone (LH), which is released in response to Gn-RH; conversely, gonadotropin release increases after castration. While this type of relationship is invariably seen in male mammals, it is not always so in females. Rising levels of oestradiol, produced by maturing Graafian follicles, are responsible for triggering a massive release of LH – the 'LH surge' – and hence ovulation: *positive feedback* is an important component of ovarian cyclicity. This difference between the male and female hypothalamus is an interesting example of sexual differentiation, and we now know that it is programmed very early in life, either before birth or in the neonatal period. Under the influence of testicular steroids from the foetus, the positive feedback component is lost and a male-type hypothalamus/pituitary develops.

The hypothalamus also receives important nervous input. For example, vision and olfaction are important in co-ordinating reproductive function via the hypothalamus. In many species of mammal and bird, a change in photoperiod triggers increased hypothalamic release of Gn-RH, and thus the onset of a 'breeding season'. Olfaction also plays a role in several aspects of reproductive behaviour in mammals: in many species the male will initiate sexual activity with a female in oestrus only after detecting the characteristic odour of a pheromone from the female reproductive tract. Similarly, male pheromones are often deposited in the urine and co-ordinate female sexual behaviour.

The hypothalamus responds also to neural inputs coming via the spinal cord. If the response provoked is hormonal, the sequence can be described as a *neuro-endocrine reflex*; the afferent arc is nervous, while the efferent arc is hormonal. There are numerous examples of such reflexes in mammalian reproductive physiology. In lactating females, oxytocin is released from the posterior pituitary as an acute response to stimulation of the nipple during suckling (the 'milk-ejection reflex'). This is an example of an extremely rapid neuro-endocrine reflex. Hormonal output from the anterior pituitary can also be modulated by neuro-endocrine reflexes. In those female mammals which are 'induced ovulators', such as the rabbit and cat, an ovulatory surge of LH is released in response to cervical stimulations from copulation. The duration of this reflex is relatively long; in the rabbit, for example, ovulation occurs about twelve hours after mating.

The role of the hypothalamus in the co-ordination of many physiological processes is truly remarkable in an organ which, in the human, is only about 0.004 per cent of the mass of the whole brain!

How does the arrival of the stimulus at the cell membrane of an endocrine cell initiate release of hormone?

Precisely how an increase in intracellular $Ca^{2+}$ initiates release of insulin is unknown but, by analogy with the role of $Ca^{2+}$ in muscle, it is possible that it stimulates the contraction of the microfilament–microtubular complex. A further role of $Ca^{2+}$ in excitation–secretion coupling is to cause fusion of granule and plasma membranes; subsequent rupture of the membrane releases the contents of the granule into the extracellular space.

**Insulin.** The difference in mechanism of stimulus–secretion coupling between glucagon and glucose raises several important questions. Several intracellular signals or *second messengers* (the first messenger being the external signal) are now recognized. This may prove to be a gross under-statement; although we know something of the roles of cAMP, cyclic guanosine monophosphate (cGMP), and $Ca^{2+}$ as second messengers, many inorganic ions could function in this manner. The nature of the interactions between second messengers is important, however. In the pancreatic β-cell, glucagon or cAMP alone (in the absence of extracellular glucose) will not cause hormone release. In fact, glucagon does not stimulate insulin release in the presence of 3.3 mmol $dm^{-3}$ glucose but does so in the presence of 16.5 mmol $dm^{-3}$ glucose. At concentrations of glucose below a critical level, cAMP concentrations increase, but insulin is not released. It would seem that in this system, cAMP plays a permissive role in secretion by potentiating the effects of glucose. This may have two important consequences. During starvation, when levels of blood glucose may fall below the critical value, glucagon will not cause an inappropriate release of insulin: and in the fed animal, glucagon release from the α-cells may prime the neighbouring β-cells to respond to the predictable increase in blood glucose by releasing insulin. This would represent a fine adjustment of the feedback mechanisms operating in the control of blood glucose levels.

There is much evidence to suggest that insulin release is also under considerable nervous influence via the splanchnic and vagus nerves. In particular, catecholamines inhibit insulin release and this appears to be related to a decrease in adenylate cyclase activity. Adrenaline will also inhibit the glucose-stimulated release of insulin, however, but the mechanisms of these interactions are not understood fully.

## 24.3 Dose–response relationships

*Principle*

1 Responses to hormones change with the titre or amount of hormone (the dose).

### STUDY ITEM

24.31 Dose–response curves to hormones

*Questions and answers*

a ***From* figure [S]224 *suggest which of the effects of insulin are operating in different tissues under the following conditions:***
***1 shortly after a normal meal;***

At the titres of insulin present shortly after a meal, the hormone stimulates glycogen synthesis in the liver, and triacylglycerol synthesis in the adipose tissue. It may (indirectly) stimulate glycogen synthesis in the muscle by its effect on the increased entry of glucose. Lipolysis (in the adipose tissue) and muscle proteolysis will be fully inhibited by the titres of insulin after a meal.

***2 three hours after a meal (called normal fasting);***

Some hours after a meal, insulin titres fall to levels at which lipolysis in adipose tissue may not be fully inhibited. Thus increased supplies of fatty acids may contribute to the energy requirements of the tissues, since glucose entry into the tissues is reduced at low insulin titres. Glycogen synthesis in muscle and liver will be minimal.

***3 after two days of continuous fasting or starvation.***

During starvation, the titres of insulin will be sufficiently low to allow muscle proteolysis to begin. Thus all the effects of low titres of insulin during normal fasting will be reinforced, and extra carbohydrate will be synthesized in the liver (gluconeogenesis) from amino acids liberated by muscle proteolysis.

b ***Suggest reasons why the level of insulin in the hepatic portal vein is so much higher than in the general circulation.***

The level of insulin in the hepatic portal vein is higher than in any other part of the blood circulation because it is this vein which carries freshly released insulin from the endocrine pancreas. However, about half of the insulin reaching the liver in this vein is destroyed by the liver and never reaches the general circulation. The peripheral levels of insulin are therefore always lower than those in the portal vein.

c ***What other hormones may be involved (in this case released) to ensure survival during prolonged fasting?***

While all the major metabolic events involved in survival during fasting can be explained to some extent by the reduction in insulin titres, both glucagon and glucocorticoids are also important. Glucagon (released in response to low levels of blood glucose) activates

(via stimulation of the enzyme glycogen phosphorylase) the release of glucose from glycogen stores in liver, and also stimulates lipolysis in adipose tissue.
Glucocorticoids release amino acids from muscle by stimulating muscle proteolysis. In the liver, and also in the kidney, glucagon and glucocorticoids stimulate gluconeogenesis.
The control of metabolism in the fed and fasting state is an excellent example of the way in which hormones act together rather than as individual isolated agents (that is, of hormone integration).

## 24.4 The fate of the hormone in the blood

*Principles*

1. Once a hormone has reached the blood, there are several possible processes that may play a part in determining its activity.
2. Mechanisms must exist for the removal of hormones from the blood.

## 24.5 Mechanisms of hormone action

*Principles*

1. Hormone action is specific, that is, although many hormones are present in the blood only the target tissues will respond to a particular hormone.
2. Specificity of hormone action depends on the distribution of appropriate receptors in the target cells.

## 24.6 Hormones in the control of growth and development

*Assumptions*

1. Some knowledge of insect life histories.
2. A knowledge of the anatomical and morphological changes associated with metamorphosis in amphibians.

*Principles*

1. The changes which occur in animals that undergo metamorphosis during their development are controlled by hormones in response to signals from the nervous system when the environmental conditions are appropriate.
2. Hormones often interact to control metamorphosis in insects and amphibians.

### The control of insect metamorphosis

**Insect moulting.** Much is known of the physiology of moulting. Its firs event, as far as changes in the epidermis or cuticle are concerned, i apolysis. Although this is the first recognizable effect of ecdysone *in viv* it is its least studied effect, and the exact mechanisms by which it occur are still not understood fully. Once apolysis has taken place the insect i said to be *pharate*; for example, once the pupa has apolysed it is said t be a pharate adult even though it has not yet ecdysed. The significance c this is that the epidermis has begun to secrete cuticle appropriate to th next state – in this case, the adult – but this is not apparent until ecdysis; developmentally, the animal is committed to becoming an adult.

**Tanning of the cuticle.** A necessary process, once ecdysis has occurred, is tanning. This is the name given to the hardening – often, although not necessarily always, accompanied by darkening – of the new cuticle. This is important structurally, since the cuticle acts as an exoskeleton, and also physiologically, in reducing further water loss through the cuticle. Tanning is a common phenomenon, not only in the leather industry, but also in living systems; it always involves the stabilization of protein by the introduction of cross-bridges between constituent chains. For example, keratin in human skin is a tanned protein in which the cross-bridges are formed by disulphide bonds. In insects and Crustacea, the cross-bridges are formed by phenolic links. Tanning of insect cuticle is under the control of a peptide hormone called *bursicon*, which is produced in neurosecretory cells in different parts of the CNS and released after ecdysis. Many older textbooks describe a role for ecdysone in stimulating tanning of insect cuticle but we know now that, while this is true for the particular insect and developmental stage studied by earlier workers (the tanning of the blowfly puparium), it is certainly not the general case.

It is worth noting that Crustacea moult and undergo changes virtually identical to those described in insects, although in these animals the tanned cuticle is reinforced by deposition of calcium carbonate crystals in the interstices of the protein chains and cross-bridges. Perhaps even more interestingly, they employ exactly the same hormone, β-ecdysone, to bring about the initiation of moulting. Moulting in Crustacea is less well understood than in insects, however, especially with respect to the control of differentiation. Certainly, no hormone equivalent to the juvenile hormone of insects has been described in Crustacea.

**The timing of ecdysis.** We have only a few sketchy ideas as to how the timing of major events such as apolysis and ecdysis are controlled. One important feature is that their timing can be altered independently; the time spent as a pharate individual varies greatly both between species and between individuals. We do know that feeding is an important process in determining the timing of release of hormones. For example, in *Rhodnius*, Wigglesworth found that the single blood meal which a larva takes during each stadium (the period between ecdyses) is the signal for release of the brain hormone (which stimulates the prothoracic glands to produce α-ecdysone); moulting thus occurred a fixed time after the blood meal. In continuous feeders like caterpillars or locusts, for example, it seems that an obligatory period of feeding is required before brain hormone stimulates prothoracic gland activity, but the precise timing of the release of ecdysone to initiate apolysis is not understood. Furthermore, there is now quite intriguing evidence that the timing of a moult is controlled partly by exteroceptive signals (for example, by photoperiod in Lepidoptera) and partly by interoceptive stimuli; in the tobacco hornworm moth, *Nanduca sexta*, the last-stage larva needs to reach a certain mass before it can pupate. How it 'knows' its mass is not known.

**Insect growth.** To say that arthropods cannot grow in size during a stadium because of the rigidity and 'unstretchability' of the cuticle is too gross a generalization. Certainly, for most Crustacea, and for those insects with a relatively hard (well-tanned) cuticle, this is a reasonable view. However, many insect larvae have rather softer and more 'stretchable' cuticles; caterpillars, maggots (which have cuticles with hardly any tanning at all!), and even locust hoppers are all examples where it is very easy to measure some increase in physical size during each stadium. Growth in cuticular area occurs also in adult locusts.

### Why do insects lose mass at ecdysis?

The reduced feeding activity at and around ecdysis may result partly from mechanical difficulties associated with attachment of the muscles to the old 'apolysed' cuticle. Insect muscles do not have tendons; the muscles are attached directly to the cuticle against which they pull.

## STUDY ITEM

24.61 Which hormones are involved in insect metamorphosis?

*Questions and answers*

**a** ***Does the timing of the peak of β-ecdysone concentration (or titre) shown in* figure [S]231 *suggest it has a role as a growth hormone?***

If you examine the timing of the peak of β-ecdysone concentration shown in figure (S)231 (late in the stadium) it should become clear that growth, at least as it is assessed in figure (S)227, occurs in the absence of α- or β-ecdysone (that is, early in the stadium). Although ecdysone initiates synthesis (growth) of the new cuticle, recent evidence suggests it is the release of α-ecdysone that is the signal for the cessation of feeding in some insects. It is therefore responsible for the slowing of body growth in general. β-ecdysone does not appear to be a growth hormone in the sense that we use this term to describe such hormones in other animals (for example, growth hormone in mammals).

**b** ***Does this ['stationary moults'] suggest that differentiation is an obligatory consequence of moulting?***

In this case β-ecdysone causes a moult, but there is no associated change in body form and there is no further differentiation. In other words, differentiation is not an obligatory consequence of moulting.

**c** ***What conclusions can be drawn from the results of these experiments?***

In the case of epidermal cell differentiation in insects, the process is at least partially reversible. De-differentiation can occur, but what is perhaps more important is that in young larvae there appears to be some factor (not β-ecdysone) which can influence differentiation.

**d** ***How could you demonstrate experimentally that JH is influencing differentiation?***

You could inject JH, or you could implant actively secreting corpora allata in the adults, together with β-ecdysone.

e *From the information shown in* **figure [S]232** *do you consider JH stops development towards the adult? Give reasons for your answer.*

In one sense this appears to be a good description of what we see. JH is present in large amounts in young larval insects and then it declines in amount in those stages where metamorphosis occurs. Recent measurements of the titre of JH in the blood of insects confirm this (figure (S)232). Thus precocious adult development is prevented in young larvae by the presence of relatively large amounts of JH. On the other hand, in the experiments we discussed earlier, where JH could be shown to cause de-differentiation of adult epidermal cells, the action of the hormone is clearly much more than one of simply blocking adult development.

f *Does the evidence in* **figure [S]232** *and* **figure [S]233** *suggest that JH maintains the existing stage? Explain your answer.*

The experiments which we discussed earlier, in which adult insects are made to moult, argue against JH maintaining the existing stage as JH can cause de-differentiation. More importantly, when you examine the changes in the titre of JH shown in figure (S)232, it should be apparent that the slight forward development of differentiation which occurs during larval moults in hemimetabolous insects (see figure (S)233) take place when JH is present, albeit in decreasing amounts as the larvae age.

g *Is there any evidence from* **figure [S]233** *and* **figure [S]234** *to suggest that JH promotes larval characters?*

This is perhaps the most reasonable of the three concepts. But here, we must introduce the idea that it is the amount of JH present which determines the degree of effect. Large amounts of hormone promote immature characters, whereas smaller amounts allow progressively more mature characters to be expressed (see figure (S)233); finally, at metamorphosis when no JH is present, the adult characters are allowed expression (see figure (S)234). □

**Growth hormones.** Some invertebrates possess growth hormones. The molluscs are a particularly clear example; a group of neurosecretory cells on either side of the brain (in each cerebral ganglion) produce a proteinaceous hormone which stimulates body and shell growth. It is, however, in the mammals, and especially in humans, that growth hormones are well known. But even the action of human growth hormone (GH) is not entirely clear. While the effects of excess or deficient levels of GH in adolescence are well known, and covered in most textbooks, the situation in later life is different: we do not stop growing as adults simply because we stop releasing GH. Rather, for some unknown reason the cells, which in the neonate responded so vigorously to GH, simply fail to respond further (indeed, adults continue to release GH during short fasts because it has very important actions in controlling metabolism, which are often quite unrelated to growth). In other words, mammalian GH could be described as a hormone which stimulates growth in those tissues that can respond to it, by stimulating them to perform the functions detailed in their genome; once the inbuilt programme of growth has been allowed phenotypic expression, they show no further growth response to the hormone. In this sense, growth is an intrinsic property of such cells and the hormone plays a more passive role than its name suggests in allowing the cells to proceed along their particular developmental path. A similar argument can be made for ecdysone except, of course, that the degree of progress along that developmental path is determined by JH.

**Juvenile hormone.** The secretion from the corpora allata is, in fact, one or more (according to species) of three closely related terpenoids (figure (S)235). The hormones are usually referred to as JH-I (C-18 juvenile hormone), JH-II (C-17 juvenile hormone) and JH-III (C-16 juvenile hormone). Our knowledge of possible differences in the roles of these three hormones is incomplete at present, but it is quite likely that a particular juvenile hormone, or a particular ratio of one hormone to another, may be characteristic of larval or adult stages. It should not be thought of as strange that insects possess such a 'family' of closely relate hormones. This is a well-known phenomenon in vertebrates (adrena steroids and pituitary peptides are well-documented examples).

**Chromosome puffing.** Polytene chromosomes are found in cells of th salivary glands, intestine, and Malpighian tubules – tissues that are al metabolically active, and that grow by an increase in cell size rather tha numbers. Polytene chromosomes consist of many chromatin filament fastened at the chromomeres (the bands, which are clearly seen alon their length) and may be up to a hundred times thicker and ten time longer than the chromosomes of typical cells. The pattern of bandin corresponds to the pattern of DNA concentration along eac chromosome; it is specific for each chromosome of a cell set, and (whe visible) appears to be exactly the same for all cells of the organism. Puf represent the sites along the chromosome where the adherence of th filaments is reduced and the mRNA synthesis (transcription) i occurring.

It has been shown that β-ecdysone becomes associated directly wit the particular primary puffs that the hormone induces. This was feasib because the ketone group in the B-ring of ecdysone (see figure (S)23 makes it possible, using intense irradiation with light, to cross-link th steroid with the proteins to which it binds. In this way ecdysone whic had entered the salivary gland nuclei was 'fixed' to its binding protei and then located histologically using rabbit antibodies raised again ecdysone. These antibodies had been 'tagged' with a fluorescent mark so that when the polytene chromosomes were examined, the fluorescen of the antibody showed the site of the ecdysone. This corresponded wit the location of the puffs. Such evidence agrees well with the scheme f steroid action shown in figure (S)226.

The chromosome puffing that occurs in the salivary glands under t action of ecdysone and JH is relevant to our understanding of the acti of ecdysone in controlling moulting, even though the salivary glands are not obvious target organs, It is unfortunate that, ordinarily, we cannot see the chromosomes of epidermal cells, but they share a common embryological origin with the salivary glands: in insects, both are ectodermal. It is not unreasonable, therefore, to infer that ecdysone and JH may influence gene activity in epidermal cells in a similar manner to that studied in salivary gland cells.

**Synergism between ecdysone and juvenile hormone.** This can perhaps be illustrated by an analogy. A moult may be thought of as a developmental gate which is split horizontally into several hinged sections, each unlocked by different titres of JH, the lowest section requiring high titres while the highest require little or no hormone. Thus, as the JH titre decreases, progressively higher sections of the gate open to reveal more of the adult pattern. However, no section of the gate can open if the master lock is closed; ecdysone opens this master lock at each moult. In this way, if JH 'turns the handle' of each section in our imaginary gate, no matter how much hormone is present (no matter how much the handle is rattled), when ecdysone is absent (when the gate remains locked and no section opens) there is no development. Similarly, if ecdysone is present, although the insect can moult, if the JH titre is not changed there is no development (the lock is open, but the handle is not turned and no gate section opens). Such an analogy does not explain the mechanism involved but does emphasize that ecdysone and JH act in concert to control metamorphosis. Ecdysone can be thought of as playing a passive role, by timing events to indicate to the tissues that they should proceed in their development. JH indicates how far (determined by its titre) the cells should progress along a particular morphogenetic programme, by switching on genes concerned with larval pattern and suppressing genes for adult characters.

In the schemes shown in most textbooks (and in figure (S)234), juvenile hormone and ecdysone are depicted as acting together at the same time. In fact, they act at completely different times. Clearly, ecdysone acts late in the instar to initiate apolysis but, surprisingly, the morphogenetic effect of juvenile hormone is exerted very early in the instar. The epidermal cells are sensitive to juvenile hormone for only a very short period at the beginning of the instar, although there is some evidence that different parts of the body and different characters are to some extent sensitive at different times. In this sense the insect body is a mosaic of sensitivity to the juvenile hormone, rather like the situation in amphibian metamorphosis.

Thus, the titre of juvenile hormone at the sensitive period for each tissue (at the beginning of the instar) determines the commitment of the cells, which is in some way 'remembered' until later when β-ecdysone is secreted and initiates a moult.

## STUDY ITEM

### 24.62 Insect metamorphosis

*Part A. Hormones and moulting*

*Questions and answers*

a ***What conclusions can you make from this observation?***

There is a causal relationship between feeding and moulting. Feeding is the signal for the moulting process to be initiated. The fact that moulting occurs at a fixed time after feeding is also important because it implies a definite causal relationship between the taking of a blood meal and moulting.

b ***What conclusions can be drawn from these observations about the importance of the head and the time of decapitation?***

The head is necessary for the initiation of moulting, at least for the first four days after feeding. In other words, during this critical period of four days the head is necessary for the feeding stimulus to be effective.

c ***What conclusions can be drawn from this experiment?***

Sectioning of the ventral nerve cord does not prevent digestion of the blood meal. Thus moulting does not occur in response to absorbed nutrients. Presumably the presence of the blood meal allows signals to be transmitted to the head (brain) and the transmission of this information is essential for the activity of the head. (We know now that stretch receptors in the abdominal wall signal the presence of the meal to the brain via the nerve cord.)

d ***What further conclusions can be made about the role of the head in moulting?***

The head produces a blood-borne factor which causes moulting.

e ***What is the source of the control exerted by the brain over moulting?***

The brain secretes a hormone which originates in the neurosecretory cells of the pars intercerebralis.

f ***Is the importance of the head in moulting the same in silkworms as in* Rhodnius*?***

Yes.

g ***What conclusion can be drawn concerning the possible role of the thorax in moulting?***

The abdomen clearly requires some factor from the thorax. The head alone (although necessary, as in *Rhodnius*) is insufficient. The ligature prevents the influence of the thoracic factor from causing moulting in the abdomen. It is presumed that the factor from the thorax must normally be carried in the blood. This supposed hormone from the thorax must be distinct from that from the brain.

**h** ***What is the role of the brain in moulting?***

The brain must activate some centre in the thorax to produce the supposed hormone (in **g**, above) which initiates moulting.

**i** ***What is the role of the prothoracic glands?***

and

**j** ***What is the relationship between the length of feeding (associated with the critical period) and the effectiveness of the prothoracic glands in promoting a moult?***

Feeding (as in *Rhodnius*) in some way activates the brain. The brain in turn activates the prothoracic glands. The prothoracic glands then produce a hormone (ecdysone) which causes moulting. The critical period represents the time taken to activate the prothoracic glands fully.

*Part B. Hormones and metamorphosis*

*Questions and answers*

**k** ***What conclusions can be drawn from these experiments concerning the control of metamorphosis?***

In the first experiment the metamorphosis of the fourth-instar larva is accelerated. Presumably the influence of the endocrine system from the fifth-instar larva is insufficient to control normal development (that is, to suppress adult development). In the second experiment, the normal metamorphosis of the fifth-instar larva is suppressed to some extent by the influence of the endocrine system of the fourth-instar larva. Thus the endocrine system of the fourth-instar larva has some inhibitory influence on metamorphosis; that of the fifth-instar larva has not.

**l** ***What conclusions can be made from this experiment? Relate your answer to your conclusions from the other experiments described in this section.***

These results are qualitatively similar to those of the first two experiments, except that instead of assuming that the difference in endocrine activity between the instars is effectively the presence or absence of inhibition of development (that is, the presence or absence of a hormone from the corpus allatum), here it can be argued that there is more inhibition from the corpus allatum of the third-instar larva than there is from that of the fourth-instar larva, but that there is inhibition from the corpus allatum in both larvae. Wigglesworth was able to suggest that the inhibitory influence of the corpus allatum decreases in progressive instars, that is, III > IV > V. Since in the fifth instar the corpus allatum appears to have no effect, it is the absence of the corpus allatum hormone (juvenile hormone) which allows the final stage of metamorphosis to be expressed in the full development of adult form and function. □

## The control of amphibian metamorphosis

**Thyroid hormone and amphibian metamorphosis.** Both $T_3$ and $T_4$ are present in tadpoles, but it is generally assumed that $T_4$ is essentially a prohormone which is converted to the active hormone, $T_3$, in peripheral tissues. This is analogous to the situation in the insects, in which the prothoracic glands release α-ecdysone which is later converted to the active or 'real' hormone, β-ecdysone.

Surprisingly, thyroid hormone (TH) is synthesized very early in development even though, classically, TH is characterized as having no positive role in regulating the growth and development of larval structures. There is, unfortunately, directly conflicting evidence as to whether the TH levels in premetamorphic tadpoles are inhibitory or (at the low levels found) stimulatory, perhaps acting synergistically with prolactin (PRL). A synergism between TH and a growth hormone (PRL in the larva is effectively a growth hormone) would be consistent with our knowledge of the interactions between such hormones in adult vertebrates.

Recent studies have forced a re-examination of some central ideas concerning TH titres and the control of TH secretion. First, there is some evidence that tissues change quite dramatically in their sensitivity to TH. Almost certainly, this reflects changes in the populations of binding sites (receptors) for $T_3$ in the nucleus and/or mitochondria of target tissues. Consequently, at certain times accelerated development could occur without corresponding changes in the TH secretion, simply through increased sensitivity of the tissues to the extant titre of hormone. It is likely that metamorphosis is controlled by a combination of such effects and changes in TH levels. Finally, there is abundant evidence that various tissues gain or lose sensitivity to TH at different times. Again, we can draw a parallel here with the situation in insects, where effectively throughout each instar the larval tissues are a mosaic of different sensitivities to JH.

A second problem concerns the control of TH secretion. The theory that is propounded most widely envisages that TH exerts a positive feedback on thyroid releasing hormone (TRH) release during climax. The concept was proposed originally by Etkin because TH stimulates the development of the hypothalamic–pituitary axis (through its stimulation of the development of the median eminence) and it therefore brings about the release of TRH. In the premetamorphic tadpole, the pituitary receives no stimulation from the hypothalamus and consequently is very sensitive to negative feedback by TH. But as the development of hypothalamic control over pituitary activity proceeds under the influence of TH, stimulation of TSH release from the pituitary by TRH would increase TH levels. Increased levels of TH would favour greater development of the median eminence and increase the hypothalamic stimulation of pituitary function. Etkin argued that this is a 'self-accelerating' process, hence the term 'positive feedback'. This is quite separate and distinct from the idea that TH exerts a direct positive feedback on the release of TRH. Indeed, most studies point to a negative feedback of TH on TRH release in tadpoles, just as in the adult. On the other hand, two recent reports do provide evidence for a positive feedback effect of TH on TRH release. Perhaps both positive and negative feedback loops exist, as they do in the hypothalamic-pituitary control of ovarian function in mammals.

**Prolactin and amphibian metamorphosis.** Prolactin (PRL) belongs to a family of pituitary (and placental) hormones. PRL and growth hormones are synthesized in similar cell types in the anterior pituitary and have similar molecular structures, and often their biological effects show considerable overlap. Thus, while, as stated in the *Study guide*, we should not be too surprised if PRL has a growth hormone function in larvae, it is strange that PRL does not appear to have any of the metabolic actions shown by mammalian growth hormone. It does not, for example, cause mobilization of lipid reserves in order to provide the essential energy substrates to support growth. Similarly it is remarkable that some mammalian growth hormones (prepared from different species) have growth-promoting activity in tadpoles, but show no antimetamorphic properties. Thus, it is clear that growth promotion and metamorphic suppression are not linked as intimately as we might expect from the fact that metamorphosis proceeds at the expense of growth.

**Conclusions.** The observations on the effects of thyroxine pellets implanted in the opercular region offer a vital insight into the way hormones control development. Clearly, individual target cells or tissues are programmed to give precise responses on a signal from the hormone. It is a fact of fundamental importance that the cell's inbuilt programme, not the hormone, determines its response.

One further point can be ascertained from the results of these experiments: because all the cells of the epidermis in the opercular region carry an identical genotype, those which respond to TH by thinning (to produce the window) must have received some additional instructions to determine their particular and individual responses. It is believed that epidermal cells receive positional information: their response is determined by their position. Our understanding of how cells 'appreciate' their position relative to other cells is fairly rudimentary, but obviously it is of crucial importance in the control of development. Pattern formation of limbs is discussed in Chapter 23, section 23.8.

## 24.7 Communication and control in plant growth

*Assumption*

1 An understanding of the action of animal hormones given in the earlier part of the chapter.

*Principles*

1 Plants produce chemical substances which act as 'messengers' in a similar way to the hormones found in animals.
2 Several groups of plant growth regulators exist which have specific effects in the development of plants.

### Do plants really need hormones?

In the last few years there has been mounting criticism of traditional views concerning plant hormones – such as those set out in this revised text. This is largely motivated by the sense of frustration occasioned by the lack of understanding of how plant hormones work.

Critical assessment of the experimental support for the traditional view reveals many inconsistencies and much conflicting evidence that has been conveniently ignored. On the other hand, the critics are not yet able to replace the 'old' system of ideas with any 'new' one that has much more consistency. There are snags in both old and new; perhaps the basic ideas were wrong and progress will only be made when we re-think the direction of new research.

Discussions of this kind have been confined mainly to research journals and specialist meetings on plant growth regulators. However, wider coverage has been afforded by the popular article 'Do plants really need hormones?' by Jonathan Weyers in *New Scientist* (17.5.84). It dealt only briefly with the main controversy and tended to give the impression that anyone who considered that the traditional ideas still had merit was rather outmoded and certainly in the minority. It is very difficult to cast aside as misguided fifty years' research and teaching when there is, as yet, no useful alternative which has better credentials. Of course, demolition does precede rebuilding – though it may be possible to modify the old structure whilst still retaining some of its charm, if we are not too hasty. Hence the fairly traditional views, found in all similar texts at the time of writing, have been retained in this second edition.

The gist of the main criticisms and proposed changes is as follows:
**1** The idea of plant hormones derives directly from the earlier discovery of animal hormones. It ignores the vastly different life styles and developmental patterns in plants as compared with animals. For instance, plants are generally non-mobile and have 'open' growth which continues throughout life, using terminal and lateral meristem. Animals are mobile, heterotrophic, and have a 'closed' type of growth in which differentiation is achieved fairly early in life and meristems cease to function. They have well defined circulatory systems and highly differentiated 'target' organs for any hormone produced elsewhere. In plants, the circulatory system is very ill-defined and 'target' tissues are often not clearly distant from the tissues that are supposed to have produced the hormone.
**2** In animals, quite small changes in hormone concentration (of, say, 50 per cent) have marked effects on the magnitude of response, whereas in plants changes in concentration need to be ten- to a hundredfold in order to produce clear effects. (This generally refers to external concentrations applied to the tissue. It may be relevant that it normally takes a tenfold increase in external salt concentration to double the internal concentration.)
**3** When attempts have been made to assess the internal concentration of a plant hormone in the responding tissue and to compare this with the magnitude of response, then what little evidence there is indicates that there is virtually no connection between internal concentration and magnitude of response.

This has led to the new postulate that changes in hormone concentration in plants are virtually irrelevant as a means of controlling plant response. In fact, it may be that the main purpose of the plant growth substances is to provide a fairly *stable* signal – that everything is normal, in spite of the changing water potentials, nutrient concentrations, and so on, which reflect a fluctuating environment. This view is contributed by Trewavas (see below).

**4** Most of the evidence about the chemical identity and concentration of plant hormones was obtained using very crude and unreliable techniques, and could lead to misinterpretation of data. Newer techniques will allow definitive results to be obtained, although only for much greater expenditure of time and money. They may well resolve some of the inconsistencies that exist at present, but the question of whether the extracted growth substance was spread uniformly through the cells or concentrated in some local area of action will remain unsolved.

Whatever one's bias, the criticisms are justified and healthy. Present views depend on rather shaky evidence.

The main thrust of Trewavas' contribution is to insist that any mechanism of hormone action is likely to depend on the following interaction:

growth substance + receptor ⇌ growth substance/receptor complex
↓
effect on growth etc.

Thus not only is the concentration of growth substance critical; so also is the availability of the receptor. Trewavas believes that such a receptor is likely to be membrane-bound protein. The affinity of this protein for the growth substance can change – and this means that the 'sensitivity' of the system to a given concentration of growth substance can change. Trewavas considers that it is this change in sensitivity which is normally the limiting factor in the response of a tissue and suggests experiments to follow through this hypothesis. This may be polarizing the argument too strongly, however; it is possible that changes in sensitivity of the receptor will explain those cases in which a response did not appear to correspond with changes in concentration of growth substance – though there is no evidence that it does. On the other hand, the sensitivity of any receptor has been conveniently ignored up to now, mainly because we cannot yet characterize the receptors very effectively. The next few years are likely to see substantial progress towards resolving these problems.

In the meantime, teachers and students who wish to learn more about these ideas should read some, or all, of the following papers; most non-specialists are not in a good position to evaluate the arguments, however, so readers should not be too easily persuaded that everything that is traditional is wrong.

TREWAVAS, A. 'How do plant growth substances work?' *Plant cell and environment*. **4**, 1981, pp. 203–28. (A major paper, fairly heavy going but worth some perseverance.)

TREWAVAS, A. 'Growth substance sensitivity: the limiting factor in plant development'. *Physiologia plantarum* **55**, 1982, pp. 60–72. (A useful review of the main points in the 1981 paper and an interesting challenge to the traditional belief that gibberellins control amylase synthesis in barley aleurone cells.)

FIRN, R. D., MACLEOD, K. and PARSONS, A. Phototropism: patterns of growth and gradients of light'. *What's new in Plant physiology*. **14**, 1983, pp. 29–32. (A readable discussion of the evidence for and against the two main theories of phototropism, suggesting some new experiments to differentiate between the two.)

*Questions and answers*

**a** ***Why is it necessary (apart from providing a means of storage) to have a system which inactivates or destroys a hormone within the plant?***

If synthesis occurred without removal, the substance would tend to accumulate at its site of action and this would then saturate the system and prevent any further response.

**b** ***Why does water enter the cell now, when prior to the increase in plasticity there was equilibrium at full turgor?***

Water potential of a cell = water potential of vacuolar sap and cytoplasm + turgor pressure

$$\psi_{cell} = \psi_s + \psi_p$$

At full turgor the water potential ($\psi_{cell}$) = 0 and $\psi_s = \psi_p$. If now $\psi_p$ decreases (because the plasticity of the wall increases and elasticity decreases) then as it does so $\psi_{cell}$ becomes more negative, so water now enters the cell until a new equilibrium occurs (or the cell bursts!).

**Practical investigation. *Practical guide 6*, investigation 24B, 'The effect of IAA on the growth of coleoptiles and radicles'.**

## STUDY ITEM

**24.71** Interaction between hormones (*J.M.B.*)

*Questions and answers*

**a** ***From the graphs in figure [S] 242 summarize the effect of IAA and $GA_3$ on elongation growth in pea internode segments.***

$GA_3$ alone increased elongation by about 25 per cent, whereas IAA alone increased elongation by about 18 per cent and the two together increased it by about 220 per cent.

**b** ***How is this similar to the way in which animal hormones act together?***

This is an example of potentiative synergism which is also seen in the action of animal hormones, when the combined response of two hormones acting together is greater than the sum of the individual responses to these hormones.

**c** ***Suggest a possible reason why the two hormones act together to promote internode extension.***

Possibly the two substances act at different points in a reaction sequence to reduce the resistance at two separate 'bottlenecks', thus leading to a synergistic or cooperative effect – or any other sensible reason.

**d** ***How would you test the hypothesis suggested in* c *above?***

First determine the concentration of one of the growth regulators (either will do) which will give maximum growth of the pea stem sections. Then, in a second experiment, find out what happens to the stem sections when the second regulator is added, at one or more concentrations, to this optimal concentration of the first. If both regulators act at the same site in an additive way, no further increase in growth would be expected on addition of the second regulator, since the site is already fully activated. If there is synergistic action by cooperation between two sites, however, addition of the second regulator will release a 'blockage' at the second site and a marked increase in growth will be observed. □

### Other effects of Gibberellins: promotion of flowering

*Questions and answers*

**c** ***What does this suggest about long days and the promotion of flowering?***

This suggests that transfer from short-day to long-day conditions in some way promotes gibberellin synthesis (or release from inactive stores).

**Practical investigation. *Practical guide 6*, investigation 24C, 'Stimulation of amylase production in germinating barley grains'.**

**d** ***Why is it advantageous to store insoluble starch as a food reserve?***

Starch grains form compact stores and have no osmotic effect on cell contents.

**e** ***Why should an inhibitor of protein synthesis affect amylase production?***

Enzymes are proteins; so amylase cannot be made *de novo* if no proteins are being made.

## STUDY ITEM

### 24.72 Some effects of gibberellic acid on potato tubers

*Questions and answers*

a **_How might bacterial and fungal contamination of the cultures and tissues have affected the results?_**

The 'background' carbon dioxide output could have been increased, they may have produced amylases or other enzymes, and they may have produced gibberellic acid-like substances which would mimic $GA_3$.

b **_How would the controls have been treated in groups one and two?_**

The controls would have been treated with water in equal volumes to the other solutions used.

c **_Explain briefly how malonic acid interferes with the Krebs (citric acid) cycle._**

Malonic acid is closely similar to succinic acid, one of the intermediates in the Krebs cycle. Malonic acid acts as a competitive inhibitor to the enzyme succinic dehydrogenase so the reaction succinic acid → fumaric acid is stopped.

d **_Relate the data in_ figure [S]246a _to those in_ [S]246b.**

For the disbudded controls the values of mean concentration of reducing sugar are similar, although not identical.

$GA_3$ applied to the disbudded region appears to restore the hydrolytic capacity of a normal tissue with buds.

e **_Account for the data obtained from the malonic acid treatment._**

Since the Krebs cycle is inhibited there is very little energy available. Reducing sugar could 'leak' out to the external medium as there is no energy to prevent leakage. Also there could be utilization of the initial low level of reducing sugar by glycolysis.

f **_How would you expect the mean carbon dioxide production to vary if cylinders with buds in group two had been treated with malonic acid solution? Use a graph to help explain your answer._** *(J.M.B.)*

□ See *figure 42*.

### Cytokinins

*Question and answer*

f **_What connection is there between adenine and nucleic acids?_**

Adenine is one of the two purine bases which are constituents of all nucleic acids.

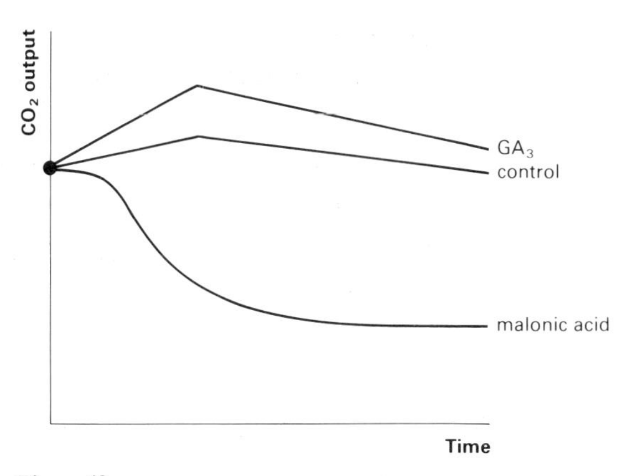


**Figure 42**
Inhibition of carbon dioxide output by malonic acid.

**Practical investigation. *Practical guide 6*, investigation 24D, 'The effect of plant hormones on seed germination'.**

## 24.8 The immune response

*Principles*

1 An immune response exists in vertebrates which eliminates foreign material from the body.

2 There are two main forms of immunity: natural and acquired.

### Natural or non-specific immunity

Activation of the terminal components of complement through activation of the first component (for example, by an antigen–antibody complex) is known as the *classical pathway* of complement activation. The order of complement activation is C142356789. Some agents, such as bacterial polysaccharides, can activate $C_3$ directly without involvement of $C_1$, $C_4$ and $C_2$; this is referred to as the *alternate pathway*, and is a more primitive system. Lysis of the target is effected by $C_8$, its activity being enhanced by $C_9$. The presence in the blood of certain serum proteins which are inactivators of various complement components provides central complement activation.

*Question and answer*

a ***Would you anticipate that individuals who are deficient in the last four components of complement would be more susceptible to microbial infection than people with a deficiency of the first component?***

Subjects who are permanently deficient in any of the last four components in the complement sequence remain healthy and are not particularly susceptible to infection, suggesting that the direct lytic activity of complement is not very important in combating microbial infection. Deficiency of $C_3$ and $C_5$ markedly predisposes people to recurrent infections; this implies that it is the generation of the fragments of these components which augments polymorphonuclear cell function and is therefore important for survival in our environment. Though individuals with deficiencies of the first three components are prone to infection, bacterial polysaccharides, as we have said, can activate complement directly via $C_3$. Complement deficiencies would not be expected to alter susceptibility to viral infection.

An important component of non-specific immunity to viruses is interferon, which is released by lymphoid and non-lymphoid cells. One of its main effects is to inhibit viral mRNA transcription in other cells. Interferon can also enhance the activity of certain cytotoxic lymphocytes.

Several pharmaceutical companies have recently invested considerable finance in constructing plants for the bulk production of interferons to be used for the treatment of particular viral infections and cancers.

### Acquired or adaptive immunity

**Humoral immunity.** There are five main classes of antibody: IgM, IgG, IgA, IgE, and IgD. They differ from one another in their relative molecular mass, amino acid composition, ability to activate complement, distribution throughout the body, and degrees of binding to different cells. IgM and IgG are the most common antibodies in our body fluids. IgM occurs as a pentameric molecule due to the binding of five single units by a polypeptide. Its ten possible antigen-binding sites thus make it a very good bacterial agglutinator. Both IgM and IgG are very effective in activating complement. IgG, unlike other antibodies, is able to cross the placenta, thus conferring immune protection on a baby during the first few weeks of its life. IgA, which is located in mucous and serous secretions, is present as a dimeric molecule. One of its main functions is to inhibit the adherence of micro-organisms to mucosal cells. IgD appears on the surface of B lymphocytes, often in association with IgM; it seems to have a role in B cell development. IgE plays an important part in immediate hypersensitivity or anaphylaxis (see page 269).

In order to respond to antigens lymphocytes often need the assistance of macrophages. This help, in antigen presentation, may be in the form of simple binding of the antigen to surface receptors or 'processing' by the macrophage's intracellular enzymes, so that its 'antigenicity' is increased.

### STUDY ITEM
### 24.81 Response to an antigen

*Questions and answers*

**a** ***Why do you think there was an improved antibody response following the second contact with antigen?***

The first antibody response to antigen is known as the *primary antibody response* and most of this interaction of antigen with B lymphocytes takes place in lymphoid tissue (see page 273). The antibody is IgM. During the primary response there is, in addition to the proliferation of B cells and secretion of antibody, a generation of B cells which retain the capability to release specific antibody. These cells circulate in the body, often for years, so that when antigens gain future entry they will soon encounter these memory cells which in turn quickly signal the recruitment of similar B cells so that a rapid antibody response is elicited. IgG is the main class of antibody generated during the reaction.

**b** ***What is the benefit of such a response?***

Acquisition of maximum immune responsiveness on contact with antigen. This is the basis of immunization.
The advantage of memory cells which induce clonal expansion of identical B cells is that large numbers of specific lymphocytes do not have to be stored in the immune system.

**c** ***What does the response shown in* figure [S]257 *tell you about the role of lymphocytes?***

This experiment shows clearly that B lymphocytes carry immunological memory. □

**Cell-mediated immunity (CMI)**

*Questions and answers*

**b** ***Offer an explanation for this observation.***

This experiment illustrates the importance of the thymus gland as a lymphoid organ in the development of mature lymphocytes.

**c** ***What conclusions would you draw from the results of this experiment?***

The experiment demonstrates that the thymus bestows immunological competence on lymphocytes.

## Hypersensitivity

A synonym for immediate hypersensitivity (or anaphylaxis) is *allergy*, a term introduced at the beginning of this century to denote altered host reactivity to an antigen. Approximately ten per cent of the population suffer from some form of allergy which is generally inherited. The antibodies to the allergic-inducing antigens, called allergens, are mainly of the IgE class. Cross-linking of the mast cell-bound IgE antibodies by the allergens results in the release from these cells of chemical mediators such as the vasoactive amines histamine and bradykinin; these induce vasodilatation and contraction of smooth muscle, particularly of the bronchioles in asthmatic sufferers. The involvement of histamine in immediate hypersensitivity explains the effectiveness of using anti-histamine drugs in this disorder.

The anaphylactic reaction is rapid because of the large density of mast cells at the sites of antigen entry – for example, the mucous membranes in the lungs, nose and throat. Allergic reactions are often associated with an increased number of eosinophils in the blood; these are attracted to sites of immediate hypersensitivity by chemotactic factors released from the mast cells. On reaching these sites the eosinophils release a number of chemical mediators which antagonize the action of the factors secreted by the mast cells.

Unlike acute inflammation, which may start and subside in a matter of minutes or a few hours, the inflammation associated with delayed hypersensitivity is not apparent until after 6–8 hours. This is because of the time taken for sensitized T lymphocytes to respond and migrate to the antigen and synthesise non-antibody proteins (lymphokines) which maintain a CMI reaction. Such lymphokines recruit and hold other leucocytes to the inflammatory site. They also enhance the function of macrophages e.g. phagocytosis and stimulate other lymphocytes to generate more lymphokines. Inflammation continues until a peak level is reached 24–48 hours after the start of the reaction. After this time the antigen is removed from the site and the inflammatory reaction subsides. However, should the antigen persist, because it cannot be degraded, e.g. asbestos, macrophage colonization of that site will occur (granuloma).

**d** ***In the experiments shown in* figure [S] 258, *state why the skin reaction in guinea pig B appeared before that of guinea pig A, and why no skin reaction was induced in guinea pig C.***

The experiment demonstrates that delayed hypersensitivity can only be transferred to non-sensitized animals by lymphoid cells and not by serum (antibodies). The reaction is thus cell-mediated.

**e** ***What additional information does this experiment give you regarding the mediation of delayed hypersensitivity?***

Delayed hypersensitivity can be transferred either by viable lymphocytes, particularly T cells, or by a soluble product extracted from them (transfer factor). Initial claims that this factor would transfer specific immunocompetence to patients whose delayed hypersensitivity was impaired have not received uniform support.

Delayed hypersensitivity skin reactions to bacterial antigens are important for the identification of certain infections and for assessing a subject's cell-mediated immune responsiveness. For example, when a small amount of antigen prepared from the *Mycobacterium* tuber-culosis is injected into the skin, the vast majority of people tested will develop, 24–28 hours later, the characteristic inflammatory reaction of delayed hypersensitivity. This is called the Mantoux test.

**Additional information on human leucocyte antigen (HLA) system.** Several diseases show a good correlation with particular HLAs, suggesting that certain individuals may have a genetic predisposition to disease. Consequently this field of immunology, at the time of writing, is receiving a great deal of interest. Human leucocyte antigens may not only be diagnostic markers of diseases but are valuable pointers to our understanding of their aetiology and pathogenesis. Several possible theories have been advanced to explain these disease associations.

**1** The surface antigens of the invading micro-organism are similar to the human leucocyte antigens and so avoid recognition by the host's immune system and escape elimination.

**2** The human leucocyte antigens act as receptors for the foreign agent.

**3** The disease is related to a gene that is linked to the HLA gene as a result of linkage disequilibrium (that is, there has not been a dissociation of closely linked genes due to genetic randomization) which may result in a poor T cell response to certain virus-HLA combinations.

## STUDY ITEM
## 24.82 Skin grafts

*Questions and answers*

**a** ***Why does the white mouse in* figure [S]259a *reject the skin graft from a black mouse (unrelated donor), yet if a newborn white mouse* [figure (S)259b] *is injected with mature lymphocytes from a black mouse it will later accept the skin graft from a black mouse?***

The immune system of the white mouse recognizes cells of the skin graft as foreign and will attack them in the same way that it attacks, for example, a virally infected cell. The newborn mouse's immune system has not yet fully developed and foreign antigens which it encounters at this stage escape elimination and in some way induce in the host a future suppression of its immune responsiveness to that antigen. The white mouse has now developed specific non-reactivity to the antigen of the black mouse, a phenomenon known as immunological tolerance. This tolerance enables the immune response to distinguish the body's own constituents from foreign antigens.

**b** ***If a white mouse* [figure (S)259b] *has accepted a skin graft from a black mouse and is later injected with mature lymphocytes from a closely related white mouse, the graft is rejected. Explain why.***

The state of tolerance has now been abolished by the injection of lymphoid cells from a closely related donor which recognizes cells of the black mouse as foreign.

**c** ***Do you find it surprising that the young mouse in* [figure (S)259c] *which is injected with mature lymphocytes from a black mouse later rejects a skin graft from a black mouse?***

The immune system of the young mouse developed so quickly (within seven days) that by the time it received the lymphocytes from the black mouse its own lymphocytes had been programmed to recognize antigens, other than its own, as foreign. Consequently the skin graft from the black mouse was rejected. This experiment shows that it was the immature and not the mature lymphocytes of the white mouse which were susceptible to tolerance induction.

**d** ***Offer an explanation why a mouse that has several months previously rejected a graft will expel a similar second graft more rapidly than the first.***

Secondary rejection is more rapid because of the memory retained by primed lymphoid cells following first antigenic challenge. Humoral antibodies with specificity against the grafted cells may also be generated following the first rejection. □

The immune system is very efficient in responding to and eliminating foreign antigens, but why does it not respond to the multitude of self antigens that it frequently encounters? This ability to discriminate self from non-self illustrates both the subtlety and complexity of the immune system. Fortunately for us, the lymphoid system has acquired the ability *not* to respond to an antigen to which it would otherwise respond (immunological tolerance). Without this ill-understood mechanism the consequences to ourselves would be catastrophic.

Tolerance can be induced in the neonate (that is, during the first few weeks after birth) by antigens interacting with immature lymphocytes and preventing their development, and in adults by applying either a very high concentration or several low doses of persistent antigen. With low doses of antigen it is the T cells which are made unresponsive, whereas high doses of antigen affect both the T and the B cells.

In general, tolerance is easier to induce in T cells than in B cells, though the susceptibility of these two cell types depends upon the concentration of antigen and the time necessary for induction of tolerance. Mechanisms of tolerance induction appear to involve the early deletion of immature lymphocytes which possess self-reactive properties. If, however, these self-reactive lymphocytes develop into fully mature cells, their antagonism by another group of lymphocytes known as suppressor cells (see page 274) is also involved.

Antibodies which combine with specific sites of the antigen-binding region of another antibody are called anti-idiotypic antibodies. (An idiotype is a unique determinant present on homogeneous antibody.) Such antibodies may react with antibodies expressed on auto-reactive B cells and render them inactive.

Loss of immunological tolerance results in immune reactions to self components (autoimmunity), which can be mediated by either T or B lymphocytes. Most autoimmune reactions, however, are due to the generation of autoantibodies which induce tissue damage either by directly activating complement or by combining with the autoantigen of the tissue to form immune complexes which in turn attract phagocytes that attempt to phagocytose and in so doing release lysosomal enzymes which destroy adjacent tissue cells.

Autoantibodies are often directed against intracellular materials such as DNA, mitochondria, and contractile proteins which, though normally sequestered within the cell, are released following tissue damage. It is conceivable that these 'antigens' are treated as foreign by the immune system because they were hidden from lymphocyte surveillance during the critical tolerance-inducing stage of the perinatal period. In other words, there is a stage during our development when certain lymphocytes closely inspect and retain a memory of the components that constitute our internal environment so that they can subsequently distinguish self from non-self. Should the lymphocytes later encounter any material that was not available for this early processing, it will be regarded as hostile and the relevant immunological response elicited.

Autoantibodies may also be directed against a person's own cells as a consequence of the attachment and/or modification of cell surface determinants by drugs or viruses, or the sharing of cross-reacting antigens between host and invader.

Several diseases are characterized by circulating autoantibodies, which probably play an important role in their pathogenesis. For example, autoantibodies are directed against DNA in systemic lupus erythematosus (a disorder of connective tissue), against thyroid proteins in chronic thyroiditis, against postsynaptic acetylcholine receptors in myasthenia gravis, and against red blood cells in autoimmune haemolytic anaemia.

Most patients with classical rheumatoid arthritis have antibodies which are mainly IgM (rheumatoid factor) directed against the Fc fragments of other IgG antibodies. These antibodies readily self-associate with one another to form immune complexes which are believed to mediate the inflammatory reaction in joint cavities. If B lymphocytes from healthy subjects are activated by a polyclonal stimulator *in vitro* they will produce significant quantities of rheumatoid factor, demonstrating that normal B lymphocytes have the potential to release autoantibodies. Thus autoantibody products may arise from excessive B cell stimulation *in vivo* due to persistent activation or a failure of the suppressor T lymphocytes to abrogate this proliferation.

### Helper cells and suppressor cells

*Questions and answers*

**f** ***What does this experiment tell you about the role of T cells in the humoral response?***

The source of B cells was either the bone marrow or spleen. When cells from either a normal or an immune spleen were transferred to irradiated mice, an adequate humoral response was seen when these animals were later challenged with sheep red blood cells. However, if only bone marrow cells were transferred a weak antibody response was obtained, even though B cells were present in similar numbers to those in the spleen. The difference between the two lymphoid organs is that, in contrast to the bone marrow, the spleen contains T lymphocytes, suggesting that T cell 'help' is necessary for an efficient humoral response.

This inference is borne out by the demonstration that when the irradiated mice received a combined transfer of bone marrow cells and thymic cells, which themselves had no effect on the antibody response, there was a restoration of humoral competence. This experiment demonstrates that a subpopulation of T lymphocytes, termed *helper cells*, are essential for antibody production. Antibody responses which require the co-operation of helper T cells with B cells and macrophages, such as the secondary immune response, are referred to as T-dependent, to distinguish them from reactions in which antigen-induced B cells secrete IgM (and not IgG) immunoglobulins in the absence of helper T cells (T-cell-independent). Most of the T-cell-independent antigens are polysaccharides which are not readily degraded by the body.

g ***Comment on the role of the splenic cells in modifying the humoral response of the recipient mice in* figure [S]260.**

The antibody response of mice to sheep red blood cells was impaired when they received a transfer of spleen cells from mice which had been made tolerant to sheep red blood cells. This depressed humoral response was not seen when the transferred spleen cells were depleted of T cells, demonstrating that the induced suppression was mediated by T lymphocytes. Thus as well as T cells helping B cell activity, there is also a subpopulation of T cells able to inhibit this function; these cells are known as *suppressor cells*. The experiment also shows that tolerance to sheep red blood cells can be transferred from tolerant mice to normal recipients by spleen cells.
There are two types of suppressor cells: those that suppress all antibody responses (non-specific) and those that suppress antibody formation induced by specific antigen (specific). In addition to promoting B cell function, helper T cells can also enhance cell-mediated immunity (see below) and generate suppressor cells, which in turn nullifies T helper cell activity. The helper and suppressor cells exemplify the balance of opposing and antagonizing forces required by most biological systems.

Helper and suppressor T cell functions are mediated by the release of biologically active soluble proteins, the biochemical characterization of which is currently being intensively studied. These helper and suppressor factors are related to the major histocompatibility complex (MHC) gene products. In addition to functional assays, helper and suppressor cells are commonly identified by antisera directed against distinct glycoproteins on their membranes.

## AIDS (acquired immune deficiency syndrome)

This disease has become an epidemic in the U.S.A., almost certainly because of the promiscuity inherent in a proportion of the homosexual population. The available evidence suggests that the disease is initiated by an infection with a human T cell leukaemia/lymphoma virus type III (HTLV III), which is a member of the retroviruses (a family of RNA tumour viruses). The virus is transmitted via blood – hence the appearance of the disease in a very small proportion of individuals who have received blood transfusions containing samples taken from AIDS sufferers. A decrease in the number of blood T helper lymphocytes is believed to be due to their selective infection by the virus.

Because of the impaired function and deficiency of blood

lymphocytes, there appears to be a requisite for reconstructing the cell-mediated immunity effector arm of the immune response. Transfer of T cells is not practical because of induction of mixed lymphocyte reaction (MLR), and hence a rational approach would be the administration of T cell products such as interferon and lymphokines.

Bone-marrow transplantation is theoretically the most appropriate therapy though this, in view of the large patient numbers, would encounter tremendous practical problems. In all probability the only satisfactory treatment will emerge with the development of a successful vaccine.

## Immunization (vaccination)

*Questions and answers*

h ***Why is the immunity acquired by inactivated vaccine not as efficient as that with the attenuated strain?***

Microbial replication presents the host with a larger and more sustained concentration of antigen.
The live virus, unlike a dead virus, can infect a host's cell and induce budding of viral antigens on its surface to activate cytotoxic lymphocytes. It also has the advantage of migrating to its normal target area where it will induce a local immune reaction; for example, cholera is dealt with more efficiently by antibodies produced locally by the gut wall than by antibodies manufactured at some distant lymphoid organ.

i ***For several months after birth measles and mumps vaccines are often ineffective in infants. Why do you think this is so?***

Even though there is satisfactory development of the infant's immune system, circulating maternally-derived antibodies may still be present which could either interfere with antigen processing or result in an adverse immunological reaction. Thus infants are not routinely immunized with these vaccines until 12 months after birth.

j ***Children with hypogammaglobulinaemia, in whom antibodies are absent or nearly so, are immune to measles virus infection following recovery from the initial infection. Suggest how this non-antibody-dependent immunity is mediated.***

For many years immunity to measles was thought to be due to the persistence of antibodies and antibody-producing cells. From studying measles infection in children with hypogammaglobulinaemia and other disorders, we now know that immunity may also be conferred by sensitized T lymphocytes which are cytotoxic for measles-infected cells. Following entry via the respiratory tract, the measles virus proliferates in the cells of local lymph nodes for ten to twelve days. During this period, in which there are few disease symptoms, T and B cells are rapidly proliferating in response to viral infection. Suddenly, the explosive release of infected cells from the lymph nodes, which disperse to lodge in capillary beds throughout the body, is the signal for attack by cytotoxic T lymphocytes. Free virus particles and soluble antigens that are shed from the lysed cells are neutralized by antibodies. This period of intense immunological reactivity is responsible for the characteristic rash of measles, which is a consequence of numerous local inflammatory reactions in the capillaries of the skin that are mediated by lymphokines, one of whose actions is to increase capillary permeability. Within a few days of the appearance of the rash the infection is eliminated.

## Tumour immunology

*Questions and answers*

k ***Why do you think tumour cells escape immune recognition? Release of antigens from the tumour surface may contribute to this phenomenon. Suggest why.***

Tumor-specific antigens are relatively poor at inducing an immune response (that is, their antigenicity is weak) compared with foreign transplantation antigens. Antigen release (shedding), either in large amounts for a short time or slowly and continuously over a long period, may induce a state of specific tolerance so that the tumour avoids immune surveillance.

l ***In which group of individuals would you expect to observe an inefficient immunological response?***

The incidence of malignant disease is sometimes increased in subjects receiving immunosuppressive drugs (such as organ transplant patients) and in children with acquired immune deficiencies. Deterioration of the immune response with age also coincides with an elevation in malignancy.

m ***In 3, it is probable that antibodies directed against the tumour augment its growth. Suggest how this could occur. Postulate any other mechanisms that might be involved.***

By binding to tumour cells antibodies may prevent the cytotoxic cells from recognizing the tumour antigens. Alternatively, if antibodies combine with antigens liberated from the tumour cells, these complexes could bind to the antigen recognition sites on the cytotoxic T cell or natural killer (NK) cell surface, or to the Fc receptors on the killer (K) cells, and thus distract them from coming into contact with the tumour cells. Studies have shown that certain lymphocytes from patients with tumours inhibit the action of cytotoxic lymphocytes, suggesting that tumours induce the generation of suppressor cells.

The type of immunotherapy referred to in the *Study guide* is non-specific. A more rapid immune response is generated by specific immunotherapy. Where a tumour is related to a particular cancer-inducing virus, administration of the relevant vaccine would be of benefit in limiting tumour development at other sites.

Specific immunotherapy, in which the host is immunized by his own inactivated tumour cells or their extracts, may augment a tumour-specific immune response. In animals, such vaccinations are successful when given before tumour induction. Alternatively, it may be possible to generate tumour-specific immunized lymphocytes when the patient's tumour cells are cultured with his lymphocytes in the laboratory. Reinfusion of these lymphocytes might benefit patients who had recently undergone surgery, though expense and laboratory practicalities may detract from this approach. Passive transfer of sensitized lymphocytes between two patients bearing similar tumours has been unsuccessful because of the problem of introducing foreign antigens on the incompatible lymphocytes and of subsequent rejection. Attempts to transfer a soluble product (transfer factor) of tumour-sensitized lymphocytes, which would appear to overcome this problem of foreign antigenicity, have not been successful.

### 24.9 Resistance to disease in plants

*Principle*

1 Plants appear to have a means of resisting disease by producing biochemicals.

## PART II The *Practical guide*

The investigations related to this chapter appear in *Practical guide 6, Development, control, and integration.*

### INVESTIGATION 24A The microscopic structure of endocrine glands

ITEMS NEEDED

Microscope, monocular 1/1
Microscope slides prepared *and/or* 35 mm transparencies of sections of adrenal medulla, pancreas, and thyroid gland
Projector, 35 mm slide 1/class

(*Study guide* 24.1 'Endocrine communication and control in animals'; 24.2 'Hormone synthesis and release'.)

*Principle*

1 The microscopic structure of endocrine cells is related to the type of hormone released.

*Questions and answers*

a ***What characteristic features would you expect endocrine cells to have?***

A characteristic feature of all endocrine glands is a very rich blood supply – thus blood capillaries (in those animals with closed circulatory systems) should be numerous. Other characteristics are prominent nuclei, numerous mitochondria, endoplasmic reticulum. Golgi bodies, and secretory vesicles.

**b** ***Are the characteristic features predicted for endocrine glands in your answer to question* a *present in the glands you have examined?***

Numerous blood capillaries will almost certainly be seen in any slide. The other features will be visible only in electronmicrographs.

**c** ***In what ways are the glands you studied dissimilar?***

The pancreas differs from the other two glands studied, in having two types of tissue, this reflects its two functions. There are the enzyme-secreting cells (acini), and the hormone-secreting cells of the islets of Langerhans, with which we are concerned here. There are two types of cell in the islets of Langerhans – α-cells which secrete glucagon and β-cells which secrete insulin – but they are usually indistinguishable from one another unless special staining is used. The β-cells are the more numerous; the α-cells are the larger.

The other and more significant difference to notice is the relationship between the hormone-secreting cells and the blood system.

The α- and β-cells of the islets of Langerhans secrete hormones directly into the blood capillaries, The adrenal medulla secretes the catecholamine hormones noradrenaline and adrenaline; they are not secreted continuously, however, but are stored as granules in the cytoplasm of the cells. Their release into the blood capillaries is under the control of the sympathetic nervous system. Because the stored catecholamine granules are oxidized to a brown colour when fixed in chrome salts the secretory cells of the adrenal medulla are often called chromaffin cells.

The thyroid gland is unique in that the hormones it secretes ($T_3$ and $T_4$) are stored in follicles in an inactive form called thyroglobulin. Cuboidal epithelial cells in the follicle walls synthesize and secrete the hormones. The release of the hormones into the blood vessels is under the control of thyroid stimulating hormone (TSH) which is secreted by the anterior pituitary gland.

*Figure 43* shows the relationship between the secretory cells of these glands and blood vessels.

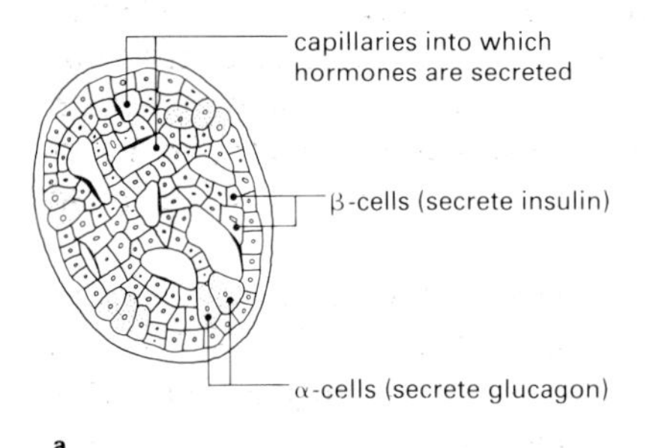


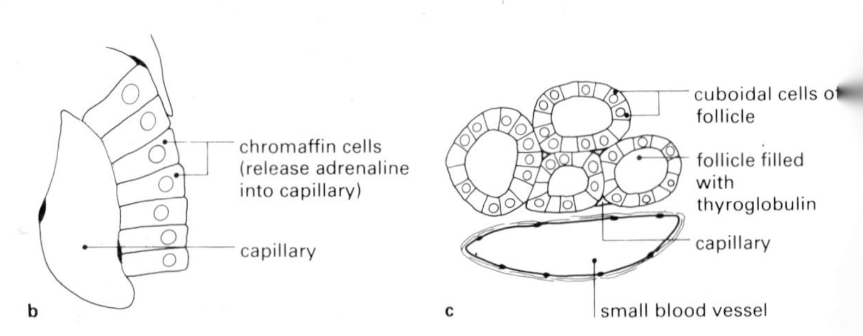


**Figure 43**
The relationship between hormone-secreting cells and blood vessels.
**a** Islet of Langerhans.
**b** Adrenal medulla.
**c** Thyroid gland.
*Based on M. B. V. Roberts,* Biology: A Functional Approach. Students' Manual. *Nelson, 1974.*

d ***In the electronmicrograph shown in figure [P]27, the rough endoplasmic reticulum is a prominent feature. You can also see a Golgi complex and mitochondria.***
***How does the presence of these structures relate to the hormone-producing function of the cell?***

The presence of the rough endoplasmic reticulum suggests a site of protein synthesis. The Golgi complex is usually associated with the way in which secretory products are packaged within membrane-bound vesicles. The presence of mitochondria, which are the sites of respiration, suggest that processes requiring energy input are taking place.

## INVESTIGATION
## 24B The effect of IAA on the growth of coleoptiles and radicles

### ITEMS NEEDED

Germinating cereal grain, with coleoptile and germinating legume seed, with radicle, 18–20 mm long 25 of each/group

IAA solution in sucrose buffer, 0.01 mol $dm^{-3}$ (1.75 g IAA $dm^{-3}$)
Sucrose buffer
Water, distilled

Beakers, 100 $cm^3$
Incubator at 25 °C 1/class
Measuring cylinder, 100 $cm^3$ 1/group
Microscope slide 1/group
Mounted razor blades, pairs held 10 mm apart
Petri dish with lid 5/group
Pipettes or syringes, 1 $cm^3$
Ruler and hand lens or vernier calipers 1/group

(*Study guide* 24.7 'Communication and control in plant growth'.)

**Sucrose buffer**
Sucrose, 20 g
Potassium hydrogen phosphate, 1.74 g
Citric acid, 0.96 g in distilled water, 1 $dm^3$

*Assumptions*
1 An elementary understanding of statistics, in particular the standard error of difference.
2 The ability to relate growth of isolated plant sections to tropic responses.

*Principles*
1 IAA significantly affects the increase in length of coleoptile segments and detached radicles.
2 IAA is effective in extremely low concentrations.
3 A given concentration of IAA may affect coleoptile and radicle tissues in different ways.

The increase in length of the plant sections is almost entirely due to cell extension. It follows that sections of different sizes will have different potentials for growth and so it is essential to cut sections of the same initial length. To ensure an adequate number of standard length coleoptiles and radicles, it is usually necessary to germinate about four times that number of seeds. Soak the seeds in water for 24 hours and incubate in the dark at 25 °C.

Pin the legume seeds to a vertical board as described in *Teachers' guide I*, page 310 to produce straight radicles. The buffer solution, pH approximately 5, ensures that the IAA is only slightly ionized; IAA enters cells more easily in its un-ionized form.

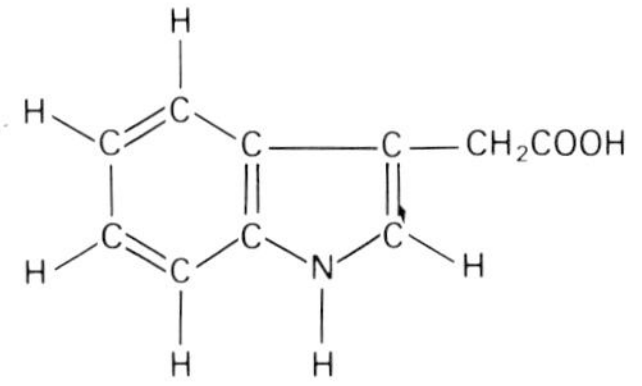

**Figure 44**
Indole-3-ethanoic acid (un-ionized form).

*Questions and answers*

a **_From the graphical analysis, does IAA appear to affect the coleoptile and radicle sections? If so, is the effect roughly proportional to the concentration of IAA?_**

Yes, though the response of the radicles is likely to be less than that of the coleoptiles. There is no obvious proportionality to IAA concentration except, perhaps, that radicle sections show an inverse relationship. Using other concentrations of IAA ($10^{-3}$, $10^{-5}$, $10^{-7}$ and $10^{-9}$ mol dm$^{-3}$, for instance) will give a clearer picture (see *figure 45*).

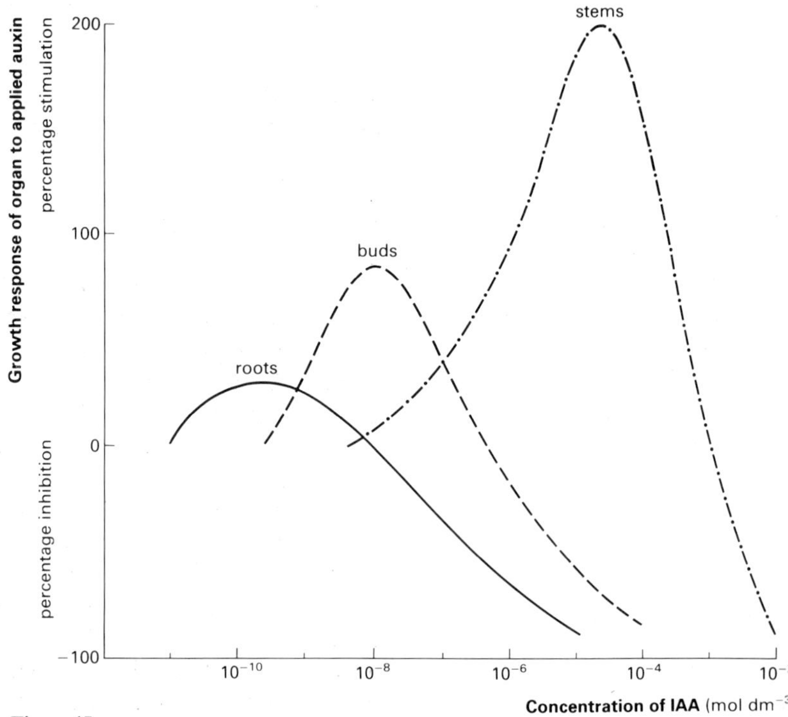


**Figure 45**
The growth responses of roots, buds, and stems in relation to externally applied IAA.

b **_In what ways do different concentrations of IAA affect the coleoptile and radicle sections?_**

See *figure 45*. The responses of the radicle sections are completely different from those of the coleoptile sections.

c **_From the statistical analysis, which concentrations of IAA appear to have a significant effect on the sections?_**

All the concentrations are likely to have a significant effect, apart from $10^{-8}$ mol dm$^{-3}$.

d **_In what ways do the results obtained with IAA aid the understanding of tropic responses?_**

Tropic responses are brought about by the differential extension of cells. IAA is one of the substances that affect cell size. Phototropism can be accounted for by supposing that light causes accumulation of IAA on the dark side of the coleoptile. This results in greater elongation on the dark side and consequent bending towards the light. Gravitropism could be explained by supposing that IAA accumulates on the lower side of a horizontal root, inhibiting elongation on this side so that the root bends down; evidence for the accumulation of IAA on the lower surface of roots is inconclusive, however.

e ***Why do you discard the 3 mm at the tip of the section?***

To exclude the endogenous supply of IAA.

## INVESTIGATION
## 24C Stimulation of amylase production in germinating barley grains

### ITEMS NEEDED

Barley (e.g. Var. Proctor)
Agar, sterile Petri dishes containing:
starch/agar 2/group
starch/agar to which gibberellic acid has been added 1/group
starch/agar into which a 'boat' made of dialysis tubing containing suitable agar has been placed 1/2 group

Gibberellic acid
Iodine in aqueous potassium iodide solution
Sodium hypochlorite (3–5 per cent available chlorine)
Starch soluble (or ordinary potato starch may be used)
Water, sterile

Adhesive tape
Dialysis membrane (Visking tubing)
Forceps 1/group
Incubator (25–30 °C) 1/class
Muslin
Petri dishes 3 or 4/group
Scalpels (or razor blades) 1/group
Tiles 1/group

(*Study guide* 24.7 'Communication and control in plant growth'.)

**Agar.** Dissolve 2 g of agar in 100 $cm^3$ of distilled water. Heat until nearly boiling.
**Gibberellic acid.** Dissove 0.1 g of gibberellic acid in 2 $cm^3$ of ethanol and dilute to 100 $cm^3$ with water. Add 1 $cm^3$ of this solution to 100 $cm^3$ of starch agar to give a final concentration 10 p.p.m.
**Iodine solution.** Dissolve 6 g of potassium iodide in about 200 $cm^3$ of distilled water and then add 3 g of iodine crystals. Make the solution up to 1 $dm^3$ with distilled water. It is essential to prepare this solution 24 hours before it is required, as iodine is slow to dissolve.
**Starch agar.** Add 2 g of agar to 100 $cm^3$ of cold 1 per cent starch suspension and allow to stand for 10 minutes. Heat the mixture slowly until it just boils, then pour it into the Petri dishes.
**Starch suspension 1 %.** 10 g of soluble starch (or ordinary potato starch) are put into 1 $dm^3$ of distilled water. Mix the starch with a little of the water, cold, and boil the rest, then pour the starch–water mixture into the boiling water. Alternatively, mix the starch with a little cold water and heat carefully until the mixture froths up; dilute this with cold water, stirring well continuously, and make the final suspension up to 1 $dm^3$.

*Principles*

1 Developing embryos produce gibberellic acid which diffuses back into the aleurone layer and stimulates amylase activity which converts insoluble starch into soluble sugars for transport.
2 The large protein molecules of the enzyme, amylase, will not diffuse through a dialysis membrane; the smaller gibberellic acid molecules will get through.

*Procedure*

It is essential to employ a good sterile technique in the preparation of the sterile Petri dishes, leaving the dishes uncovered for the shortest possible time even when pouring the agar. This will reduce the risk of infection by various micro-organisms that also produce amylases. It is, however, often possible to discriminate between the amylases produced by the grain and by micro-organisms: that produced by the grain will form a concentric ring around the grain, that from micro-organisms in the grain an eccentric ring. Small, clear patches scattered all over the surface of the agar are also the result of amylases produced by micro-organisms which have contaminated the plate.

Each group needs two Petri dishes containing starch agar and one containing starch agar plus gibberellic acid. In addition they may need a further Petri dish of starch agar into which a 'boat' made of dialysis membrane has been set. Prepare the 'boat' as follows: Fold the dialysis membrane (Visking tubing) round a book of indicator papers or something of similar shape and size. Fold up the ends and seal with adhesive tape to make a 'boat'. Pour starch agar into a Petri dish and hold the 'boat' gently in the molten agar until it sets in position (if the molten agar is not very hot when it is poured it will set quite quickly). Then pour some plain agar into the 'boat' and leave it to set. Embryo-containing halves of barley grains are placed on the plain agar in the 'boat'. If there is any doubt that gibberellic acid can diffuse out of the boat, it can be settled by putting agar containing gibberellic acid inside the boat and omitting the embryo halves of the barley grains.

*Questions and answers*

a ***Under which conditions is a starch-degrading enzyme (amylase) produced?***

Embryo halves on starch agar and non-embryo halves on starch agar containing gibberellic acid should both show evidence of amylase production.

b ***Suggest a hypothesis for the control of amylase production in germinating barley grains.***

These results should suggest that when the grains absorb moisture the embryos produce gibberellic acid, which stimulates amylase production.

c ***What can you deduce from these results?***

Expected results are shown in *figure 46*. They suggest that something passes from the embryo halves of the barley grains through the dialysis membrane into the non-embryo halves of the grains which stimulates amylase production in them. From the previous investigations students should suggest that it is gibberellic acid which, being a smaller molecule than amylase, can pass through the membrane and diffuse into the non-embryo halves.

d ***What part of the barley grain would you suggest produces the amylase? Look carefully at the base of the grains that have produced amylase to see if their appearance gives a clue.***

Students may not have much knowledge of the structure of a barley grain, but after studying *figure (P)10.7* in the *Practical guide* they might suggest the aleurone layer. Often the starch can be seen to be digested from the outside, which goes translucent whilst the centre remains opalescent and granular.

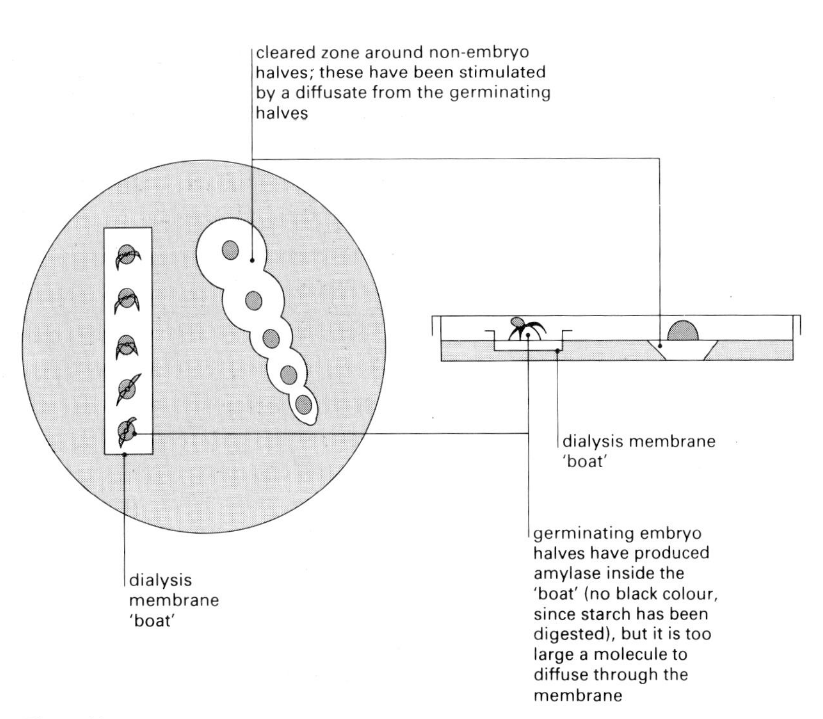


**Figure 46**
Amylase production in germinating barley grains: expected results.

e ***What is the significance of your findings in seed germination and the control of dormancy?***

Food reserves are usually stored in insoluble form, which avoids the complications which high solute concentration would produce osmotically. They must be soluble for transport, however, as otherwise they cannot pass through membranes. If starch is to be stored, then amylase activity must be minimized until hydrolysis is required. The embryo controls the process by producing the gibberellic acid which is the 'signal' to the aleurone layer to produce amylase. If the embryo is dormant it will not produce gibberellic acid, so there will be no 'signal' for amylase production in the aleurone layer; in this way, the food reserve is protected against mobilization when the grain is first moistened. Of course, it is also possible that other hormones are involved in germination and the control of dormancy as well as gibberellic acid.

## INVESTIGATION
## 24D The effect of plant hormones on seed germination

**TEMS NEEDED**

Lettuce seeds – Dandie, Kloek, or other light-sensitive variety 50/group
Abscisic acid, 0.002 g $dm^{-3}$
Water, distilled *(continued)*

(*Study guide* 24.7 'Communication and control in plant growth'.)

*Principles*

1 Plant hormones can act as germination inhibitors.
2 One plant hormone may counteract the effect of another.

Results obtained in March 1982 are shown in table 22.

## ITEMS NEEDED *(continued)*

Filter paper
Incubator and/or environmental chamber (25 °C), (as a means of maintaining constant temperature in light and dark)
Labels or wax pencils
Petri dishes 8/group
Syringes or graduated pipettes ($5.0\ cm^3$)

| | percentage germination | |
|---|---|---|
| | *light* | *dark* |
| Water only | 95 | 53 |
| $0.001\ g\ dm^{-3}$ ABA | 0 | 0 |
| $0.01\ g\ dm^{-3}$ KIN | 90 | 89 |
| $0.001\ g\ dm^{-3}$ ABA + $0.01\ g\ dm^{-3}$ KIN | 90 | 87 |

**Table 22**

*Questions and answers*

**a** ***What effect does each hormone have on the germination of lettuce seeds?***

Abscisic acid imposes dormancy in the light and dark. Kinetin raises germination in the dark.

**b** ***What is the effect when both hormones are present?***

The dormancy imposed by abscisic acid is counteracted by kinetin.

**c** ***What evidence do you have from your results that these plant hormones have a role in the control of dormancy in plants?***

The results may demonstrate that abscisic acid acts as a germination inhibitor, but more evidence is needed about its presence in seeds and the way it acts. The antagonism between abscisic acid and kinetic might feature in the control of dormancy and germination, but again more evidence is needed before one can be sure of their significance in the processes.

# PART III Bibliography


GOLDSWORTHY, G. J., ROBINSON, J., and MORDUE, W. *Endocrinology.* Blackie, 1981. (A short book which includes details of cellular aspects of hormones.)

RANDLE, P. J., and DENTON, R. M. Carolina Biology Readers No. 79, *Hormones and cell metabolism.* Carolina Biological Supply Company, distributed by Packard Publishing. 2nd edn. 1983.

TATA, J. R. Carolina Biology Readers No. 46, *Metamorphosis.* Carolina Biological Supply Company, distributed by Packard Publishing. 2nd edn. 1983.

WIGGLESWORTH, V. B. Carolina Biology Readers No. 70, *Insect hormones.* Carolina Biological Supply Company, distributed by Packard Publishing. 2nd edn. 1983.

# CHAPTER 25 DEVELOPMENT AND THE EXTERNAL ENVIRONMENT

*A review of the chapter's aims and content*

1. In this chapter the effects of factors in the external environment on the rate and pattern of development are considered.
2. The phenomenon of dormancy in seeds is discussed.
3. Photomorphogenic effects in plants are described.
4. The need for a timing mechanism and a light and a dark detector in a photoperiodic response is established.
5. The evidence for the existence of a flowering hormone is examined.
6. Some practical uses of the photoperiodic control of flowering are considered.
7. The economics of increasing crop yield by controlling the external environment are studied.
8. The effect of environment on the behavioural development of animals is discussed.
9. A method of differentiating whether maturation or learning is responsible for changes in behaviour is considered.
10. The development of behaviour is illustrated from a study of bird song.
11. The importance of early experiences in the development of behaviour is examined.
12. Experimental investigations on mother–infant relationships in primates are described and discussed, including their relevance to human behaviour.
13. It is emphasized that experiments may not be performed on humans, but that observational studies are possible.

# PART I The *Study guide*

## 25.1 Introduction

*Assumption*

1. That the work in earlier chapters concerned with development has been studied.

*Principle*

1. Factors in the environment contribute to the development of organisms.

## 25.2 The external environment in relation to growth and development

**Practical investigation. *Practical guide 6*, investigation 25A, 'The effect of temperature on root growth'.**

*Principles*

1. A seed is a resistant stage in the life history of a plant.
2. Many seeds have a dormancy system which has a survival value.
3. The time of germination may be controlled by such factors as low temperature or light, as they are required to break dormancy.
4. The morphology of seedlings may be determined by conditions in the environment.
5. Temperature affects the rate at which metabolic processes take place and this has an effect on the rate of growth and development of organisms.

## STUDY ITEM

### 25.21 Control of the time of seed germination

*Questions and answers*

**a** ***Why do you think that a subzero temperature was ineffective?***

Two possibilities may occur to students:
**1** Damage caused by freezing; but this answer is inappropriate, since the intracellular fluids of the seeds will not freeze at −1 °C, because their freezing-point is depressed well below that of pure water by the dissolved solids (mainly sugars and mineral salts) that they contain. The seeds that have been stored at this temperature can subsequently be made to break their dormancy in the usual way and will then germinate normally, showing them to have survived the chilling without any injury.
**2** Cessation of metabolism, which will occur at 0 °C. This hypothesis suggests that enzyme-controlled reactions may be necessary to remove the dormancy factor.

**b** ***If the seeds were kept dry at these temperatures, dormancy remained intact. Does this reinforce your answer to a?***

Yes, as lack of water will also impede metabolism. This could be shown by measuring carbon dioxide output as a measure of respiratory rate from dry and wet seeds. (If ice damage was suspected in the answer to **a**, then lack of water would be expected to restrict ice formation too and would also reinforce this wrong answer.)

**c** ***At 3 °C it took nine weeks chilling to get 50 per cent germination. Do you think there is any advantage in a requirement for such prolonged chilling?***

Yes, because unless there is a long requirement for chilling seeds may germinate after a few early frosts in November and December.
□ Dormancy must be maintained until the hazards have passed.

**STUDY ITEM**

## 25.22 Controlling the quality of growth of the germinated seedling

*Questions and answers*

a ***How does growth in darkness achieve protection of the shoot apex?***

The shoot apex is protected in this case between the cotyledons and is, in effect, dragged up backwards through the soil so that abrasion is kept to a minimum.

b ***What is the evidence that the cells are more elongated in the seedling grown in darkness? What further information do you need to confirm this?***

The longer hypocotyl (the region between the cotyledons and the radicle) suggests *either* more cells *or* longer cells. In order to confirm which is the case a few cell lengths would have to be measured under a microscope. This is quite easy to do as the epidermal cells can be measured; there is no need to cut sections.

c ***In what ways is the light-grown seedling adapted for life above ground?***

**1** The cotyledons have unfolded and are greatly expanded, thus providing good light interception for photosynthesis.
**2** The short sturdy stem gives support.

d ***How could you test that phytochrome is responsible for the changed morphology of a seedling as a response to light?***

You would have to test whether the light effects can be reproduced by using small doses of red light. If they can, then a similar dose of far-red light should be given immediately following the red light to see if the dark type development is restored. (In both cases the result is as anticipated.)

e ***How do these two facts relate to the survival of seedlings?***

They have a better transport system for water and food and more support. □

**STUDY ITEM**

## 25.23 Hatching trout eggs

*Principles*

1. The rate of development of trout eggs is dependent, among other factors, on temperature.
2. The production of large numbers of young compensates for the high mortality rate.

*Questions and answers*

**a** ***From these data, make an equation to relate incubation period and temperature as exactly as possible.***

Using the data given, the temperature multiplied by the number of days of the incubation period gives the figure 410.
Let $T$ = temperature in degrees Celsius
$I$ = incubation period in days

and the equation would be

$T \times I = 410$

that is,

$T \times I = K$ (a constant)

**b** ***From your general biological knowledge, or from your equation, determine what you could expect the incubation period to be at −1 °C, +3 °C, +8 °C, and +15 °C.***

$I = \frac{410}{T}$, so the incubation period at the temperatures given could be expected to be

| $T$ | $I$ |
|---|---|
| − 1 °C | – |
| + 3 °C | 137 days |
| + 8 °C | 51 days |
| +15 °C | 27 days |

If the temperature falls below freezing the unprotected eggs are killed. The lowest temperature at which the eggs will hatch is about +1 °C. At +15 °C the amount of oxygen dissolved in the water becomes the limiting factor. Because of this, up to +21 °C it is possible to hatch trout eggs only if the flow of water is sufficient to provide the necessary amount of oxygen.

**c** ***Can you explain how trout eggs fertilized throughout late winter and early spring tend to hatch at about the same time?***

The water temperature rises gradually from winter to spring; later-fertilized eggs therefore hatch more quickly.

**d** ***What effect, if any, might this mass trout hatching have on the trout population?***

There is no parental care of eggs or young trout. A large population of newly hatched trout (alevins) is therefore at risk from many possible dangers. These include bruising, overheating by day and chilling at night, fungal and bacterial diseases, drying up, buffeting from fast currents, and predation by such animals as water birds, newts, other fish (stickleback, eel, older trout), water beetles, and dragonfly nymphs. Competition for food is not a cause of death in the alevins as the supply of food in the yolk sac lasts about six weeks. □

## 25.3 Light and plant growth

**Practical investigation. *Practical guide 6*, investigation 25B, 'Effects of light on the germination of lettuce seeds'.**

*Principle*

1 Unlike photosynthesis, for which prolonged light of high intensity is required, photomorphogenesis only requires short exposure to light of low intensity.

### STUDY ITEM

25.31 The significance of action spectra and absorption spectra

*Questions and answers*

a ***How can the amount of photosynthesis be measured? What units would it have?***

There are several possibilities, but usually, in the short term, carbon dioxide uptake or oxygen evolution are measured. The units would be mass (or volume) of gas per unit leaf area (or per unit chlorophyll) per unit time: for example, mg carbon dioxide $m^{-2}$ leaf area $h^{-1}$, or $cm^3$ oxygen $g^{-1}$ chlorophyll $minute^{-1}$.

b ***Suggest why more photosynthesis results from light in the green region than you might predict from the absorption spectrum of chlorophyll.***

The presence of accessory pigments such as carotene and xanthophylls extends the light absorption towards the green region.

c ***Do you think the absorption spectrum of a crude extract of chloroplasts would be a good match to the action spectrum of photosynthesis?***

One would expect it to be better than chlorophyll alone. *Figure 47* illustrates this, using dye reduction as a measure of photosynthetic potential. □

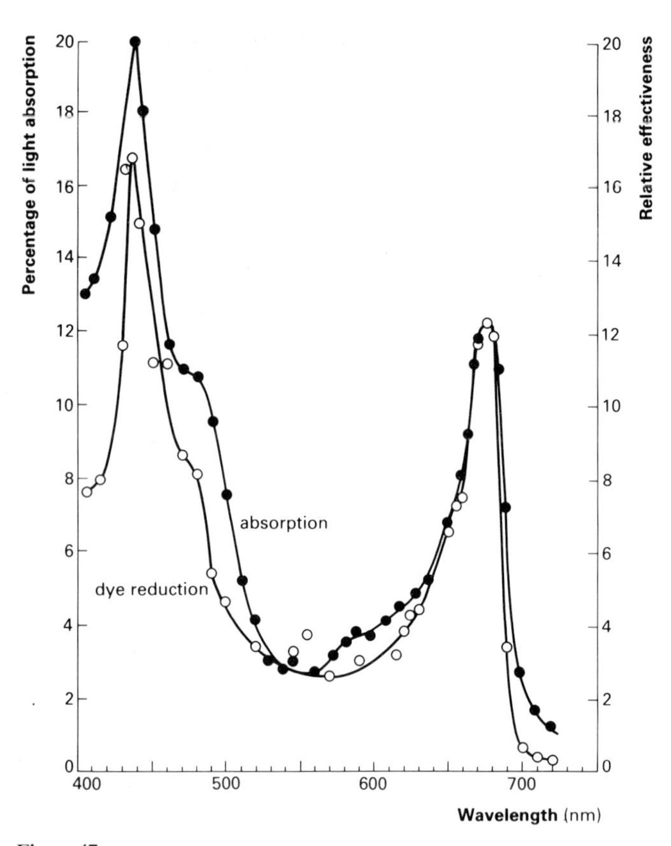


**Figure 47**
Absorption spectrum (●——●) of extracted chloroplasts, compared with the effectiveness of light of different wavelengths to cause the photosynthetic reduction of a dye (○——○), using the same chloroplasts.
*Based on Chen, S. L., 'The action spectrum for the photochemical evolution of oxygen by isolated chloroplasts'*, Plant Physiology, **27**, *1952, 35.*

## STUDY ITEM

## 25.32 Action spectra and photomorphogenic responses

*Questions and answers*

a ***How many different pigments do you think are involved? Could any of them be chlorophyll?***

Two, one of which is phytochrome which can exist in two forms. Neither corresponds to chlorophyll action.

b ***The absorption spectra of four plant pigments (A, B, C, and D) are shown in* figure [S]263b. *Select the pigment which you believe best fits the action spectrum for each of the different responses shown in* figure [S]263a.**

From figure (S)263a, A is clearly the pigment involved with phototropism; the other three curves in that diagram concern responses to C and D – the two forms of phytochrome. Pigment A = carotene, B = chlorophyll, C and D = $P_{660}$ and $P_{735}$ respectively c ***What colour pigment would you expect to be involved in a process for which you obtained the action spectrum shown in figure [S]263c?***

Red or blue would be good at absorbing green, which is what figure (S)263c represents. This is the most effective colour for human vision and the pigment is visual purple (rhodopsin). □

## Photoperiodism

*Questions and answers*

a ***On what dates are day and night of equal duration?***

March 21st and September 21st, the spring and summer equinoxes.

b ***Predict what changes in day length might induce a native British tree to go dormant.***

Shortening days from summer to autumn.

c ***Suggest what day length might induce a perennial plant from a hot climate to go dormant.***

Lengthening days which will correspond to the onset of the hot, dry period.

d ***Can you think of any phenomena in humans which are connected with the biological clock?***

One is the jet lag that affects travellers making journeys where there are considerable differences in local times between their departure and their destination (over and above the time taken to make the journey). They tend to want to eat and sleep at the local times of their journey's origin, which may be very different from those prevailing at their destination; for example, if passengers leave London for Vancouver at 10 a.m. London time, after a flight of around ten hours they will arrive in Vancouver at about midday local time. They may have eaten lunch and dinner on the plane and arrive in time for another early lunch. By teatime their local clock will be telling them that it is well past bedtime. Even the following day they will still feel uneasy, and it will take them about two days to reset the clock to local time.

Many bodily functions such as urination, defecation and body temperature are also in phase with the body clock.

e ***What does this suggest about the importance of photosynthesis in the photoperiodic response?***

Dim light was effective, so photosynthesis is unimportant; in treatment D even one hour of dim light completely altered the effect of the eight hours daylight whereas in E this same total illumination produced an opposite effect – this could not be photosynthesis.

f ***Suggest an experiment to check that the continuity of the light phase is not very critical.***

Insert ten minutes of darkness into the middle of the day (by comparison with ten minutes light in the middle of the night, as in regime D in figure (S)264). If continuity is not important you would expect a minimal response, and this is what happens – nothing!

g ***Can you deduce whether the timing mechanism acts like an egg-timer or like an oscillator (such as a pendulum)?***

An egg-timer will only time a single event; once the sand is in the bottom of the glass the timing is complete, until someone turns it upside down and starts it off again. In the treatments in which eight hours of light are followed by four hours of dark the timer is continuing for at least forty-eight hours. This would not be possible if the equivalent of setting off the sand in the egg-timer was either the light coming *on* at the beginning of the 'day', or the light going *off* at its end. In either case, the timer would run out after twenty-four hours and then stop. It seems more likely that the clock is an oscillator of some kind so that, like a pendulum that has been given a push, it continues to go on swinging for several cycles, but it slowly runs down if it is not restimulated. This model would fit the evidence if we think in terms of a pendulum oscillating once per day: it might run for several days following the single 'push' provided by the 'light on' or 'light off' trigger.
The egg-timer could be an analogy of a biochemical reaction which produced (or destroyed) a certain amount of metabolite. Such a biochemical reaction would also tend to go faster or slower as the temperature varied. A clock that ran faster when it was warm and slower when it was cool ($Q_{10}$ about 2) would simply negate the advantage of the reliability of seasonal daylength. If the biological clock is to be any good it must be virtually unaffected by temperature over the physiological range between about 5° and 35 °C – and it is (though examples from photoperiodic responses are hard to find because there are much better systems with which to demonstrate temperature insensitivity of the biological clock).

h ***How would you proceed to do this?[That is, to find what pigment is responding to light.]***

You would use coloured filters to select different regions of the spectrum and determine the effect of ten minutes of equal energies of the various colours.

i ***What colour would you expect the pigment phytochrome to be?***

Blue or blue green to complement red.

j ***If ten minutes of red light is equivalent to ten minutes of white light, how would you check that phytochrome was the receptor pigment?***

Follow the ten minutes of red light immediately with twenty minutes of far-red light – this should nullify the effect of the red light. (Students will probably suggest ten minutes of far-red light but it is common to use about twice the length of time of red light for the far-red. This is to be on the safe side since the $P_R \rightarrow P_{FR}$ transformation proceeds more readily than $P_{FR} \rightarrow P_R$.)

**STUDY ITEM**

25.33 Control of flowering

*Questions and answers*

a ***What do you conclude about the site of perception of the photoperiodic stimulus? How might this fit in with the idea of a flowering hormone?***

The leaves, not the apex, must perceive the stimulus. Since the apex is where the flower is formed a 'message' must pass from leaf to apex – this would fit in with the idea of a hormone provided the 'message' acted at great dilution.

b ***What does this tell us about the production of 'florigen' in relation to photoperiod being received by the leaf?***

Florigen production appears to continue once it has been 'triggered' by the appropriate day length, even when the plant is transferred to a non-inductive day length.

c ***Can you suggest the appropriate conditions, and result, of the reciprocal experiment in which a flowering shoot of tobacco is grafted to non-flowering* Hyoscyamus*?***

The reciprocal experiment would be to maintain both plants in short days. The tobacco plant will flower and the *Hyoscyamus* will remain vegetative. If a flowering shoot of tobacco is grafted to non-flowering *Hyoscyamus*, both plants still being maintained under short days, the *Hyoscyamus* will begin flowering even though none of its leaves have received an inductive long day.
It is usual to remove all leaves above the graft to ensure that the florigen from the graft will reach the apex. If other younger leaves are left above the graft then they may supply the nutrients to the apex, with the possible result that no florigen arrives there since it has been transported down to the roots.

d ***What do you conclude about the flower-inducing hormone in long-day versus short-day plants?***

Florigen appears to be the same in both long- and short-day plants, since it will induce flowering in both types of plant. □

## STUDY ITEM

### 25.34 The economics of growing a glasshouse crop using a controlled environment

*Assumptions*

1 An understanding of some aspects of photosynthesis.
2 Some knowledge of mineral nutrition.

*Principles*

1 Schemes for altering environmental factors may result in greater yields of crops.
2 Unless these schemes are economically viable it is not sensible for a producer to operate them.

The purpose of this exercise is to relate scientific discovery to its practical application and to evaluate the economics. Many good ideas are scientifically feasible but economically unsupportable. The development of microbial protein for animal feed from certain oil fractions is a case which highlighted the difficulties. When the price of oil suddenly rose in the mid-1970s, a process that may till then have been marginally economic immediately ceased to be so.

The understanding of plant nutrition led to great increases in crop yields. Remarkable increases in yield were achieved in underdeveloped countries in the early stages of the 'Green Revolution', demonstrating how the use of good varieties, plus irrigation, fertilizers, and pest control, could increase yield (*in the same climate*) by factors of four to five for maize, wheat, and rice in countries as diverse as Mexico, the Philippines, and India.

In glasshouse cultivation, adequate water and nutrients have always been supplied and with extra heating as well many crops can be grown out of season. Only since the mid 1960s has the benefit of extra carbon dioxide in the atmosphere been commercially exploited. The principle applies not only to tomatoes but also to winter lettuce, chrysanthemums, and other winter glasshouse crops. The result of applying this knowledge can be a huge increase in productivity per unit area of land – but at a cost. The last section is concerned with evaluation of the financial benefits of the innovations. Along the way there will be useful revision of mineral nutrition and some aspects of photosynthesis.

In 1980–82 the value of the tomato crop was over £50 million per annum; we imported nearly twice the bulk and rather more than twice this value of tomatoes into the United Kingdom from Holland, the Canary Islands, the Channel Islands, and Spain. The Canaries and Spain mainly provide for the winter months but Holland and the Channel Islands imports are in direct competition during the United Kingdom season. It will be interesting to see how the United Kingdom industry develops.

*Questions and answers*

a ***Why is it more useful to compare dry masses rather than fresh harvest mass?***

**1** Superficial water on the plants is very variable in quantity and is irrelevant.

**2** Internal water content is variable and depends largely on variables in the environment.

**3** Internal water makes up a high proportion of fresh harvest mass and growth is best measured in terms of the products of anabolism. Only insignificant quantities of these are lost by heating quickly to constant dry mass.

**b** ***Suggest why there is less variation in the yields of tomatoes from different growers than there is in the wheat yields of different farmers.***

Apart from the hours of sunshine, tomato growers maintain a close control of the environment in which their crop grows. Farmers have no such control, hence wheat yields vary much more widely, reflecting the variations in the environment.

**c** ***Suggest two major factors which contribute to the higher yield of tomatoes as compared to wheat.***

The two major factors which improve tomato yield are, firstly, a greatly extended growing period, allowing a longer duration of photosynthesis, and, secondly, the provision of extra warmth which promotes more rapid growth.

**d** ***Why does the addition of carbon dioxide to the atmosphere in which plants are growing give an increased yield?***

Extra carbon dioxide is beneficial because it improves net photosynthesis. In bright sunlight carbon dioxide is often the main limitation; it is being used faster than it can get to the site of utilization in the chloroplast. Increasing the external concentration directly increases the concentration of dissolved carbon dioxide around the photosynthetic cells, which in turn increases the rate of movement towards the site of depletion in the chloroplasts.

**e** ***What are the reasons for these measures?***

The effectiveness of extra carbon dioxide depends on optimizing photosynthesis. If the greenhouse is not well sealed carbon dioxide will escape to the exterior; dirty glass and glazing bars will cut down light.

**f** ***Suggest why some of the early workers found injury to plants when they supplied extra carbon dioxide.***

Industrial flue gases contain sulphur dioxide as well as carbon dioxide. Sulphur dioxide is a very toxic pollutant and could easily override any potential benefit from the accompanying carbon dioxide. The usual method of producing carbon dioxide these days is to burn propane or butane, or low-sulphur kerosene or North Sea gas (methane). (The stipulation about low sulphur content is to ensure that sulphur dioxide is not produced during combustion.) In retrospect it is easy to understand the reason for the confusion in interpreting the benefits. The plants receiving relatively pure carbon dioxide benefited from their treatment; those given carbon dioxide plus pollutants were damaged.

g ***Why is the supply of carbon dioxide stopped during the summer months?***

This is because the extra sunshine produces extra warmth and unless the ventilators are opened the temperatures inside the greenhouse become supra-optimal and the crop suffers. When the ventilators are open most of the carbon dioxide would escape and be wasted.

h ***At what time during a twenty-four-hour period should the extra carbon dioxide be supplied for it to be beneficial?***

If the effect is on photosynthesis there is no need to enrich the atmosphere with carbon dioxide at night. Enrichment from soon after sunrise until just before sunset would be reasonable.

i ***Why is the maintenance of a thin 'film' an important consideration?***

A thin 'film' not only ensures good aeration, but keeps the volume of nutrient solution to a minimum. This is important because it has to be collected in a tank and recirculated. Large volumes would require a massive tank and extra energy for pumping.

j ***List the elements that would be needed in such a nutrient solution.***

The essential elements are the macronutrients nitrogen, potassium, phosphorus, sulphur, calcium, and magnesium, plus the micronutrients iron, manganese, zinc, copper, boron, molybdenum, and chlorine. Generally only nitrogen, potassium, and phosphorus are routinely monitored and replenished. If the crop begins to show signs of magnesium deficiency then extra magnesium is also added.
The element that tends to become unavailable is iron. To prevent this iron is added as an organic complex which withstands the tendency of inorganic iron to go out of solution as insoluble hydroxide/phosphate. It was characteristic of all early solution cultures (such as those of Knop and Sachs) that they showed iron-deficiency chlorosis unless tiny amounts of iron salts were added at regular intervals.

k ***Suggest reasons for the large difference between the cost of seed for a commercial variety and that for a domestic variety.***

The difference in cost pays for the extensive breeding work which goes into a commercial cultivar to make it the most disease-resistant, highest-yielding, best-quality variety available. Disease resistance can be critical for a commercial grower. About 7 g of seed are required to produce 1000 plants and planting is around 22 000 $ha^{-1}$, so the seed alone can cost aroung £650 per hectare of tomato plants. Since the cost of other inputs could be around £180 000 $ha^{-1}$, only the best seed is good enough. These financial considerations do not concern the domestic grower who is content to grow the familiar, well-tried varieties.

**l** ***The cost of producing carbon dioxide for enrichment is about £3800 ha$^{-1}$. Can you estimate its net benefit in crop value, assuming it improves overall yield by 30 per cent?***

If the total yield is 278 tonnes ha$^{-1}$ (as in the table) then the yield without any carbon dioxide enrichment would have been

$\left(\frac{278 \times 100}{130}\right) = 214$ tonnes ha$^{-1}$, and the increase in yield per hectare is then $(278 - 214) = 64$ tonnes ha$^{-1}$.

Value of 64 tonnes at an average of £667 tonne$^{-1}$ = £42 688.

Net benefit = 42 688 − 3800 = £38 888.

Thus the grower cannot afford *not* to use carbon dioxide enrichment. (Of course, this is an overestimate of the benefit because the extra fruit has to be picked, packed, transported, and so on, but the broad conclusion is still the same.)

**m** ***In fact, it is likely to raise yields up to the end of June by 40 per cent and thereafter to make little difference to the yield from July to October. Does this make much difference to the net value of the crop? Can you explain why its effects are minimal after June?***

Up to the end of June the yield was 149 tonnes ha$^{-1}$. If this was a 40 per cent increase 'unenriched' yield would be
$\frac{149 \times 100}{140} = 106$ tonnes ha$^{-1}$, and the increase in yield due to enrichment = 43 tonnes ha$^{-1}$.
The value of the crop to the end of June was £128 770 for the 149 tonnes, so over this period the average price per tonne = £864. Then the value of the extra 43 tonnes at £864 per tonne = £42 336, or virtually the same as the result of the first calculation.
The reason why the effects of carbon dioxide enrichment were small after the end of June is that the enrichment ceased in mid-May – six weeks previously – so that most of the growth which produced fruit from July onwards would not have benefited from enrichment.

At this point teachers may like to discuss photorespiration in the context of carbon dioxide enrichment and the growth of glasshouse crops.

If carbon dioxide can be supplied only when sunshine is low then its effect must be considerably less because light is then the main limitation. However, it is now realized that photorespiration occurs, and that oxygen and carbon dioxide are competing for what appears to be the same active site on the enzyme ribulose bisphosphate carboxylase. (See *Study guide I*, Chapter 6.)

The net effect of photorespiration is to reduce the efficiency of carbon dioxide fixation. But if the external carbon dioxide concentration is raised whilst the oxygen level stays constant, then photorespiration is suppressed because the competitive ability of carbon dioxide against oxygen has been enhanced. This is a major reason why carbon dioxide enrichment is useful even in poorish light. It is known that if the oxygen concentration is lowered from the 21 per cent level in air to, say, 5 per cent, then the efficiency of photosynthesis is greatly enhanced even at air levels of carbon dioxide. This supports the evidence for competitiveness with carbon dioxide. On the other hand, if oxygen levels are raised beyond 21 per cent, even up to pure oxygen (100 per cent) then the efficiency of photosynthesis falls as the amount of carbon dioxide evolved in photorespiration increases.

Thus carbon dioxide enrichment in the winter and early spring is worth while (as the figures show) but the effect would be even greater from the same level of enrichment in the brighter summer months.

Many older textbooks produced graphs to illustrate the idea of limiting factors in photosynthesis derived from F. F. Blackman's ideas (*figure 48*a and *b*). If *figure 48a* were a true representation, then carbon dioxide enrichment would not be of any advantage in poor light (below the point A). The real situation is that there is no sharp discontinuity from a straight line when carbon dioxide becomes limiting at points A and B. Rather they are two curves, as shown in *figure 48b*, as many workers found even before the discovery of photorespiration and the benefit from carbon dioxide enrichment in poor light.

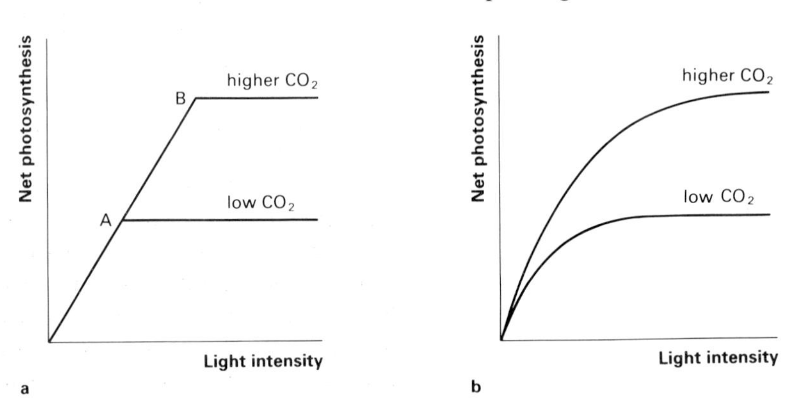


**Figure 48**
Two interpretations of the interactions between light intensity and carbon dioxide supply in photosynthesis.

## 25.4 The study of ontogeny

*Assumption*

1 Some general awareness of the nature/nurture debate, and its educational and philosophical importance.

*Principles*

1 There is a distinction between maturation and learning which has enormous relevance to practices of child rearing and education. In the interests of fostering a critical approach to current wisdom and practices, it is worth pointing out that phrases like 'learning to walk'

and 'potty training' suggest learning, probably quite unjustifiably. How about 'learning' to ride a bicycle, or play a musical instrument, or speak a foreign language?

2 To describe behaviour as being due to 'instinct' does not usefully add to the understanding of the causes or development of the behaviour concerned.

3 Deprivation experiments have very limited usefulness in trying to tease apart the respective roles of genes and environment in the development of behaviour.

4 Learning is very variable in many respects, including what is learned, the time during development when it is learned, and the speed and durability of the learning process. The key point which should become clear is that learning is geared to the special needs of the species concerned.

### Pecking in chicks

*Questions and answers*

**a** ***How would you design an experiment to distinguish between these possibilities?***

Students might suggest some form of deprivation experiment, whereby chicks are denied the opportunity to learn through trial and error while they are maturing. It would be reasonable to keep them in a container of uniform background, without features to focus on, and to feed them with forceps or a pipette. It is worth pointing out that this would not be easy and might be considered unacceptable. The other possible approach would be to provide chicks with unusually rich learning opportunities well in advance of the usual appearance of the accurate pecking behaviour. Again, this would not be at all easy. If learning is important in the development of the behaviour, the pecking should *not* improve in accuracy in the first situation but should in the second, in advance of the normal developmental stage. But if maturation is the main influence and learning is unimportant, the behaviour should improve in accuracy normally in the first case, and not be accelerated in the second.

**b** ***Are there any other possible explanations you can think of for the improved accuracy of pecking?***

As learning and maturation are both such broad terms it is difficult to think of other possibilities which fit into neither category.

### The ontogeny of birdsong

This section demonstrates the complex interaction between inherited and environmental factors during development, and illustrates the dangers of emphasizing one or the other exclusively. Thorpe's classic experiments are a good example of a situation where understanding could not have progressed without new technology: in this case, the sound spectrograph. Without this, detailed and objective analysis of sounds would have been impossible.

**STUDY ITEM**

## 25.41 Mother–infant relationships

**Practical investigation.** ***Practical guide 6*****, investigation 25C, 'Early environment and later behaviour of mice'.**

It is worth prefacing this section with some words of warning about the dangers of adopting a deterministic approach to such a sensitive issue. Clearly, one would not wish to convey to a student with a disturbed family background the notion that his or her own past experiences will inevitably lead to unsatisfactory relationships in the future. This interpretation would also denote a misunderstanding of what is being said in this section. Whereas it is undoubtedly true that different early relationships lead to differences in later behaviour, it is much less clear that any one kind of early relationship is any better than another. Neither is it the case that the influences of early behaviour produce effects that are subsequently fixed.

*Questions and answers*

**a** ***Do you think that it is legitimate to extrapolate conclusions from studies of non-human primates to humans?***

No – or, at least in the field of behaviour, only with great caution. All the information presented in the last two sections indicates that the developmental pattern of an individual species is finely adapted to the particular needs of that species in its own environment. Although we are fairly closely related to rhesus monkeys (though less closely than to many other monkeys, and apes) their social organization is very different from ours. Nevertheless, rhesus monkeys have many advantages as experimental animals, and it is legal to conduct experiments on them that could never be carried out on humans. (These reservations do not necessarily apply equally to physiological and other kinds of biological investigation.)

**b** ***Do you think that it is any more ethical to interfere with the relationship between a mother rhesus monkey and her infant than it is to do the same kind of experiment with humans?***

This is a question for general discussion.

**c** ***What implications does this passage have for those concerned with the care of laboratory and domestic animals?***

Isolated animals may be emotionally unstable even if physically sound. The behavioural needs of an animal should be considered as an important aspect of husbandry.

**d** ***What do you suppose is the meaning of the statement: '... investigations of social and behavioural development are long-term. In a sense, they can never end, because the problems of one generation must be traced into the next'?***

Such studies are necessarily long-term, because social and behavioural development must be studied throughout the life of an animal. The problems of one generation are traced into the next in the sense, for example, that the emotional stability of parents will influence that of their children. Thus a chain reaction through several generations is set up when one generation is influenced by experimental circumstances.

e ***To what extent are the cloth mothers adequate substitutes for real mothers?***

They are very poor substitutes if we consider the monkeys' emotional development in later years. As the authors mention, the behaviour of monkeys raised with cloth mothers was indistinguishable from that of monkeys raised in isolation. Only in stress situations, especially in their younger years, did the monkeys derive a sense of security from their cloth surrogates.

f ***Professor Harlow once wrote that 'the close behavioural resemblance of our disturbed infants to disturbed human beings gives us the confidence that we are working with significant variables...'. Is this statement justified?***

From observations in children's homes and such places it is clear that a child's relations with his/her peers and mother (whether a real or a foster mother) are important. These are the variables being studied with the rhesus monkeys.

g ***In what respects are the conclusions you would draw from the results of the investigations adequate to explain the way psychological development of rhesus monkeys takes place in the wild?***

It is not possible to give a full answer. In the wild, more complex social relations are found. In laboratory studies, this complexity is deliberately reduced. Thus careful investigation of behaviour in the wild is necessary before laboratory findings can be interpreted in terms of the wild life of these animals.

h ***What psychological factors other than those mentioned in the quotations may affect the development of young rhesus monkeys? Through further reading find out to what extent your answer is borne out by research findings that followed those mentioned here.***

**1** Play within peer groups. Play itself has many important functions, perhaps associated with infant–infant affectional bonds.
**2** Mother–infant relationships where the mother was raised in isolation or partial isolation. (H. F. Harlow and M. K. Harlow found interesting parallels with the human situation. Such mothers paid little attention to offspring and were abusive, while the offspring exhibited extremes of sexuality and aggressiveness.)
**3** Possibly infant–father relationships.

i ***In what way could the behaviour of mothers be responsible for the increasing independence of infants as they get older?***

By rejecting her infant, a mother is likely to encourage it to be less dependent on her. By increasing her rejection rate as the infant develops, the mother ensures that this is an accelerating process.

j ***How do the relative responsibilities of the mother and of the infant for maintaining their relationship change with time?***

The graphs suggest that, initially, contact is maintained by the mother but that her role diminishes with time, whereas the infant, gradually becoming more mobile, plays an increasing part in seeking contact.

k ***What do you think is the significance of the variation in the contact index shown in* figure [S]280d*? What features of the behaviour of individual infants and mothers could account for this variation?***

There are two kinds of variation to be considered: variation over time, and variation between mother–infant pairs, as shown by the dotted lines on each side of the median line (these represent 25 and 75 percentiles). The variation with time is covered in the answer to question **j**. Variation between mother–infant pairs is more interesting. Since the contact index is a ratio comparing maternal and infant behaviour, variation in the index could be due to variation between mothers, between infants, or both. Hinde classifies mothers as being 'rejecting', 'possessive', or 'controlling'. Rejecting mothers often reject the infants' attempts to cling on them and rarely initiate contacts. Possessive mothers rarely reject their infants and initiate contacts frequently. Controlling mothers reject a high proportion of their infants' advances and initiate a high proportion themselves. This is summarized in *figure 49*.

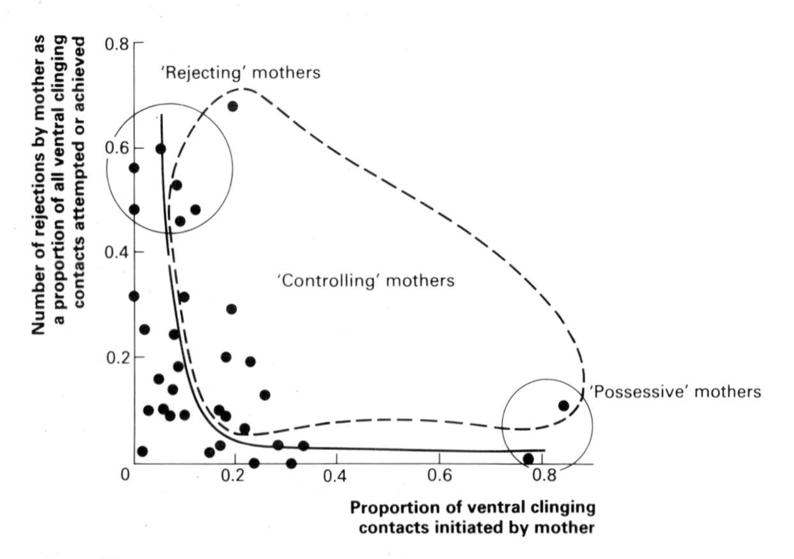


**Figure 49**
Mother–infant relationships: variations in mothers' behaviour.

# PART II The *Practical guide*

The investigations related to this chapter appear in *Practical guide 6, Development, control, and integration.*

## INVESTIGATION
## 25A The effect of temperature on root growth

### ITEMS NEEDED

Pea seedlings with 2–3 cm roots

Water: sources of hot and cold
Beaker, 600 $cm^3$ 1/group
Boiling tube with one-holed cork to fit 1/group
Glass rod, approximately 18 cm long, which moves fairly easily through cork 1/group
Graph paper
Pins
Ruler 1/group
Stopclock/watch 1/group
Thermometer 1/group

(*Study guide* 25.2 'The external environment in relation to growth and development'.)

*Assumptions*

1 An appreciation of the fact that the growth process involves chemical reactions mediated by enzymes.
2 A knowledge of the effect of temperature on enzyme activity.
3 An understanding of the graphical presentation of results and its interpretation.

*Principles*

1 By changing common environmental variables in a controlled way, it is possible to illustrate their effect on growth and development.
2 Threads of work involving other investigations, such as those concerned with the effect of temperature on metabolism, should be drawn together.

Temperature coefficient ($Q_{10}$) will probably have been met when considering enzyme-controlled reactions earlier in the course. Most chemical reactions, including those mediated by enzymes, have a $Q_{10}$ of about 2, at least within the range 5° to 35 °C.

It is easy, using elementary geometry, to convert the arbitrary growth units of '*H*' into absolute units, such as mm $hour^{-1}$. The relationship depends upon the relative radii of the boiling-tube and the glass rod. Thus a value of *y* mm 'change in *H*' $= y \times \frac{r^2}{R^2}$ mm (where $r$ = radius of the rod and $R$ = radius of the tube).

It will be seen that the rod measurements are about 30–40 times amplified compared with the actual change in length of the root.

It is possible to use this method to assess changes in root growth caused by other variants, such as lack of nutrients (nitrogen deficiency, for instance) or the presence of heavy metal contaminants such as lead, copper, or cadmium, or the effect of herbicides.

It would be feasible to replace the sliding glass rod with a disposable 2 $cm^3$ narrow-bore (or similar) syringe and a long needle fitted firmly through the cork so that it plunges well below the water level at the root tip. A measured volume of water could be withdrawn until the meniscus just 'jumped' back on to the root apex: the volume withdrawn per unit time $\propto$ root growth. Using a 12 mm diameter boiling-tube, a fall in height of liquid of 1 mm is equivalent to a liquid volume of 450 $mm^3$ = 0.45 $cm^3$ water withdrawn into the syringe. If 0.5 mm growth per hour is expected at around 25 °C, volume changes of at most 0.1 $cm^3$, and preferably of 0.05 $cm^3$, will have to be measured.

*Questions and answers*

a **_What information do the slopes of the graphs give you about the rate of root growth?_**

If the graph of growth against time is a straight line, then the rate of growth is constant. Fluctuations of temperature, either at the initial phase of equilibration or during the running of the experiment, could cause anomalous results.

b **_What effect would you expect further increases of 10 °C to have on the rate of root growth? Give your reasons._**

If $Q_{10}$ doubled between 15° and 25 °C, or between 25° and 35 °C, it might seem reasonable to suppose it would double between 30° and 40 °C, between 35° and 45 °C, and so on. In fact $Q_{10}$ decreases, and becomes negative as the temperature rises above 35 °C, largely because of the effect of higher temperatures on enzyme activity. The enzymes tend to be denatured at the higher temperatures.

## INVESTIGATION
## 25B Effects of light on the germination of lettuce seeds

### ITEMS NEEDED

Large green leaf (runner or French bean, sycamore) 1/group
Lettuce seeds – one sensitive to light (Dandie or Kloek), one not sensitive (Great Lakes, Webb's Wonderful, Avoncrisp) (for sources see page 305)

Black polythene or dark area
Filters, Cinemoid: No. 14 ruby and No. 20 deep primary blue plus any others (for source see page 306)
*Filter paper, 9 cm Whatman No. 1
Incubator and/or environmental chamber (as a means of maintaining constant temperature in light and dark) 1/class
Labels or wax pencils
Light boxes (home-made)
Light sources (incandescent, or incandescent and fluorescent)

Petri dishes
Stopclock/watch

*If different papers are used, adjust the amount of water so that there is approximately 1 cm³ 'free' water when moistened dish is tipped sideways. If there is any more than this the seeds will float into clumps; if there is less they may dry out.

(*Study guide* 25.3 'Light and plant growth'.)

*Assumption*

1 Some knowledge of the nature of light, which is necessary for a full understanding of this investigation.

*Principles*

1 The effect of light on the germination of seeds is peculiar to certain varieties.
2 To be effective, the duration of the exposure to light can be quite brief, and the light intensity fairly low.
3 A photochemical reaction depends on *total* energy, that is, duration of illumination × intensity of light.
4 Certain wavelengths of light inhibit germination while others promote it.
5 The effects of inhibition and promotion are reversible.
6 The process has an important ecological significance.
7 The investigations are essential to a rational explanation of the role of phytochrome in germination.

*Procedure*

Four points to ensure a successful outcome of these investigations must be stressed.

1 The seeds must have *complete darkness* while they are soaking, and those having continuous dark treatment *must* remain in complete darkness throughout the experiment.
2 After treatment with various light regimes the seeds must be placed in a constant temperature (25 °C if Dandie is used), that is, in an incubator or environmental chamber unless otherwise stated or unless temperature is the variable being investigated.

3 Old lettuce seed should not be used as it gives unreliable germination in this experiment.

4 Fluorescent light alone should not be used since it contains very little red light and no far-red. Daylight or incandescent light is required.

*Varieties of lettuce seed*

A most useful variety is the winter-forcing variety Dandie, but an older variety Kloek would do instead. Unfortunately lettuce varieties change as the breeders improve their characteristics, so Dandie may eventually disappear. Possible successors are Diplomat and Freeman, which may need an incubator temperature of 23–24° rather than 25 °C, and Baronet which may need 26 °C. Sources of seed are:

National Seed Development Organization (N.S.D.O.), Newton Hall, Newton, Cambridge.

Sutton's Seeds, Hale Road, Torquay, Devon, TQ2 7QJ.

E. W. King and Co. Ltd, Coggeshall, Essex.

*Procedure A*

Besides demonstrating the general effect of light on germination, this investigation illustrates the need for a specific constant temperature. It also shows that quite brief exposure to fairly dim light is effective.

*Questions and answers*

a ***What is the effect of light on the germination of the two varieties of lettuce?***

The results should show that light promotes the germination of Dandie or Kloek lettuce seed but has little or no effect on Great Lakes, Webb's Wonderful, or Avoncrisp, all of which germinate well in darkness.

b ***Does the length of time for which the seeds are exposed to light have any effect?***

The results should show that 30s or 300s give the same percentage germination as the control in continuous light, but it is possible that 30 s may be less efficient than 300 s.

c ***Is temperature a critical factor?***

The effect of light on germination is best seen within a fairly restricted temperature range. The differences between the results at room temperature and at 25 °C for Dandie, for example, should illustrate the need for a specific temperature.

d ***In photosynthesis light is used as a major source of energy. Do you think this is likely to be the case in germination, or from your results can you suggest another role for it?***

Since the response is not related to the length of time of exposure to light, and since a short exposure to light is sufficient to promote germination, light appears to be acting as an initiator of a process – a kind of sensory 'trigger'.

*Results obtained in November 1981 using the variety Dandie*

| *Temperature (°C)* | *Light % germination* | *(S.E.)* | *Dark % germination* | *(S.E.)* |
|---|---|---|---|---|
| 18 | 86.5 | (9.9) | 93.0 | (3.1) |
| 25 | 91.5 | (5.4) | 26.0 | (8.7) |
| 27 | 88.5 | (6.6) | 6.5 | (3.9) |
| 30 | 44.0 | (12.2) | 0 | (0) |

(S.E. = standard error of mean)

**Table 23**
Effect of different incubation temperatures
(var. Dandie; imbibition time 1½ hours; 10 minutes exposure to tungsten lamp; six replicates per treatment).

| | *Complete darkness* | *Duration of exposure*(s) *1* | *3* | *10* | *30* | *100* | *300* |
|---|---|---|---|---|---|---|---|
| *% germination* | 26 | 40 | 54.5 | 67.5 | 90.5 | 95.5 | 96.5 |
| *S.E.* ($n = 4$) | 8.7 | 5.9 | 1.1 | 6.8 | 3.4 | 2.4 | 1.1 |

**Table 24**
Effect of duration of exposure to Microlite lamp at 25 cm distance from Petri dish (approximately 1100 lux) (imbibition time 1½ hours; incubation at 25 °C).

| | *Imbibition time* (minutes): *0* | *15* | *30* | *60* | *120* |
|---|---|---|---|---|---|
| *% germination* | 30.4 | 65.3 | 89.3 | 93.3 | 95.2 |
| *S.E.* ($n = 6$) | 5.1 | 8.0 | 1.7 | 2.4 | 2.2 |

**Table 25**
Effect of duration of imbibition (10 minutes' exposure to light; incubation at 25 °C).

## *Procedures B and C*

These experiments test the effects of different wavelengths of light. The red and far-red filters are essential; others may also be used. The filter can be obtained from Rank Strand Electric, P.O. Box 70, Great West Road, Brentford, Middlesex TW8 9HR (01-568 9222).

| *Colour* | *Cinemoid filter No.* |
|---|---|
| red | 14 |
| deep primary blue | 20 |
| far red | 14 + 20 |
| primary green | 39 |
| yellow (light green and deep orange) | 23 + 5A |

The investigations show that red light is able to produce the same effect upon germination as white light. Far-red light appears to inhib growth, but as the percentage germination in the dark is low it is not eas to decide if this is true inhibition. You can check this by alternating re and far-red light and then, to show that the seeds are not just killed b far-red light, trying red light a second time.

Repeated alternation of red with far-red light can be tried; the seed will be found to respond only to the last quality of light given. (Be ver careful not to let white light in when the filters are being changed.) Th effect is seen more clearly if the seeds are exposed to far-red light for twic as long as to red light (600 s instead of 300 s) since $P_R \rightarrow P_{FR}$ proceeds more easily than $P_{FR} \rightarrow P_R$.

*Questions and answers*

e ***How do different wavelengths of light affect germination? Which wavelengths appear to promote and which appear to inhibit germination? Are these effects reversible?***

Red light should allow the same germination as white light; far-red light should produce a lower germination than complete darkness does. (Deep blue may also produce an inhibition because it also transmits far-red – and is used to make the far-red filter.) Red light appears to promote germination and far-red light to inhibit it. Far-red light reverses the effect of red, and *vice versa*.

f ***Why would fluorescent lamps be useless as the sole light source for these experiments?***

Light from these lamps contains virtually no far-red and very little red light.

g ***From these experiments formulate a hypothesis involving a pigment to explain the mechanism concerned in influencing seed germination. What colour would you expect the pigment to be?***

When light affects an organism there must *always* be a pigment that absorbs it. Students should be able to postulate the existence of a light-absorbing pigment (or pigments) inside the plant. It is easier to explain the reversibility of the system by postulating one pigment existing in two readily interchangeable forms, rather than two separate pigments (*figure(S)267*). The change from one to the other could be caused by different wavelengths of light, one form ($P_R$) absorbing red light and changing to the other ($P_{FR}$), which absorbs in the far-red region of the spectrum and reverts to $P_R$. Germination is promoted by $P_{FR}$ but not by $P_R$.

The colour of a pigment is due to the light it reflects. Reflected light is photochemically useless, however; only the absorbed light is effective. A pigment which responds to red and far-red light must be absorbing it, and reflecting other light. It should therefore look blue or blue–green. In fact, a blue light-absorbing pigment, phytochrome, does exist in minute amounts in plants and was first extracted in 1959 from corn seedlings. Although phytochrome may be of limited significance in seed germination it has a ubiquitous role in effecting the germinated seedling's response to light as it emerges from darkness in the soil. Then the stem growth slows to a sturdier type, while the epicotyl (or hypocotyl) hook – which protected the apex as it passed through the abrasive soil – unfurls and the leaves (or cotyledons) unfold and expand to optimize photosynthesis, which is needed to supplement the rapidly exhausting seed reserves. Without phytochrome the plant would continue the weak, undeveloped growth of the etiolated seedling – and die. Phytochrome is also the light detector for photoperiodic phenomena such as control of flowering, leaf fall and the onset of

dormancy in deciduous trees. Some of the most rapid responses to a change in phytochrome occur within minutes of stimulation, and suggest that a primary effect is a change in permeability of the cells, with all the varied results that might ensue. There is also considerable evidence that phytochrome is itself bound to cell membranes, so it would be well placed to effect changes in their permeability.
The conversion process is thus reversible by red and far-red light and can be repeated many times *in vitro*. In darkness the reversion of $P_{FR}$ to $P_R$ can take many hours. It is slowed down by a low temperature, and does not occur at all in the absence of oxygen, which suggests that it may be brought about by an enzyme. This suggestion is consistent with evidence that the dark reversion does not occur in extracted and purified phytochrome, which still responds to red and far-red light.

**h** ***In view of the inhibiting effect of one wavelength of light, suggest a reason why light-sensitive seeds will germinate in white light, which contains all wavelengths.***

Red light is about three times more effective than far-red in phytochrome interconversion. If equal amounts of red and far-red ligh are given simultaneously, most of the pigment is converted to $P_{FR}$. (This is why we suggest that for this experiment the exposure time in far-red light should be twice that given in red light.)

*Procedure D*
This should demonstrate the ecological significance of the inhibition o germination by far-red light, since the green leaf will behave like a filte that transmits in the far-red region. (Chlorophyll absorbs red light fo photosynthesis but lets the far-red through.) It should also show tha inhibition with prolonged illumination via a leaf is not dependent on a critical temperature of 25 °C but also occurs at room temperature.

You can show that a green leaf filter is not equivalent to darkness since if a copper(II) chloride filter is placed on top of the leaf – thu absorbing most of the far-red light that the leaf would otherwise hav transmitted – the seeds will germinate.

*Results obtained in March 1981 with ambient light and temperature (variety Dandie)*

| | *Red* | *Far-red* | *Leaf 1* | *Leaf 2* | *Leaf 3* | *Leaf 4 +* $CuCl_2$ |
|---|---|---|---|---|---|---|
| *% germination* | 70 | 2 | 0 | 0 | 0 | 68 |

*Questions and answers*

**i** ***What is the effect of a green leaf filter on the germination of the two varieties?***

The leaf behaves like a far-red-transmitting filter and causes inhibitio of germination.

**j** ***Is temperature a critical factor for this effect?***

No, because the effect is seen at room temperature, which fluctuates.

**k** ***What is the ecological significance of the effect of a green leaf filter?***

Light which has filtered through a leaf canopy has an inhibiting effect on seed germination. The sharp temperature sensitivity required for stimulation of germination by light is not required for inhibition, although it does require a long exposure. This makes sense, as far-red inhibition of germination due to leaf shading would need to be independent of temperature to be of much value, and exposure would be fairly prolonged. The message to the seed is 'don't germinate in "secondhand" light – all the photosynthetically useful light has gone and you will starve to death'.
The ecological significance of this effect is important. Light-controlled seed dormancy is not of such obvious significance as dormancy controlled by temperature, but light effects seem to provide an extra factor extending the period over which seeds can germinate, an important attribute in an annual. If all the seeds germinated at the same time, there would be no reservoir of ungerminated seed in the soil. In addition, the canopy of leaves in a wood and the carpet of leaves in the herb layer act as filters to daylight, excluding red light and letting in only far-red. If a tree is felled or the carpet cleared, any dormant seeds inhibited by far-red light will respond and germinate. Also, when a light-requiring seed is buried in soil or covered by litter, it may not get sufficient light to promote germination – and will stay dormant until conditions become more favourable for seedling photosynthesis. The mechanism thus plays a significant part in the lives of such plants, preventing germination when competition for light or space is too intense for survival.

## STUDY ITEM

### 25B.1 The effect of light on the germination of *Lycopus europaeus* seeds

(*J.M.B.*)

An experiment was set up to investigate the effect of light on the germination of the seeds of the plant *Lycopus europaeus* (gipsy-wort). Five batches of seeds were initially soaked for two hours in the dark at the temperatures given in table 26. Each batch was then exposed to a different cycle of temperature fluctuation. At the same time, within each batch, half of the seeds were exposed to alternating light and dark and the other half were kept in continuous darkness. The treatments and results are shown in the table.

| *Batch treated* | *Temperature during preliminary soaking (°C)* | *Temperature cycles (°C)* | *Percentage germination: in cycles of dark/light* | *in continuous darkness* |
|---|---|---|---|---|
| 1 | 25 | 25/25 | 0 | 0 |
| 2 | 25 | 15/25 | 64 | 0 |
| 3 | 25 | 10/25 | 88 | 0 |
| 4 | 15 | 15/15 | 0 | 0 |
| 5 | 10 | 10/10 | 0 | 0 |

**Table 26**

**a** ***Give two reasons why the seeds were soaked in the dark for two hours before the experiment.***

Firstly, imbibition is required to sensitize the seeds to light; secondly, if they had been soaked in the light, then dormancy may be broken during the soaking period.

**b** ***Suggest a possible reason why there is no result showing 100 per cent germination.***

Some seeds may be dead or the conditions may still not be optimal for instance, the temperature regime could perhaps be improved, or continuous light may be better.

**c** ***Using the table of results, give two conditions which are shown to be necessary for germination.***

Alternating temperature, and light.

**d** ***No experiments were carried out in this case on the wavelength of light involved in germination. From your knowledge of other work on the effect of light wavelength on the physiology of germination, outline how you would test the hypothesis that the production of far red phytochrome ($P_{735}$) is necessary for germination.***

Alternate 10/15° temperatures, and use dim red light in place of white light. If this works well try short exposure to red light; if it still works try far-red illumination to negate the benefit of red light, by following short exposure to red light with short exposure to far-red. □

## INVESTIGATION
## 25C Early environment and later behaviour of mice

### ITEMS NEEDED

Mouse litter, newly born 2/class

Clock or watch 1/class
Lamp (40 W, 250 V) 1/class
Means of cleaning box
Mouse cage 2/class
Open field apparatus 1/class
(see *figure 50*).

(*Study guide* 25.4 'The study of ontogeny'.)

*Principles*

1 The early environment of mammals has a considerable influence on the development of behavioural patterns.
2 Studies with animals can provide starting points for a discussion of the effect of environment on child development.

Advance preparation for this practical consists of obtaining at least two litters of mice at about the same time. If possible four litters should be used, giving two control and two experimental litters. This can be accomplished if several female mice can be mated on the same night. To ensure this, use should be made of the 'Whitten effect'. It has been found that maintaining female mice in a cage together results in synchronization of their oestrous cycles, so that when they are subsequently paired, mating tends to take place on the third night after the introduction of the male. If females are exposed in a stock box to a caged male for two days prior to pairing, they will tend to mate on the first and second nights. To ensure that several are mated on the same

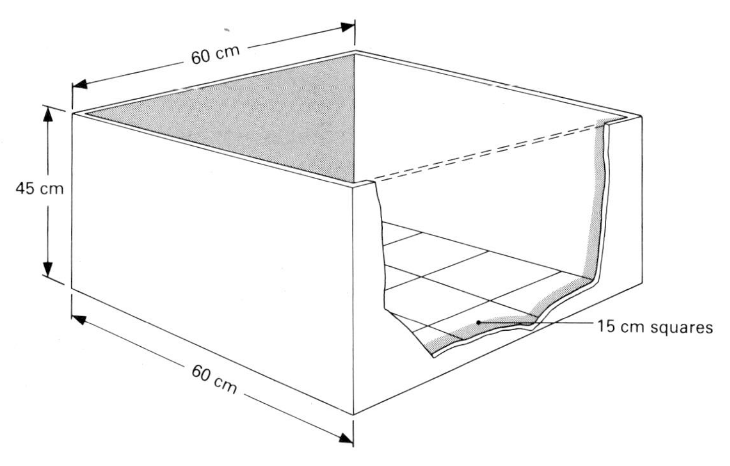


**Figure 50**
An open field apparatus. The apparatus may be built to the dimensions shown using cardboard, sheet hardboard, or plywood; the squares of the grid are numbered from 1 to 16.

night it may be wise to pair more than the number of females to be used. A mated female mouse can be recognized by the vaginal plug which is present for some hours afterwards. The parents used in the investigation must be either all of the same inbred strain or able to produce an $F_1$ litter. The female mice should have been well handled previously, to reduce the possibility of their destroying the subsequent litter.

The behaviour of the animals is assessed in terms of the amount of exploration they carry out in a limited time, and of their emotionality as measured by the amount of defecation in the same period. Emotionality is a term used by psychologists to cover such concepts as excitability, instability, timidity, and fearfulness. It is really a blanket term describing the various reactions of an animal to strange situations.

The results of this type of investigation are usually quite clear-cut. However, sometimes there is no significant difference in the defecation scores of treated and untreated mice. Fairly clear evidence is now appearing that the effects of handling mice operate through the mother. The important aspect of the experiment is that the young are removed for a time. Whether they are handled as suggested, or just isolated individually in a box, does not seem to affect the result. The hypothesis is that such mice elicit more maternal behaviour from the mother. Evidence to support this could be obtained by observing the general behaviour of mothers with both litters and measuring the amount of licking by the mother. It should be found that this is a significantly higher level in the handled litter.

*Questions and answers*

**a** ***Why are inbred strains or $F_1$ litters used?***

To reduce the amount of genetic variability to the minimum.

**b** ***Why are animals removed in a random order when repeating the 'open field' test?***

To eliminate the effect of an animal being always either the first or the last in a test series. In either event this might have significant effects upon the animal's activity.

**c** ***Why was it necessary to illuminate the arena as uniformly as possible and to reduce the noise level during testing?***

Uniform illumination ensures that the animals' behaviour is not complicated by a response to light or dark patches in the arena. Low noise levels are necessary because the animals may not respond adequately in the presence of a lot of extraneous noise.

**d** ***Apart from differences in the amount of movement of the animals, were any other aspects of their activity noticeable?***

One difference that may be observed is that some animals tend to remain at the periphery of the square, often in contact with the wall.

**e** ***What, if any, was the relationship between the 'exploratory behaviour' and 'emotionality' of the animals as measured in the investigation?***

The animals with the highest activity should give the lowest defecation scores.

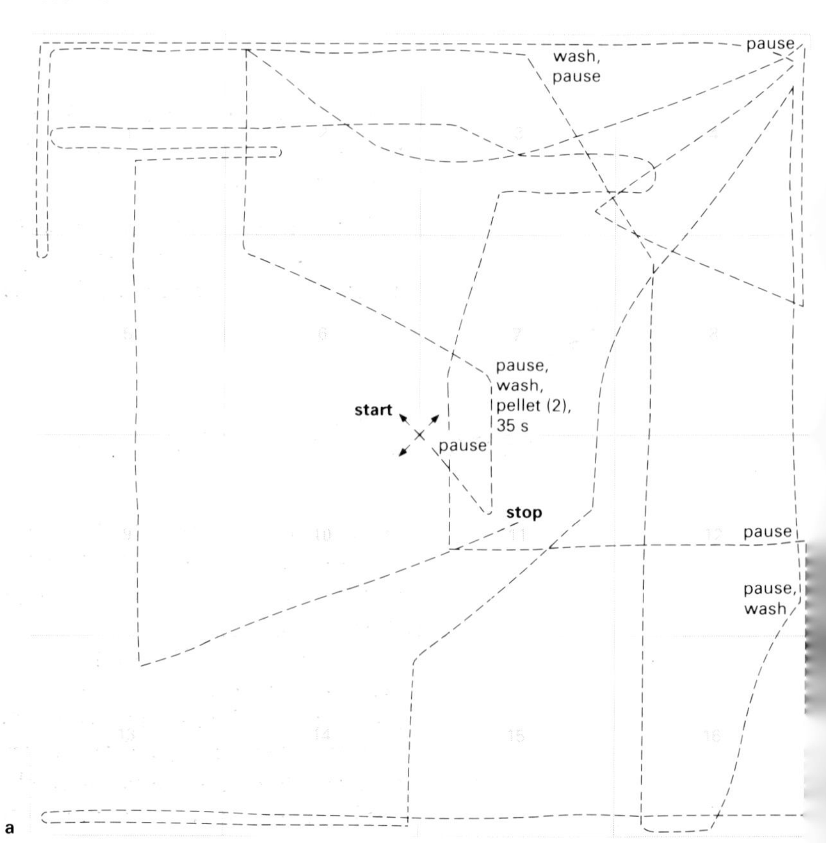


**Figure 51**
The behaviour of mice in an open field apparatus: an example of results.
**a** A mouse from a handled litter.

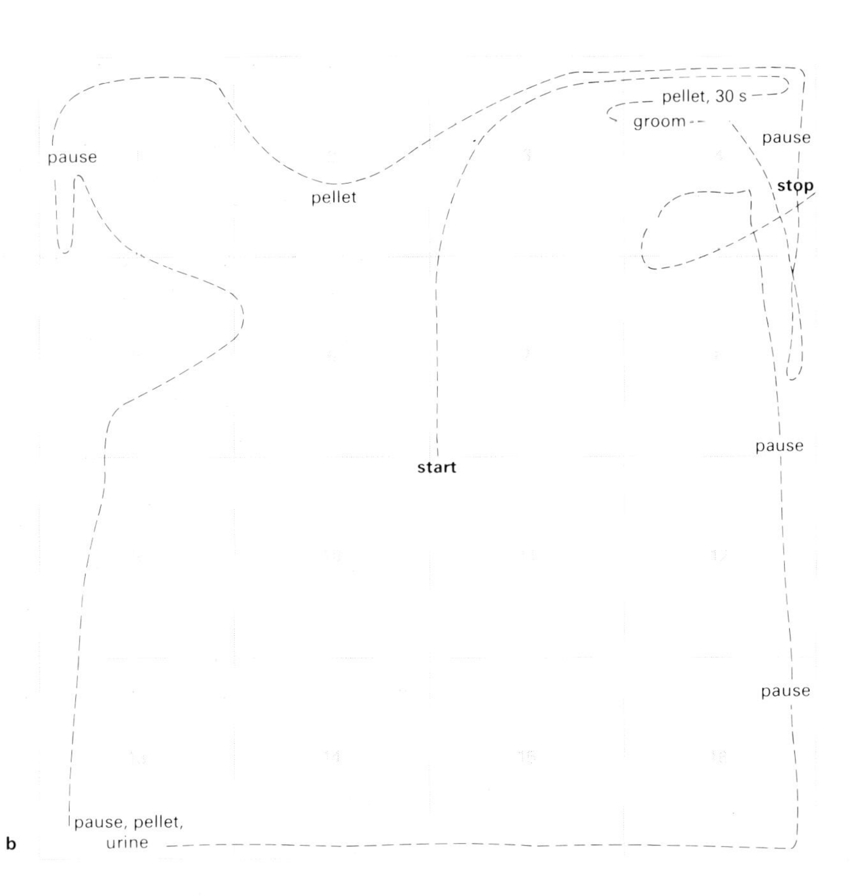


**b** An unhandled mouse.

**f** ***If you found a significant difference in the behaviour of the two groups, how easy was it to assign an individual animal to one or other category?***

This will depend on how different the behaviour of the two groups was found to be. Often it is possible to place an animal in one or the other group merely by observing it in its cage.

# PART III Bibliography

BUTLER, W. L. and DOWNS, R. J. 'Light and plant development'. *Scientific American* offprint No. 107. (An account of the discovery of phytochrome and its role in the response of plants to a change in the length of night.)

DENENBERG, V. H. 'Early experience and emotional development'. *Scientific American* offprint No. 478. (An account of the author's investigation with infant rats on early social experiences and later emotional behaviour, and on mother–infant relationship.)

HARLOW, H. F. 'Love in infant monkeys'. *Scientific American*, **200** (6), 1959. (An account of the author's studies of young rhesus monkeys.)

HARLOW, H. F. and HARLOW, M. K. 'Social deprivation in monkeys'. *Scientific American*, **207** (5), 1962. (This looks at the way peer relations influence the development of infant monkeys.)

HINDE, R. A. and HINDE, J. S. Carolina Biology Readers No. 63, *Instinct and intelligence*. 2nd edn. Carolina Biological Supply Company, distributed by Packard Publishing Ltd., 1980.

# PART FOUR ECOLOGY AND EVOLUTION

In references to figures and tables, **'S'** denotes the **Study guide**.
**'P'** refers to the **Practical guide**.
Example: 'figure (S)2'.

□ denotes the end of a Study item.

# CHAPTER 26 THE ORGANISM AND ITS ENVIRONMENT

*A review of the chapter's aims and contents*

1 The chapter introduces Chapters 26–29, which are concerned with the ecology of organisms. Some terms are defined.
2 The complexity of interactions between species is emphasized.
3 Modern ecology is an experimental subject. Ecologists often perform field and laboratory experiments to test hypotheses.
4 The factors which influence the local and geographical distribution patterns of organisms are introduced. They are loosely classified as physical and chemical on the one hand, and biotic on the other.
5 This chapter concentrates upon physical and chemical factors in water, soil, and air.
6 The distribution patterns of organisms in water are influenced in particular by oxygen concentration and temperature.
7 Soils are formed by the breakdown of rock and are utilized by plant roots for ion and water uptake. Earthworms can enhance soil fertility. Ions in the soil have a major influence upon plant distribution patterns.
8 Relative humidity and temperature may influence the distribution patterns of animals and plants.
9 The interactions of physical and biotic factors are illustrated by their effects on the zonation on rocky shores.

## PART I The *Study guide*

### 26.1 Introduction to ecology

**Practical investigation. *Practical guide* 7, investigation 26A, 'A qualitative study of a community'.**

*Principles*

1 Ecologists do not merely identify organisms and study their natural history, although a background knowledge of natural history is essential. Ecologists try to detect patterns in nature. They ask questions and test their ideas by controlled experiment.
2 Ecology has considerable economic importance.
3 The interactions between species in nature are complex.

## STUDY ITEM

### 26.11 The chestnut-headed oropendola bird

*Assumptions*

1 That students have the ability to calculate a $\chi^2$ value from a $2 \times 2$ contingency table.

2 An understanding of the meaning of the probability value generated by the significance test.

3 An understanding that many fly species parasitize animals. The females lay their eggs in the animal's body, and the fly larvae hatch and eat the host tissues before pupating.

*Principle*

1 The distribution patterns and behaviour of a species in nature may be influenced by the presence of other species.

This study is based on an investigation by Neal Smith (Smithsonian Tropical Research Institute, Panama Canal Zone), beautifully described by Ricklefs (1976 and 1980). In order to sample oropendola nests up to 17 m above his head, Smith stood on a 5 m ladder at night, balancing 15 m of extendable aluminium poles with instruments at the tip. He cut off each nest, lowered it to the ground, examined it, and then replaced it, using contact tape, contact glue, or rat traps to replace the nest in its former position.

*Questions and answers*

a ***Suggest why bot-flies did not occur in discriminator colonies. How could you test your hypothesis?***

The discriminator birds could have eaten them. They could have been eaten by wasps or bees. Wasps and bees could have deterred the bot-flies from entering the oropendola colonies.
It is difficult to test these hypotheses in the field because bot-flies are mobile. The faecal pellets of parent and offspring oropendolas, and the ground beneath their nests, could be examined for the remains of flies. Similarly, fly wings could be sought in areas around wasp and bee nests. At the edges of discriminator and non-discriminator colonies, counts could be made of the frequency of bot-fly flights into the area. These might confirm whether the wasps and bees deter bot-flies, or whether invading bot-flies are eaten once they invade the colony.

b ***Apply the $\chi^2$ test to the data in table 50 and work out the probability of obtaining these results by chance if the presence of bot-flies had been independent of the presence of cowbirds. Discuss these results and suggest a hypothesis to explain them.***

For a $2 \times 2$ contingency table:

$$\chi_1^2 = \frac{[(ad - bc)^2 n]}{[(a + b)(c + d)(a + c)(b + d)]}$$

where a, b, c, and d are the entries in the table. In the contingency table, a is diagonally opposite to d, b is diagonally opposite to c, and (a + d) etc. are the marginal totals. In this case:

$$\chi^2 = \frac{[(57 \times 42) - (382 \times 619)]^2 \times 1100}{439 \times 661 \times 676 \times 424} = 724.6 \; (P < 0.001)$$

The presence of cowbirds is very strongly associated with the absence of bot-flies. The absence of cowbirds is strongly associated with the presence of bot-flies. Cowbirds may eat bot-fly adults and larvae when the flies attack oropendola chicks.

**c** ***Explain why, in colonies built a long way from wasp or bee nests, it was advantageous to the oropendolas not to discriminate between their eggs and those of cowbirds.***

Oropendola colonies a long way from wasp or bee nests might be heavily infested with bot-flies. Bot-fly larvae may parasitize and kill the chicks. The cowbirds in the nest eat bot-flies. In this case it is advantageous for the oropendolas to be non-discriminators, allowing the adult cowbirds to lay eggs in their nests.

**d** ***In non-discriminator colonies the cowbirds were ignored, but in discriminator colonies they were chased away by the oropendolas. Is this consistent with your hypothesis?***

Yes. The behaviour of the oropendolas in a colony probably depends on their perception of the local abundance of bot-flies. Discriminators will only accept a cowbird egg if it is laid soon after an oropendola egg and mimics it. Female cowbirds enter the oropendola colony alone and with care. In non-discriminator colonies a female cowbird lays two to five eggs over several days. Cowbirds enter non-discriminator colonies in aggressive groups. □

## Ecological terminology

*Principle*

1 The distribution patterns of the main types of terrestrial plant communities can be related to environmental factors.

Some more ecological terms are introduced and defined because they are fundamental to Chapters 26–29.

Students should be aware of the distinctions between the words 'habitat', 'microhabitat', and 'niche'. A simple discussion of the niche concept is provided in Chapter 2 of King, 1980. (see the bibliography to this chapter).

## STUDY ITEM

### 26.12 Biomes; some examples of vegetation types which extend over large areas of the Earth's surface

*Principles*

1 A holistic view of the biome types on Earth is required, to provide a perspective, before single species are considered in Chapters 26–28.
2 Boundaries between community types are frequent.
3 Ecologists do not know all the answers!
4 Fire can be an important ecological factor.

Students should be encouraged to ask themselves how they might test the hypotheses that they suggest in answer to this Study item. Other types of biome (sea, desert, freshwater, estuaries, etc.) should be mentioned. Their distribution can be related to temperature and the availability of water (*figure 52*).

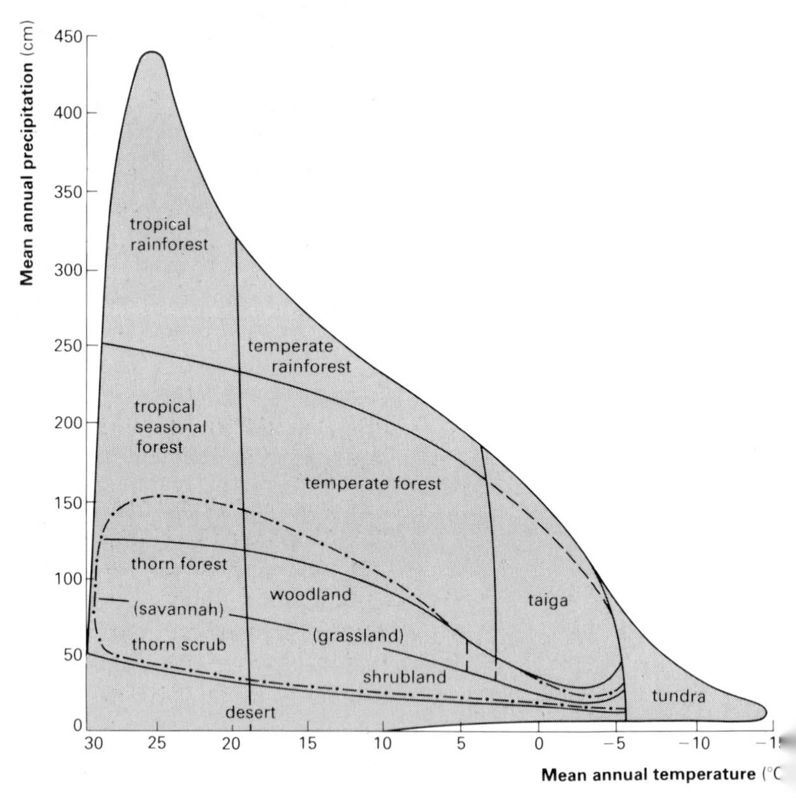


**Figure 52**
The relationships of biome types to precipitation and temperature. The boundaries are approximate since they can be altered by a maritime climate, fire, and many other environmental variables.
*Based on Whittaker, R. H.,* Communities and ecosystems, *Macmillan, 1970.*

*Questions and answers*

a **Where the boreal forest (figure [S]286) meets the tundra (figure [S]285) there is often a distinct boundary, known as the 'tree-line'. Discuss the factors which may prevent the trees from occurring in the tundra zone, and suggest why most of the tundra species do not grow beneath the trees.**

Tundra species would be shaded or covered with leaf litter beneath the trees. Amongst reasons why the trees cannot grow further north are:

**1** Their buds are above the soil surface in the coldest season and are killed by low temperatures or the ice (rime) which accumulates on them.

**2** The boundary is reinforced by the grazing of tundra animals close to the trees, preventing seedling invasion.

**3** Tundra soils are often frozen for most of the year (permafrost). Tree roots may not be able to penetrate it.
**4** Fire, peat bogs, or shallow soils may also contribute to the boundary in places.

There have been few convincing investigations of the tree line. The boundary is discussed in Pruitt, 1978.

**b** ***Discuss the various factors which cause the differences in tree frequency and species composition between tropical rain forest* (figure [S]284) *and tropical savannah* (figure [S]287).**

Tropical savannah develops in zones which have less rainfall than tropical rain forests (*figure 52*). Nevertheless, one would expect secondary succession to cover the area rapidly with trees. The invasion of shrubs and trees is prevented by grazing and fire. The sparse trees are fire-resistant. □

## 26.2 Making and testing hypotheses

*Principle*

**1** There are four types of ecological investigation. They are field observation, natural field experiments (such as the colonization of a new island), field experiments in which the treatments are decided by the investigator, and laboratory experiments. Ecologists often use a combination of these approaches to test their ideas and to establish new principles.

### STUDY ITEM

### 26.21 Determining a causal relationship (multiple choice)

*Questions and answers*

**a** ***If you observed a general agreement between the distribution of a species and a particular environmental factor, you might consider using one of the following methods to establish that there is a causal relationship between them. Which would be the best method to employ?***

***A A controlled experiment in the natural environment with all the factors held unchanged except the one under investigation.***
***B An experiment with the species living in isolation from its natural environment under controlled conditions.***
***C Both (A) and (B).***
***D Neither (A) nor (B), but applying statistical methods to a large number of observations.*** *(J.M.B.–modified)*

**b** ***Give the reasons for your choice.***

The correct answer is C. Alternative A is unsatisfactory because altering one factor on its own in the field often alters many others. For example, fencing a plot to prevent grazing may alter its microclimate. Alternative B is unsatisfactory on its own because in the laboratory biotic factors are removed and the climate and soil outside are difficult to duplicate, even with growth chambers. Statistical methods, as in D, can never establish causal connections – correlation does not imply causation. Combined laboratory and field approaches are desirable. □

## STUDY ITEM

## 26.22 The formation and testing of hypotheses

*Principles*

1 Ecologists investigate phenomena by field observations and experiments.
2 One experiment suggests another.
3 Inter-specific competition may influence the distribution patterns of a species in nature.
4 Closely-related species may utilize the environment in different ways spatially or temporally. This minimizes competition between them.

This example was selected because of its simplicity. There are four investigations, each based on the previous one and designed to answer a simple question.

*Questions and answers*

**a** ***Discuss these results. Do you think that the ants and rodents are competing for seeds? State the evidence for your answer.***

When either ants or rodents were removed, there was a statistically-significant increase in the numbers of the other group of seed 'predators'. This is expected if the ants and rodents were competing. When the rodents were removed, the ants ate as many seeds as the ant and the rodents together in the plots to which both had access. Similarly, when the ants were removed, the rodents' food intake was equivalent to that of the ants and rodents combined in the control plots. When both groups were removed, seed production increased. In the control plots, in which both ants and rodents occurred, both groups were fewer in number, and ate less, than in the plots in which one group was present and the other was absent. The ants and the rodents were competing. The results strongly suggest that they were competing for seeds.
Students may justifiably point out that the experimenters should have set out more than two plots per treatment to make the results of statistical tests more precise. On the other hand, the practicalities which often limit the scale of field experimentation (money, time, and labour) need to be stressed.

**b** ***Devise another experiment which might confirm your conclusion.***

Using the same plots, ants could be allowed to re-invade either the plot with rodents only, or the plot from which both groups had been removed. In the former case the numbers of rodents should decline, and in the latter, the seed density should decline. The equivalent experiment could be performed with rodents.

**c** ***State two conclusions that can be drawn from the data in the graph.***

**1** Different ant species tend to collect different and characteristic sizes of seeds.
**2** The longer the body of the ant species, the larger the average seeds which it collects.

The graph gives no indication of the range of seed sizes which each species carries. □

## 26.3 Factors which influence the distribution patterns of organisms

**Practical investigations. *Practical guide* 7, investigation 26B, 'The effects of different water regimes on plant growth', investigation 26C, 'The relationship between nettle distribution and soil phosphate', and investigation 26D, 'The autecology of woodlice'.**

*Principles*

1. Environmental factors which affect species distributions can be regarded as physical or chemical on the one hand, and biotic on the other.
2. The ability of an organism to tolerate certain physical and chemical factors determines its potential geographical range. Each species is usually restricted by biotic factors to only part of that range.
3. It should be possible to define one major factor which limits the distribution of each species. If this factor is altered, the range of a species should expand or contract when another factor begins to limit its geographical pattern.

### STUDY ITEM

### 26.31 Buttercups (experimental design)

*Assumption*

1. An understanding of the principles of experimental design – randomization, replication, and controls.

*Question and answer*

**a** ***It was observed that three species of* Ranunculus *(buttercups) were growing in a ridge and furrow grassland liable to flooding.* R. bulbosus *occupied the ridge tops,* R. acris *the ridge slopes, and* R. repens *the bottom of the furrows. How would you test the hypothesis that this distribution of buttercups is determined by the amount of water in the soil?*** *(J.M.B.)*

J. L. Harper investigated the patterns of these buttercup species in a laboratory experiment. He sowed seeds of the three species separately in pots in which the water table was maintained at different heights. The species responded as expected (*figure 53*). *Ranunculus repens*, for example, established best in the wetter conditions, and *Ranunculus acris* and *R. bulbosus* hardly established seedlings at all on waterlogged soil. This experiment, however, is too simplistic. Most effective

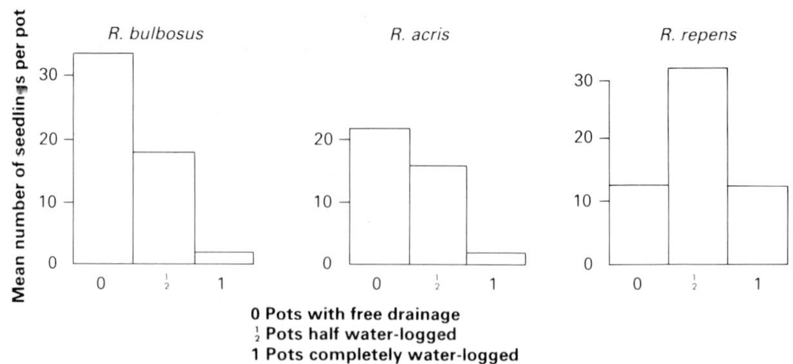


**Figure 53**
The early establishment of three species of buttercup sown in pots with the water table maintained at different levels. Fruits were sown on 26 June 1953 and counts made on 28 July 1953.
*Based on Harper, J. L., Sagar, G. R., 'Some aspects of the ecology of buttercups in permanent grassland'.* Proceedings of the British Weed Control Conference, *256, 1958.*

reproduction in *R. repens* is by stolons, not by seed. Competition between the species ought to be considered and investigated. It could be, for example, that *R. repens* could grow anywhere, but was excluded from the ridge tops and slopes by the other two species. Water *per se* may not be important, but a waterlogged soil is higher in hydrogen sulphide, iron(II), and methane, and lower in oxygen, than a dry one (see Etherington, 1983). Laboratory experiments to test each of these factors in turn will appear in a good answer. Grazing, and the appearance of worm casts fit for seed germination, may also have an influence. □

## 26.4 Aquatic factors

*Assumption*

1 A knowledge of the main properties of water, such as specific heat, evaporation, and meniscus.

*Principles*

1 Water is quite different from air as a medium for organisms to live in.
2 Oxygen is not very soluble in water, yet is essential for many aquatic organisms. Its concentration sometimes limits the distributions of water organisms.
3 Many aquatic organisms are also sensitive to temperature.

### STUDY ITEM
### 26.41 The influence of changes in river water

*Assumptions*

1 The solubility of any gas dissolved in water decreases with an increase in water temperature.
2 Fluctuations in pH in the same body of water are usually due to fluctuations in the concentration of carbon dioxide dissolved in it.

*Principles*

1 Diurnal fluctuations of oxygen and carbon dioxide are caused by daily variations in photosynthesis and respiration.

2 Environmental factors interact in subtle ways, which must be taken into account when their influence on the distribution of organisms is considered.

*Questions and answers*

a ***Explain why fluctuations in pH can be used as an indication of changes in the concentration of carbon dioxide dissolved in water.***

Dissolved carbon dioxide forms carbonic acid ($H_2CO_3$). This weak acid dissociates to form protons ($H^+$), which acidify the water, and hydrogen carbonate ions ($HCO_3^-$). The pH is $-\log_{10}[H^+]$, where $[H^+]$ represents the concentration of hydrogen ions. When $[H^+]$ increases, the pH decreases.

b ***Describe and explain the changes in graphs A and B over a 24-hour period.***

From midnight to dawn, the oxygen concentration decreases and the carbon dioxide level in the water increases. During the day the oxygen level increases and the carbon dioxide level falls. In the afternoon and evening both concentrations reach a plateau. Then, as evening falls, the oxygen level declines rapidly as the carbon dioxide increases slowly. At night both plants and animals are respiring, absorbing oxygen from the water and producing carbon dioxide. During the day the rate of plant photosynthesis exceeds the combined respiration of plants, animals, and decomposers. Plants rapidly remove carbon dioxide from the water and produce oxygen.

c ***What is the significance of the changes in graph C?***

Temperature influences the maximum concentration of the gas which can dissolve in water. The higher the temperature, the less gas can dissolve.

d ***In what way do the changes in graph C influence your interpretation of graphs A and B?***

The temperature fluctuations are small (3 °C), but it is possible that they might influence the solubility of gas in water. Temperature changes are only relevant to a gas when the water is saturated with it. Then a temperature increase might reduce gas concentration or a temperature decrease might increase it. During the day the higher water temperature may have reduced the volume of oxygen which could dissolve in the water. In this case the graph underestimates oxygen production in daylight. There is no information about saturation with carbon dioxide. However, the carbon dioxide concentration is highest between 0500 hours and 0800 hours and its decrease may partly be the result of increasing water temperature. □

## The restoration of the tidal Thames

*Principles*

1 Sewage pollution has had a considerable influence on the fauna of waterways.

2 Many fish, in order to survive, require water reasonably high in oxygen.

3 Sewage pollution reduces the oxygen levels in water.

The tidal Thames is used as an example because it is so well documented. The dissolved oxygen and the fish fauna of the Thames have been monitored since 1883 and there are many records before that time. In two cases a spell of considerable pollution was followed by a period when the river was clean. The fauna changed accordingly.

## STUDY ITEM

### 26.42 The distribution patterns of some caddis-fly nymphs in the river Usk in Wales

*Assumption*

1 An understanding that oxygen is required for the oxidation of foodstuffs in aerobic respiration. The higher the oxygen uptake, the more food is being respired.

*Principles*

1 The geographical distributions of some species can be related to their environments and their physiological responses to them.

2 Closely-related species may differ physiologically in ways which determine their relative success in different habitats.

*Questions and answers*

a ***Write down, from the respiration rate graph for each species, the temperature at which a respiratory rate of 1 mg $O_2$ $g^{-1}$ dry mass per hour occurs. What does this result indicate?***

14.5 °C for *Diplectrona felix*, 16 °C for *Hydropsyche instabilis*, and 21 °C for *Hydropsyche pellucidula*.
The part of the vertical axis below an oxygen consumption of 1 mg $g^{-1}$ dry $mass^{-1}$ $hour^{-1}$ covers the zone in which the species have particularly low respiration rates. This occurs at temperatures up to 14.5 °C in *Diplectrona felix*, up to 16 °C for *H. instabilis* and up to 21 °C for *H. pellucidula*.
We do not know how food intake varies with temperature. Assuming, however, that food intake is similar at all temperatures a slowly respiring organism devotes more of its energy to growth and less to respiration than a rapidly respiring one. Thus, according to the laboratory results, *H. instabilis* can thrive at higher temperatures than *D. felix*, and *H. pellucidula* can thrive at a higher temperature still. Their metabolic adaptations differ. The temperatures at which the species are habitually found in nature correspond – *D. felix* occupies the coldest water and *H. pellucidula* the warmest (table (S)54). These results do not prove that the metabolic adaptations of the species influence their distribution patterns, but they are suggestive.

**b** ***At what temperatures do the flattest parts of the graphs occur in 1* Diplectrona felix *and 2* Hydropsyche instabilis? *What is the importance of these flat portions on the graph?***

**1** 5–10 °C
**2** 5–15 °C.

Most enzyme-controlled reactions, such as cellular respiration, more than double with each 10 °C rise in temperature. In these species, over these limited temperature ranges, re. piration hardly increases with temperature. This might give each species a competitive advantage in its own particular river section.

**c** **Hydropsyche pellucidula** ***has a lower respiration rate at all temperatures than the other two species. Suggest what prevents it from surviving further towards the source of the river.***

**1** Temperature fluctuations are too intense.
**2** The range of potential food species is different.
**3** The presence in the upper reaches of a predator which prefers it to the other species.
**4** Competition for space or food with other species.
**5** Adults only hatch, mate, and lay eggs in the warmest air or water.
**6** *H. pellucidula* is swept away by fast-flowing streams.

**d** ***Discuss critically whether temperature is likely to be a major factor which limits the distribution patterns of these species.***

At the lower end of the temperature range (water temperatures of 5–10 °C) the respiration rates of the species do not differ much. It seems unlikely that *H. instabilis* and *H. pellucidula* are prevented from living upstream by low temperatures.
At the upper end of the temperature range (water temperatures 10–20 °C), however, *D. felix*, confined to headwater tributaries (*figure (S)293*), might not be able to live in the warmer water downstream, because its respiration rate would be so high.
Fluctuations in temperature also need to be considered.
Temperature seems unlikely to affect the distribution of *H. instabilis*. Water temperatures will only rarely reach 20 °C and below that temperature the respiration rates of the two *Hydropsyche* species hardly differ.

**e** ***Describe further experiments to test your hypotheses.***

In the field, net cages containing marked individuals of all three species alone and together could be lowered into the water in all three habitats. Each individual could be weighed at intervals. This would not test a specific hypothesis, but would indicate whether each species was able to survive in the others' habitats in the presence or absence of competition.
In the laboratory, the responses of the three species to temperature fluctuations equivalent to those in the field could be tested. The numbers of feeding episodes per hour could be recorded for each

species by placing individuals in water at known temperatures for 24 hours with a known number of prey organisms. □

## 26.5 Soil factors

*Principles*

1. Soil is a complex mixture, teeming with life, in which plants grow, with their roots taking up water and ions (see *Study guide I*, Chapter 8).
2. Plant roots grow rapidly through soil and encounter new sources of nutrient ions. Their root hairs and cortical cells absorb these from the soil solution by active transport.
3. Earthworms can enhance soil fertility.
4. Soils differ from place to place, both in their profiles, which reflect the history of the soil, and in their ionic content.
5. The ions available in the soil affect the range of plant species which it supports. Plants differ in their responses to ions and this is reflected in their local and geographical distributions.

General discussions of soil and its inhabitants appear in Russell (1957) and Leadley-Brown (1976). Etherington (1978) concisely describes the effects of soil variation on plant growth.

### Nutrient ion uptake by plants

*Assumptions*

1. Knowledge that elements often occur in nature in ionic form.
2. A knowledge of root structure and its anatomy in cross-section (see figure (S)296.

*Principles*

1. Some elements are essential to plants.
2. Some elements are taken up from the soil in much greater quantities than others.
3. Plant roots grow rapidly.
4. Ion uptake by plant roots is an active, energy-requiring process.
5. The ions enter the xylem and move up the plant in the transpiration stream.

Ion uptake is discussed by Epstein, 1972 and 1973, and Sutcliffe and Baker, 1981. Baker, 1981, describes the mechanisms of ion uptake across cell membranes. (See the bibliography.)

The mechanism of ion uptake across the cell membrane probably involves a proton pump and proteinaceous carrier molecules. It seems likely that the cell uses ATP, derived mainly from aerobic respiration, to pump protons ($H^+$ ions) from the inside to the outside of the membrane. This creates a potential difference across the membrane, with the inside of the cell being more negative than the outside. In the cells of the giant alga *Nitella*, this potential difference is about $-140$ mV. It provides the energy which can stimulate an ion to pass from the outside to the inside of the membrane.

Once the ions enter the cells they probably migrate from cell to cell via the plasmodesmata. Studies with the electron probe, which emits X-rays and records the characteristic X-ray frequency emitted by the ions of each element, suggest that ions become concentrated in the pericycle cells immediately surrounding the xylem vessels. If this is so, the active accumulation of ions by these cells creates a concentration gradient down which ions move towards the centre of the root. Presumably they move across the cell membranes of these inner stelar cells by a reverse carrier mechanism, into the walls of a xylem vessel. Then they are carried passively up the plant in the transpiration stream. Since the transport of water across the cortex is largely passive and occurs in the wet cell walls, living cells being unnecessary, water and nutrient ion uptake by plants are largely independent processes.

*Questions and answers*

a **_Discuss the assumptions underlying this experiment._**

The roots stood in the radioisotope solution long enough for an equilibrium to be reached between the concentration of the ions in the cell walls and the solution. When the roots were blotted dry any radioactive solution was removed. When the roots were placed in the de-ionized water, the radioactive ions leaked out of the cell walls but not out of the cells themselves.

b **_Explain the results in_ figure [S]297.**

In the first phase of the curve the rate of ion uptake is the same at both temperatures. This suggests that a physical process is involved (the rate is proportional to temperature above absolute zero, $Q_{10}$ = just above 1) rather than a chemical one (the rate more than doubles for each 10 °C rise in temperature). This part of the curve represents diffusion of the ions into the cell walls and inter-cellular spaces of the root cortex. Ion uptake by the cell membranes occurs rapidly at 20 °C but not at 2 °C because the respiration rate at 2 °C is so low.

## STUDY ITEM

### 26.51 Earthworms and soil fertility

*Principle*

1 Earthworms can enhance soil fertility.

*Questions and answers*

a **_Explain these three effects on the basis of the normal activities of earthworms._**

Some earthworm species eat leaves, dragging them down into their burrows. A layer of leaves reduces evaporation from the soil and may contribute to waterlogging. The soil may not drain freely without the channels created by worms. Waterlogging may reduce the rate of oxygen supply to apple tree roots and reduce their growth rate. In this case they may encounter fewer nutrients and the growth of the whole plant will slow down. Some of the essential elements may be abundant in the waterlogged leaf layer, but in the absence of earthworms they are only released slowly for plant growth.

**b** ***Discuss these results in detail. In what ways do the earthworms affect the growth of the winter wheat?***

The small number of earthworm casts in treatment C shows that fungicide killed the worms, which were sparse anyway (treatment D). The addition of earthworms obviously enhanced earthworm activity in the soil (treatments A and B).
The deep-burrowing species were most effective at removing straw from the surface. This straw, when incorporated into the soil, would enhance its content of humus and nutrient ions. The deep-burrowing worms also created channels down which roots grew. Compared to the control treatments C and D, treatment A produced more grain and four or five times the root growth. The extra root growth is a mixed blessing: the plants were better anchored and had more access to water and nutrients, but they had to support a greater mass of respiring tissue.
The shallow-burrowing earthworms were less effective. They slightly enhanced grain production and root growth.

**c** ***Why was this?***

The vertical holes had been created only at specific depths in each box. Earthworms, on the other hand, could burrow at all levels in the soil, thus improving root growth all the way down the profile. An alternative explanation is that earthworm holes, as distinct from artificial ones, were lined with available nutrients from their mucus and egesta.
The artificial vertical holes promoted barley growth more when created in the upper soil layers. This laboratory result conflicts with the results from the field experiments, in which deep-burrowing earthworms greatly enhanced root growth and grain production compared with shallow-burrowing species. The deep-burrowing species may have influenced crop growth more by the incorporation of straw into the soil, or by the burrowing more frequently near the surface (note the greater number of casts in treatment A than in treatment B) than by burrowing at depth. Notice, however, that the field experiment was performed on wheat and the laboratory experiment on barley. □

## The Broadbalk wilderness

*Assumptions*

1 An understanding of the roles of photosynthesis and respiration in the carbon balance of plants.
2 An elementary knowledge of the nitrogen cycle.

*Principle*

1 The humus in the soil contains carbon and nitrogen and is continually being replenished and broken down.

*Question and answer*

**c** ***Explain in as much detail as you can how the soil has accumulated nitrogen and carbon in such quantities.***

In the early stages of succession from bare soil to woodland net primary production by plants above and below ground exceeds the rate of decomposition. The ecosystem accumulates organic carbon and nitrogen compounds in trees, litter, and roots. Litter and humus decomposition is slow because of the toxic compounds in litter and the low temperatures at the soil surface. The quantities of litter deposited are considerable. In the 82-year-old stand of *Picea abies* mentioned in *Study guide II*, Chapter 27, Study item 27.81, for example, 2.2 tonnes $ha^{-1}$ of dry mass are being deposited each year in litter. The question above covers 83 years of the Broadbalk wilderness. Ultimately, in a mature forest, the carbon and nitrogen contents of the soil should reach a plateau.
The carbon is derived from carbon dioxide trapped in photosynthesis. The nitrogen comes from precipitation, which includes nitrates produced as a result of lightning discharges in the atmosphere, and the fixation of nitrogen gas by micro-organisms. Ultimately the trees will absorb these nitrogen atoms as nitrate or ammonium ions (see Study item 29.42).

**Practical investigations. *Practical guide* 7, investigation 26B, 'The effects of different water regimes on plant growth', and investigation 26C, 'The relationship between nettle distribution and soil phosphate'.**

### The preferences of some species for basic or acidic soils

*Assumption*

1 An understanding of the pH scale.

*Principles*

1 Many species are restricted in nature to soils of a narrow pH range.
2 These distribution patterns may be reinforced by competition.
3 Basophilous and acidophilous species are restricted to soils of a particular pH by the effect of pH on the solubility of other ions in the soil water.
4 The concentrations of various ions in the soil limit the natural distribution patterns of many plant species.

## 26.6 Air factors

*Principles*

1 Desiccation prevents many terrestrial organisms from living in habitats in which they might suffer evaporative stress.
2 Many organisms require a certain minimum temperature for reproductive development.

**Practical investigation. *Practical guide* 7, investigation 26D, 'The autecology of woodlice'.**

## STUDY ITEM

### 26.61 The distribution pattern of the stemless thistle (*Cirsium acaule*).

*Assumptions*

1 An understanding that in general the summers are colder towards the north than towards the south of Britain.

2 A realization that when water evaporates from an organism the organism cools.

*Principle*

1 Climate can affect the distribution patterns of plant species.

This example is discussed by Pigott (1969 and 1970) and is included here because of its experimental approach and its simple, clear-cut results.

*Questions and answers*

a ***Account for each of these five observations and construct a hypothesis to explain them.***

**1** The more successful fruit development in southern Britain suggests a climatic cause, unless a fruit 'predator' or parasite increases in abundance further north.

**2** South-facing slopes are warmer. North-facing slopes are colder and moister. South-facing slopes should have more pollinators, and north-facing slopes perhaps a greater incidence of fungal infection.

**3** Flowering time probably depends on the length of the night. Many species require nights longer than a critical length to flower. A certain night length will be achieved later in the season in the Pennines than in the south of Britain. The later the flowering time, the lower the average temperature at flowering, and the higher the relative humidity.

**4** The heads were artificially pollinated to ensure a high seed set. The polythene bags would have increased temperatures around the flowering heads. They might also have increased relative humidity, stopped rain from falling upon the fruits, and prevented fruit 'predators' from reaching the fruits.

**5** When solar radiation began to shine on the sprayed side its energy would initially have been used to evaporate water. The evaporation of water would have cooled the fruits. Not until all the water had evaporated would the fruits warm up on that side. On the dry side, however, the fruits would warm up much more rapidly.

*Hypothesis:* The higher the average temperatures around the fertilized ovaries as they ripen, the greater the chance that they will develop into viable fruits. Warm, dry days are more likely to occur in southern Britain in August than in the Pennines in September, and on the south-facing than the north-facing slopes in the Pennines.
Other contributing factors may be that

**1** dry, warm days favour pollinating insects and

**2** cool moist days favour fungi such as *Botrytis cinerea*, which infect the flowering heads and reduce seed set.

b ***Suggest some different experiments to test your ideas.***

Small heaters could be installed in some heads but not others and their effects on fruit set determined. Flowering stemless thistle plants, artificially-pollinated, could be potted and taken to laboratory growth rooms where different average temperatures could be simulated and fruit production monitored.
Fruit set, and the incidence of fungal infection, could be investigated on south- and north-facing slopes in southern Britain. □

## 26.7 Zonations

*Principles*

1 Zonations of plants and animals often occur in nature.
2 In some cases the zones move, and reflect a succession of species.
3 In other cases, such as a rocky shore, the zones are static.
4 The environmental fluctuations causing zonations are complex but can be investigated by experiment.

### Zonation on rocky shores

*Assumption*

1 A general understanding of tidal cycles on the coastline.

*Principle*

1 Biotic factors may considerably influence the patterns of species in nature.

Note that *Chthamalus* is a Mediterranean species, whereas the centre of distribution of the balanoids is in the arctic. In dry conditions *Chthamalus* keeps its opercular plates closed, and respires anaerobically, with the production of lactic acid.

*Questions and answers*

a ***List the differences in environment between the* C. stellatus *zone and the* B. balanoides *zone.***

The *C. stellatus* zone is further up the rock face than the *B. balanoides* zone. The *C. stellatus* zone is covered by sea water at high tide less frequently and receives less splash water. It is exposed to greater fluctuations in temperature, higher mean temperatures, higher maximum temperatures, lower relative humidities, and greater wind speeds than the *B. balanoides* zone. In the lower zone, the activity of organisms would be greater. There will be more predation, food, and fouling by algae. The pattern of predation by birds is an unknown quantity.

b ***In an experiment* Balanus *larvae were presented with slowly-rotating slates, half of which had been thoroughly cleaned and the rest of which had been treated with* Balanus *extract. One hundred larvae settled on the clean slates, but 2197 established themselves on the treated ones. What conclusions can you draw?***

*Balanus* larvae are more likely to settle on rocks where *Balanus* already lives, than on areas such as newly-bared rock, which produces no olfactory stimuli. Since the exoskeletons of crabs and cockroaches are almost as effective as *Balanus* extract in promoting settling of the *Balanus* larvae, it seems most likely that the substance which promotes settling is an arthropod cuticular protein.

## PART II The *Practical guide*

The investigations related to this chapter appear in *Practical guide* 7, *Ecology*.

### INVESTIGATION
### 26A A qualitative study of a community

**ITEMS NEEDED**

Access to a restricted habitat: pond, stream, or artificially-constructed ecosystem, such as an aquarium

Books of suitable keys for identification of specimens
Dishes, white, such as photographic dishes, for sorting specimens 1/group
Hand lenses
Microscope, binocular 1/group
Microscope, monocular 1/group
Nets 1/group
Petri dishes and other suitable containers for specimens
Pipettes (wide diameter), dropping, for collection of specimens 1/group
Spoons, plastic, for collection of organisms 1/group
Viewing tubes

(*Study guide* 26.1 'Introduction to ecology' and 26.2 'Making and testing hypotheses'.)

*Principles*

1 The investigation of organisms must begin with the observation of living things in their environments. Careful observation and recording almost always results in the formulation of questions which need to be answered by devising experiments.
2 Students must learn to name organisms and to use keys, so that they can consult books and articles to obtain more information about the habits of the organisms.

*Assumptions*

1 Ignorance of ecology, classification, and the use of dichotomous keys.
2 The ability to use the monocular microscope, binocular microscope, lens, viewing tubes, and other devices to help in the identification of the specimens.
3 The ability to recognize continuous and discontinuous variation.
4 The ability to construct plans, charts, and graphs to present the data.

If this investigation is carried out in the field, it provides an opportunity to stress the need for conservation. The students should be told to collect sparingly, in the interests of *i* maintenance of the current species composition *ii* subsequent groups of students *iii* future generations and *iv* legal requirements. As far as possible, only common species should be collected.

*Questions and answers*

a ***Did you find that the two species you selected each had a characteristic distribution pattern, occurring in some places, and being absent from others? How did these observations link up with your knowledge of the mode(s) of life of these organisms?***

It is usually possible to find species which have marked distribution patterns, related to food supply, camouflage, oxygen supply, or other environmental conditions. The students can use books on aquatic life to look up the organisms concerned. Knowledge of the habits of the species may allow them to interpret the distribution patterns they have found in relation to food supply and other environmental factors.

**b** ***Describe any problems you encountered in the identification of the animals and plants which you found.***

**c** ***What characteristics, besides structural external features showing discontinuous variation, can be used in the construction of keys?***

Behavioural, cytological, chromosomal, biochemical, and anatomical.

**d** ***What use could be made of characteristics showing continuous variation in the construction of keys?***

These are useful if the modal value and the possible spread are carefully stated in the key and these do not overlap markedly with related species. They are used quite often in classification to confirm identifications, but only in conjunction with other features.

### Notes on the investigation of habitats

The selection of habitats for study will depend on the facilities available to the school. These notes are merely intended as a very general guide to the main areas of study in a habitat. The bibliography gives more detailed guidance.

**1** Look carefully at the area and try to find vegetation types with clearly defined boundaries such as woodland plantations, seashores with distinct zones, a pond or stream. The transition zones between one community and another can give important clues to the way in which the factors influencing vegetation alter.

**2** Produce a field sketch, photograph, or plan to show the main physical features of the area and the distribution of the plants.

**3** Try to discover some background to the information you have found, such as the influences of people and their agricultural practices, industrialization, and so on.

**4** Make a simple soil survey, including the following.
*i* Origin – is it derived from the parent rock or has it been transported by wind, water, ice, or gravity?
*ii* Drainage – the water content will vary and the soil may smell sweet or sour. Sun, rain, and wind will all influence the water content.
*iii* The influence of people – ploughing, manuring, grazing, mining, and other activities, even in the long-distant past, could affect present soil structure.
*vi* Horizons – horizons in the soil can be seen in samples taken with a screw auger or core sampler (see *Study guide II*, section 26.5). Any clear boundaries should be measured and the colour noted. The type

and extent of horizon formation may influence and may be influenced by the type of vegetation which becomes established.

*v* Colour – note the colour of each horizon and whether or not it has other colours mixed with it.

*vi* pH – determined with a pH meter or a Sudbury soil test kit.

*vii* Texture – sandy soils feel gritty and silty ones feel silky. Clayey soils are sticky and can be given a polish so that they shine when damp. Loam soils break up into distinct crumbs when moist. Most soils can be roughly classified as mixtures of the above types and are named accordingly (clayey loam, sandy loam, and so on).

*viii* Organic content – some soils such as peat are almost entirely organic, while boulder clay has a very small amount of humus.

*ix* Litter – the presence of a litter of undecayed material on the surface may indicate that the rate of decay is very slow, as in a coniferous woodland.

*x* Roots – the presence of roots and their size and distribution can be noted if this does not cause too much disruption of the habitat.

*xi* Animals – the presence or absence of animals such as earthworms or ants can profoundly influence soil properties and vegetation.

If the habitat has no soil, for instance rocky and watery habitats, then select those factors which apply specially and note any new factors which may appear.

**5** Make a map or scale drawing to show the major patterns of plant communities. Divide the habitat into areas and give each species a code symbol. Do not become too involved at this stage with the identification of difficult species.

**6** Collect samples of the animals using standardized sampling methods so that a comparison of numbers may be attempted.

Sampling methods for animals in the field include the following. In the laboratory, other methods may be employed:

*i* Flying insects – kite net.

*ii* Animals on vegetation – sweep net.

*iii* Animals in litter – sieving and then collecting by pooter from a sheet of light-coloured material after allowing them to become mobile after disturbance.

*iv* Animals in water – the smaller floating animals are caught in a plankton net, while the larger ones are extracted from weed and from the softer substrates by a pond-dipping net with a coarse mesh. A careful, quiet approach to the area will reveal the larger animals such as birds and fish. The system of 'kick sampling' will remove animals from the substrate in quickly-flowing streams, but it should be used with great care, as some species, once dislodged, may not be able to re-establish themselves. Stones should be lifted towards the current, and replaced after examination.

**7** The more detailed the identifications the better, but they must not be allowed to take too long. Attention must be concentrated as much as possible on the *ecology* of the species.

8 Apart from naming the organisms, other ecologically significant characteristics can be recorded. For example, is the heather (*Calluna vulgaris*) or the bracken fern (*Pteridium aquilinum*) in the pioneer, building, mature, or degenerate stage? Parasites, saprotrophs, and epiphytes can be recorded as well as the producers and consumers. Record where each animal is found – on plants, on the ground, flying, on flowers, and so on. Try to discover the role of each animal in the habitat.

## INVESTIGATION
## 26B The effects of different water regimes on plant growth

### ITEMS NEEDED

Turves 3/group

Gauze, perforated (metal or plastic)
Seed tray 3/group
Syringe, disposable, $5 \text{ cm}^3$ 1/group
Tubing, for siphons
Taps, 3-way 3/group

(*Study guide* 26.3 'Factors which influence the distribution patterns of organisms' and 26.5 'Soil factors'.)

*Assumptions*

1 A knowledge of soil composition and structure.
2 An appreciation of the roles of water in the soil.

*Principles*

1 Alteration of a single environmental factor can cause many changes in the structure of the plant community.
2 Alteration of a single environmental factor usually alters many other factors in the factor complex. These may contribute to some of the vegetation changes seen during the experiment.

Many difficulties will arise in the interpretation of the results of this experiment if it is not replicated. Ideally, equipment permitting, there should be in the class more than one turf per treatment. Otherwise, differences in the initial species composition and relative abundance of the species in the turves may be confounded with the effects of the waterlogging regimes.

Waterlogging, and its effects on plants, are dealt with by Etherington (1983).

Students can be encouraged to follow up this experiment by performing simple analytical tests on the soils in the seed trays, and comparing the results with those for the original grassland soil (for example, for pH or oxygen content at various depths). However, effects such as the influence of waterlogging on plants should be investigated in the field not merely by measuring environmental variables, but by devising field experiments. The problems of altering a single environmental factor, without affecting the other factors, must be emphasized.

It may be possible for teachers to set up long term field experiments near the school, on a lawn, playing field or garden. The results can be monitored each year by a group of students and interpreted in relation to the treatments and the data collected in previous years.

*Questions and answers*

a **_What effects seen during the experiment do you think are caused by the different water tables in the trays? Give reasons for your answers._**

The results may differ considerably from turf to turf and from one group to another. If only one turf has been used per treatment, the results may prove difficult to interpret. Students should be encouraged to compare the rapid changes in their laboratory experiments with the persistence of most grassland communities.

When the turves were cut, roots were severed. The deep-rooted plants with severed roots may die, especially in the driest soil. Waterlogging may also promote fungal infections.

b **_What other soil factors may have changed compared with the turves in their natural environment?_**

Oxygen levels, availability of nutrient ions and toxic ions, light (if indoors), and temperature. A lowered soil oxygen level lowers the oxygen supply to roots, soil fauna, and aerobic micro-organisms. Bacterial action in anaerobic conditions may have changed sulphates and proteins to hydrogen sulphide, and Fe(III) ions to toxic Fe(II) ions (see *Study guide II*, Chapter 26, section 26.41).

c **_What are the problems in using leaf length as a measure of plant performance?_**

During the interval of three weeks, the original longest leaves may have died, leaves which were already present may have increased in length, or new leaves may have been formed which were longer than those originally present.

d **_What would be a better measure of performance of the plants than any of the other features you have measured or observed?_**

Dry mass. It is a direct measure of net primary production (Chapter 29). The dry mass has not been measured in this experiment because the individuals of a species in the original turves may differ in dry mass. There is no way of knowing whether these plants increased or decreased in dry mass during the experiment.

e **_Suggest three other observations or measurements which would be useful in estimating the performance of plants._**

Concentration of chlorophyll, sugar or starch levels, rate of photosynthesis, respiration rate, uptake of nutrient ions.

f **_What practical importance could your observations have for the gardener?_**

Different species grow better under different soil water conditions. Careful control of soil water content can encourage beneficial species and discourage harmful ones.

g **_Describe a suitable method for investigating the effect of one other_**

***soil factor, such as soil nutrient level, on a grassland community.***

There is little point in throwing quadrats, sampling the vegetation in the quadrats, and analysing the soil from the quadrats for the levels of various nutrients thought to be important for plants. This is unlikely to yield satisfactory results unless there is evidence to suggest that the level of a particular soil nutrient, such as phosphate, varies considerably over the site (Investigation 26C). Also, it does not prove cause and effect. The best way to try to establish a causal connection is through field experiments.

Students should understand the principles of field experimentation and this should come across in their answers to this question. The vegetation should be recorded both before, and a long time after, nutrient application. The treatments should be replicated and randomized, and suitable controls must be set out. There should be gaps between the plots to minimize 'edge effects'. To make the results suitable for statistical analysis, one of the experimental designs explained in *Mathematics for biologists* (such as a Latin Square) should be used, if possible.

## INVESTIGATION
## 26C The relationship between nettle distribution and soil phosphate

### ITEMS NEEDED

Ammonium para-molybdate solution
Potassium dihydrogen phosphate solution
Sodium hydrogen carbonate solution
Tin(II) chloride solution
Water, distilled

Beaker 1/group
Colorimeter and tubes 1/group
Conical flask, 250 $cm^3$, with rubber bung 1/group
Filter funnel 1/group
Filter paper, Whatman's No. 40 or similar
Flasks, 25 $cm^3$ volumetric 7/group
Glass marker
Graph paper
Pipette, dropping, 1 $cm^3$ 1/class
Specimen tubes 4/group
Syringes 5 $cm^3$, 10 $cm^3$ and 20 $cm^3$ 1/group
Trowel 1/group

(*Study guide* 26.3 'Factors which influence the distribution patterns of organisms' and 26.5 'Soil factors'.)

**Ammonium para-molybdate solution.** Dissolve 15 g ammonium para-molybdate $(NH_4)_6Mo_7O_{24}.4H_2O$ in 300 $cm^3$ warm distilled water. Filter if necessary, and allow to cool. Add 342 $cm^3$ of concentrated HCl gradually with mixing. Dilute to 1000 $cm^3$ with distilled water in a volumetric flask.

**Potassium dihydrogen phosphate solution** for the preparation of phosphate standards. Weigh 0.4393 g of potassium dihydrogen phosphate $(KH_2PO_4)$ into a 1000 $cm^3$ volumetric flask. Add 500 $cm^3$ of distilled water. Shake until the salt dissolves. Make up to the mark. This stock solution contains 0.1 mg of phosphorus per $cm^3$. Add 5 drops of toluene to reduce microbial activity. To make the dilute solution for standards (2 μg of phosphorus per $cm^3$) dilute 20 $cm^3$ stock solution to 1000 $cm^3$ with distilled water.

**Sodium hydrogen carbonate solution.** 0.5 mol $dm^{-3}$ adjusted to pH 8.5 with 1.0 mol $dm^{-3}$ NaOH.

**Tin(II) chloride (stannous chloride) solution.** Dissolve 3 g of large crystals (preferably) of tin(II) chloride $SnCl_2.2H_2O$ in 25 $cm^3$ concentrated hydrochloric acid. These crystals may take a day to dissolve. Store in a plastic bottle in a refrigerator. To produce a dilute solution of stannous chloride for the experiment, add 0.5 $cm^3$ of this stock solution to 66 $cm^3$ distilled water.

*Assumptions*

1. A knowledge of how to use a colorimeter.
2. An ability to use Student's *t*-test (see *Mathematics for biologists*).
3. Background knowledge of the roles in plants of various nutrient ions.

*Principles*

1 The distribution patterns of organisms may be controlled by a single environmental factor.
2 A statistical significance test must be used as an objective test of the existence of a relationship.
3 It is not always possible to separate cause and effect in an experiment.

A discussion of the relationship between nettle distribution patterns and soil ions appear in Pigott and Taylor, 1964 (see the bibliography).

It must be stressed by the teacher that if nettle distribution pattern is correlated with high soil phosphate levels, this does not imply that high phosphate levels determine the pattern of nettles. There is plenty of scope here for long-term field investigations, monitored by the students each year. For example, several zones of equal area could be defined near nettle clumps. Phosphate salts could be added to some of them but not to others, and the rate at which nettles spread into the treated and non-treated areas could be compared.

Although this procedure provides an exact quantitative analysis of phosphate in soil, it is time-consuming in the preparation of chemicals, and in teaching time. An alternative is to use a Sudbury soil test kit, used by gardeners to test for nutrient element deficiencies. If preliminary analysis with the Sudbury kit reveals that soil phosphate deficiency varies considerably over the nettle-infested area which you intend to use, it may be more economical, in terms of time and money, to analyse the soil in this exercise with a Sudbury kit instead of by the colorimeter method detailed in *Practical guide 7*.

*Questions and answers*

a ***Describe any visible factors which might affect nettle distribution in the area you studied.***

Nettles often grow where urination or defaecation by man and animals commonly takes place. This may be indicated by the position of the nettles with respect to hedges and walls. The dumping of farmyard rubbish has a similar effect. In fact phosphates are so insoluble that locally-high concentrations can remain in soils for thousands of years.

b ***Are the phosphate concentrations in nettle-infested areas significantly different from those in the other areas?***

Preliminary investigations suggest that the nettle-infested areas often have a higher phosphate level (significant at the 5 % level).

c ***Why are phosphates important for the growth of plants such as nettles?***

Phosphate is a constituent of nucleic acids, NAD(P) and ATP; phosphorylated carbohydrates are intermediate products in photosynthetic and respiratory pathways. The element also has structural functions. It forms part of the phospholipid molecules in cell membranes. The presence of phosphate in the soil enhances growth of seedlings, increases root development, and causes fruits and seeds to ripen earlier. It is well known that in experiments performed under standard conditions the dry mass achieved by nettles is linearly related to the phosphate concentration in the soil solution. When nettle seedlings in solution culture are deprived of phosphate, growth is greatly retarded and symptoms of phosphate deficiency appear.

**d** ***What other nutrient ions may also differ in concentration in the areas with and without nettles?***

The concentrations of nitrates and phosphates in soil are positively correlated. One reason for this is that animals egest organic matter which contains a great deal of both nitrogen and phosphates. The concentrations of total nitrogen, potassium, and calcium can be estimated approximately with a Sudbury soil test kit.

## INVESTIGATION
## 26D The autecology of woodlice

### ITEMS NEEDED

Woodlice of two different species

Drying agent (anhydrous calcium chloride or silica gel)

Balance, weighing to 0.001 g 1/class
Choice chamber 1/group
Felt-tip markers, or alternative, for marking woodlice 1/group
Fine black marker 1/group
Glass plate on blocks 1/group
Paper towel
Pooter 1/group
Sellotape
Stopwatch 1/group
Weighing containers

(*Study guide* 26.3 'Factors which influence the distribution patterns of organisms' and 26.6 'Air factors'.)

*Assumptions*

1. Some appreciation of the biology of woodlice.
2. Knowledge of the capture/recapture method for estimating population sizes (investigation 28B).
3. Ability to tabulate, record and graph data.
4. Ability to carry out the $\chi^2$ test to determine the significance of results.

*Principles*

1. The distribution patterns of organisms are determined by environmental factors.
2. The behaviour of an organism also affects its distribution pattern.
3. The overall responses of an organism may be determined by several factors acting simultaneously. A change in one factor may affect the response of the organism to another factor.
4. There are several different ways in which an organism may react to a change in its environment. Use of the different experimental procedures may help to separate these different mechanisms.

*Questions and answers*

**a** ***Do the data suggest that woodlice have a preferred shelter site?***

If the woodlice return to the same site, the percentage of the population which return will be very high at first, and will decline later. If the level of fidelity is low, there will be an initial drop, followed by a rise as animals eventually return to the original site. In a restricted area, equilibrium will be reached quickly.

**b** ***If the number of marked woodlice recovered in such a field experiment declined rapidly, what might you suspect?***

The presence of unsuspected shelter sites, or rapid hiding of the woodlice when disturbed.

c ***If there were a large number of unsuspected shelter sites in the area under investigation, and the woodlice showed no preference for the home site, what pattern of results might you expect?***

There would be no equilibrium; the number of individuals present would vary in a random manner.

d ***How would you change the experiment so that interference between individuals from different home sites was avoided?***

Choose areas with clearly defined shelter sites well away from one another; erect barriers.

e ***What data, other than those referred to in the instructions for the experiment, could be used to interpret the results of the experiment, and how would you obtain and record such data?***

Temperature and humidity, wind speed and other environmental conditions may have influenced the degree of activity of the organisms. There could also be differences caused by the breeding or feeding activities that could influence the results.

### Water loss in clustered and isolated individuals

*Questions and answers*

a ***Do your results indicate that the clustered woodlice lose mass more slowly than the isolated ones?***

A change in mass can be assumed to represent an equivalent water loss. This is not strictly true, but serves as a point of comparison. Isolated woodlice will usually lose water more rapidly than the clustered ones.

b ***Suggest a reason for the results which you have obtained.***
Reduction in the surface area to volume ratio. The relatively smaller surface area reduces water losses.

c ***Do woodlice cluster together in their natural habitat?***

Yes, although it is not clear from observation whether this phenomenon is a result of behaviour or of the restricted size of the habitat in which they are found.

d ***What bearing do your results have on your field observations?***

They confirm that clustering could have a beneficial effect when it occurs in the field.

### Isolating aspects of behaviour which lead to water loss

*Questions and answers*

a ***What preference, if any, did woodlice have in the humidity gradient***

They accumulate in the regions of the highest humidity.

**b** ***Did all the woodlice show a similar response to the same conditions? Can you explain any differences which occurred?***

It is most likely that woodlice will show the same responses. Any differences may be due to the pre-treatment, physiological differences between individuals, morbidity, breeding, and so on.

**c** ***Why is a standard pre-treatment laid down for the woodlice?***

To eliminate at least some of the variability in response between individual woodlice.

**d** ***Describe how you would determine the effect of pre-treatment on the behaviour of woodlice.***

Compare the same woodlice after different times of different pre-treatment. Compare them with other woodlice.

**e** ***Under what circumstances would you expect differences in behaviour to occur? Bear in mind how the selection of experimental animals is carried out.***

Shedding exoskeleton, bearing young, age, sex, disruption of rhythms due to disturbance of the habitat – these could all have an effect.

### Reference

Edney, 1954 (see the bibliography).

# PART III Bibliography


BAILEY, N. T. J. *Statistical methods in biology*. 2nd edn. English Universities Press, 1981.

BAKER, D. A. 'The transport of ions across plant cell membranes'. *Journal of biological education*, **15**(2), 1981. (It explores very clearly the Nernst equation and the mechanisms of ion transport across root cell membranes.)

BEEDHAM, G. E. *Identification of the British Mollusca*. Hulton Educational Publications, 1973.

BELCHER, H. and SWALE, E. *A beginner's guide to freshwater algae*. H.M.S.O., 1977.

BELCHER, H. and SWALE, E. *An illustrated guide to river phytoplankton*. H.M.S.O., 1980.

BENNETT, D. P. and HUMPHREYS, D. A. *Introduction to field biology*. 3rd edn. Edward Arnold, 1980.

CAMPBELL, R. C. *Statistical methods in biology*. 2nd edn. Cambridge University Press, 1974.

CHINERY, M. *A field guide to the insects of Britain and Northern Europe*. 2nd edn. Collins, 1976.

CLEGG, J. *The observer's book of pond life.* 3rd edn. Warne, 1980.
CLOUDSLEY-THOMPSON, J. L. and SANKEY, J. *Land invertebrates.* Methuen, 1961.
DARLINGTON, A. and LEADLEY-BROWN, A. *One approach to ecology.* Longman, 1975.
DAVIS, B. N. K. *Insects on nettles.* Naturalist's Handbooks (edited Corbet, S. A. and Disney, R. H. L.). Cambridge University Press, 1983.
DOWDESWELL, W. H. *Practical animal ecology.* Methuen, 1959.
DOWDESWELL, W. H. *Ecology: principles and practice.* Heinemann Educational, 1984.
EDNEY, E. B. 'Woodlice and the land habitat'. *New Biology,* **17**, 41, 1954.
EPSTEIN, E. *Mineral nutrition of plants: principles and perspectives.* Wiley, New York, 1972. (A valuable summary of plant nutrition in the broadest sense.)
EPSTEIN, E. 'Roots'. *Scientific American,* **228**(5), 1973. Offprint No. 1271. (A beautifully balanced summary of roots and how they take up ions.)
ETHERINGTON, J. R. Studies in Biology No. 98, *Plant physiological ecology.* Edward Arnold, 1978. (A very concise summary of the effects of climate and soil on the distribution patterns of wild plants.)
ETHERINGTON, J. R. Studies in Biology No. 154, *Wetland ecology.* Edward Arnold, 1983. (A masterly compression of a vast amount of information difficult to obtain elsewhere.)
KENT, J. W. 'Exploring the realized niche: simulated ecological mapping with a microcomputer'. *Journal of Biological Education,* **17(2)**, 131, 1983.
KING, T. J. *Ecology.* Nelson, 1980. (It provides several simple examples of the effects of the environment on the distribution patterns of organisms.)
LEADLEY BROWN, A. *Ecology of soil organisms.* Heinemann, 1978. (A solid and detailed account of the soil fauna.)
LINCOLN, R. J. and SHEALS, J. G. *Invertebrate animals: collection and preservation.* British Museum and Cambridge University Press, 1979.
LYNEBORG, L. *Field and meadow life.* 2nd edn. Blandford, 1973.
MACAN, T. T. *A guide to freshwater invertebrate animals.* Longman, 1959.
MANDAHL-BARTH, G. *Woodland life.* 2nd edn. Blandford, 1972.
MELLANBY, H. *Animal life in fresh water.* 6th edn. Chapman and Hall, 1963.
NUFFIELD ADVANCED BIOLOGICAL SCIENCE. *Key to pond organisms.* Penguin, 1970.
NUFFIELD REVISED ADVANCED BIOLOGICAL SCIENCE. *Mathematics for biologists.* Longman, 1987.
OLSEN, S. R. and DEAN, L. A. 'Phosphorus'. In Black, C. A. (editor), *Methods of soil analysis,* 1035 (this method 1046). American Society of Agronomists, Madison, 1965.
PAVIOUR-SMITH, K. and WHITTAKER, J. B. *A key to the major groups of British free-living terrestrial invertebrates.* Blackwell Scientific Publications, 1968.

PIGOTT, C. D. 'Biological flora of the British Isles. *Cirsium acaulon* (L.) Scop.' *Journal of Ecology*, **56**, 1968. (A detailed account of the biology of the stemless thistle.)
PIGOTT, C. D. and TAYLOR, K. 'The distribution of some woodland herbs in relation to the supply of nitrogen and phosphorus in the soil'. *Journal of Ecology*, **52** (Jubilee Supplement), 175, 1964.
PIGOTT, C. D. 'The response of plants to climate and climatic change' in Perring, F. (ed.) *The flora of a changing Britain.* Classey, 1970. (It describes effects of climate of several plant species, but mainly the stemless thistle.)
PRUITT, W. O. Studies in Biology No. 91, *Boreal ecology.* Edward Arnold, 1978. (An interesting summary of the effects of frost and snow on plant and animal life in the tundra.)
QUIGLEY, M. *Invertebrates of streams and rivers: a key to identification.* Edward Arnold, 1977.
RICKLEFS, R. E. *The economy of nature.* Chiron Press, Portland, Oregon, 1976. (An elegant and fascinating introduction to ecology – the best available.)
RICKLEFS, R. E. *Ecology.* 2nd edn. Nelson, 1980. (A clearly written discussion of the principles of ecology, animal behaviour, and evolution: a standard textbook.)
RUSSELL, E. J. *The world of the soil.* Collins, 1957. (A classic introduction to soils, their fauna, and the relationships between soils and agricultural production.)
SCHOOLS NATURAL SCIENCE SOCIETY. *Key to pond organisms*, 1964.
SUTCLIFFE, J. F. and BAKER, D. A. Studies in Biology No. 48, *Plants and mineral salts.* 2nd edn. Edward Arnold, 1981. (It describes the roles of the mineral elements required by plants and the mechanisms by which they are absorbed.)
SUTTON, S. *Woodlice.* Ginn, 1972.

# CHAPTER 27 ORGANISMS AND THEIR BIOTIC ENVIRONMENTS

*A review of the chapter's aims and contents*

1 The interactions between organisms in nature include competition, predation, parasitism, commensalism, mutualism, dispersal, and succession.

2 Organisms often deprive others of environmental resources which are in short supply. The growth, life span, or reproductive potential of at least one of the interacting organisms is reduced. This is known as 'competition'.

3 The existence of competition is easy to demonstrate in the laboratory, although it is often difficult to ascertain the resource(s) for which the organisms are competing.

4 Many species have feeding habits or life-cycles which depend on other specific species. Many insects depend on specific food plants. Parasites often specialize on a single host species. Two species often cooperate in mutualism, living in a long-term association which benefits both.

5 In these associations the life-cycles and physiology of the co-existing species are integrated.

6 The existence of these associations affects the geographical and local distributions of the species concerned.

7 Many species might be able to occupy a much wider geographical area if they could overcome a dispersal barrier, such as the sea, a river, or a mountain range. Humans have altered the distribution limits of many species, intentionally or unintentionally.

8 Organisms continually modify their own environments, creating conditions more or less suitable for themselves and other species. Some successions, such as the 'hydrosere', result from drastic habitat modification by the plants themselves.

## The *Study guide*

### 27.1 Competition

*Principles*

1 Competition for limited resources is frequent in nature and an important influence on the geographical and local distribution patterns of species.

2 At least in this text, the word 'competition' is used in two senses. It means the 'struggle' for resources in short supply by two organisms of about the same size when both suffer. It also means the accumulation of scarce resources by one individual, which is hardly affected, at the expense of another, which may stop growing and die.

3 Two individuals of the same species often have more similar environmental requirements than two individuals of different species.

## STUDY ITEM

## 27.11 The effect of density on the growth of monocultures

*Principle*

1 In crop plants of a single species grown in experiments at a range of densities, the dry mass and the seed production per plant are often reduced at higher densities.

Competition amongst crop plants and between crop plants and weeds is dealt with simply by Hill (1969) and in detail by Harper (1977).

*Question and answer*

a ***With reference to* figure [S]310*, explain why, with increasing numbers of plants per unit area, the rate of dry matter production by each remaining plant is reduced.***

As a plant grows, it captures resources from an increasing area with time. The greater the number of plants per unit area, the closer its neighbours will be, and the sooner its leaves and roots will compete with those of its neighbours. If one plant begins to take up limiting resources which another plant might use in its growth, such as light, carbon dioxide, water, or nutrients, the growth rate of the deprived plant will decrease. This begins to happen sooner in dense than in sparse populations. □

### The −3/2 power 'law'

*Assumptions*

1 An understanding of the difference between linear and logarithmic scales.

2 An understanding of slopes on graphs.

*Principles*

1 In a dense population of a single species, the number of plants decreases with time and the mean mass of the remaining individuals increases. The mean mass can be plotted against the remaining number of plants for different stages in the development of a single population. If logarithmic scales are used a straight line is obtained with a slope of −3/2. This relationship holds for a wide variety of plant species and is known as the '−3/2 power law'.

2 Mathematical tools are valuable in ecology.

The most accessible account of the −3/2 power law is by Hutchins (1983).

## STUDY ITEM

## 27.12 Shepherd's purse (experiment design)

*Assumption*

1 Some knowledge of experimental design including randomization and replication.

*Question and answer*

a **_Shepherd's purse,_ Capsella bursa-pastoris, _is a common annual weed of disturbed ground. By means of a field-based investigation, how would you test the hypothesis that seed production in shepherd's purse is related to its population density?_** (*J.M.B.*)

The answer should include the following points:
Clear a large area, removing all plants. Mark out the area into at least twenty plots. At least four plant densities should be used. At least five replicates of each density are required, to allow for some wasted replicates and to provide enough plots to make statistical analysis worthwhile. If the design is 4 seed densities $\times$ 5 replicates, five randomized blocks can be used, with each block across any environmental gradient. With 5 seed densities and 5 replicates a Latin square would be appropriate. Leave paths between the plots. The plots should be at least $30 \times 30$ cm to minimize edge effects.
Sow seeds at various densities. When plants are mature, count numbers of fruits on each plant, then total and record for each plot. Multiply totals by average number of seeds set per fruit, after checking whether seed production per fruit differs with plant density or with position on the fruiting inflorescence.
The major problem with shepherd's purse is that the plants continue to produce flowers for a long time. Thus, it is difficult to assess when fruit production has ceased. Note also that this species can produce many abortive or empty capsules. □

**Practical investigation. *Practical guide 7*, investigation 27A, 'Intra-specific competition in *Drosophila*'.**

**STUDY ITEM**

### 27.13 Competition between two flour beetle species in a limited environment

*Assumption*

1 A knowledge of the life cycles of beetles.

*Principles*

1 The outcome of competition between two species depends on the environment in which it occurs.

2 Competition is easy to demonstrate in the laboratory but it is difficult to ascertain the resources for which organisms are competing.

*Questions and answers*

a ***Comment on these data.***

At 34 °C and 29 °C *T. castaneum* won the majority of encounters under moist conditions, but *T. confusum* usually eliminated *T. castaneum* in moist flour. At 24 °C *T. confusum* was dominant.

b ***In what ways could the two species have interacted within the vials so that one species was placed at a disadvantage compared with the other?***

Predation of eggs, larvae, pupae, and adults.
Higher metabolic rate, allowing more rapid life-cycle, greater reproductive potential.
Loss of water might reduce growth rates of larvae and adults and hence slow down life cycle (for example, in *T. castaneum* in dry conditions at 34 and 29 °C).
Fungal infection of eggs.

c ***If the temperatures and relative humidities within a set of vials containing both species had been altered every twelve hours between hot–moist and cold–dry, to simulate the changes between night and day, what might have been the outcome of the experiment and why?***

Co-existence. In hot–moist conditions *Tribolium castaneum* would be favoured, and in cold–dry conditions *T. confusum*. The environmental fluctuations which occur in nature probably allow species which have different physiological attributes to coexist. □

**Practical investigations. *Practical guide* 7, investigation 27B, 'Inter-specific competition between two *Lemna* species' and investigation 27C, 'Competition between clover and grass'.**

**STUDY ITEM**

**27.14 Competition between two ant species**

*Principles*

1 Evidence for competition in nature can be obtained by adding or removing the competing species.
2 Competition in nature is a potent force, reinforcing the physiological differences between species to produce and maintain differences in their distribution patterns.

*Questions and answers*

a ***From the data in the table, assess the effect of* Lasius flavus *colonies on 1 other* L. flavus *colonies and 2* L. niger *colonies.***

**1** When colonies of *L. flavus* were removed, nearby *L. flavus* colonies became more vigorous. When the transplanted nest was placed near to an existing nest, the thriving colony which occupied the existing nest became much less vigorous. Competition between adjacent colonies was considerable.
**2** The colonies of *L. flavus* also affected *L. niger nests*. When *L. flavus* nests were removed, queen production by nearby *L. niger* colonies increased. When *L. flavus* was added, *L. niger* nests subsequently produced fewer queens. The effects of *L. flavus* on *L. niger* were not as dramatic as the effects of *L. flavus* on *L. flavus*.

b ***What were two adjacent colonies competing for? Explain in detail the various ways in which the vigour of a colony might be reduced by another colony nearby.***

Competition for a limited food supply is most likely. The ants live

mainly on aphids and coccids which live underground on the roots of plants. The ants suck 'honeydew' and eat young aphids as they are born. *Lasius niger* also forages on the soil surface for small living and dead insects.
Individuals from two different colonies, either of the same or different species, may fight when they meet. Presumably there is a tension zone, a sharp boundary, between two colonies. When one of the colonies is removed the potential foraging area of the neighbouring colony is increased and the colony can trap more food. Conversely, if a new colony is introduced the new individuals may take over some of the area of the surrounding colonies. The mechanism of competition could be fighting (competition for space) or the removal of food which individuals from the other colony would otherwise eat.

c ***How would you expect the foraging behaviour of* L. niger *to change when a new colony of* L. flavus *appears nearby?***

It depends partly which colony is the more aggressive. At least at the contact zone between the two colonies, the less aggressive species would begin to eat those food items avoided or overlooked by the more aggressive species. In fact *L. flavus* is a subterranean forager. *L. niger* can forage both above and below the soil surface. Where they are competing intensely, *L. flavus* forages underground and *L niger* only at the surface.

d ***What is the importance of the result that competition within one species may be more severe than competition between species?***

Competition between *L. flavus* nests certainly seems more severe than competition between adjacent *L. flavus* and *L. niger* nests. This may contribute to the coexistence of the species in the long term. For if *L. flavus* nests ever became very abundant, intra-specific competition would reduce their vigour and perhaps allow *L. niger*, with which *L. flavus* does not compete too much, to expand. Conversely, if *L. niger* colonies dominated, new colonies of *L. flavus* would be more likely to invade the area than those of *L. niger*. □

## Competition and the properties of sun and shade plants

*Assumptions*

1 An elementary knowledge of leaf anatomy.
2 A basic understanding of photosynthesis (*Study guide I*, Chapter 7) and respiration (*Study guide I*, Chapter 5).

*Principles*

1 Different species of plant and different leaves on the same plant, may differ physiologically and anatomically in their response to shading.
2 Shade leaves have a lower respiration rate than sun leaves at all temperatures and achieve maximum photosynthesis at lower light intensities and temperatures.

3 At 'compensation point', the rate of photosynthesis equals the rate of respiration. Plants, and leaves, will only increase in dry mass if photosynthesis exceeds respiration.

## 27.2 Interactions apart from competition

*Principles*

1 Some species habitually live in association with another. Sometimes these associations involve lifetime contact.
2 These associations are of four main types: predation, parasitism, commensalism, and mutualism.
3 Many associations cannot be assigned with certainty to any of these categories.

**Practical investigations. *Practical guide* 7, investigation 27D, 'The holly leaf miner *Phytomyza ilicis* and its parasitic insects' and investigation 27E, 'A study of inter-specific associations'.**

## 27.3 Predation

*Principle*

1 A predator may influence the geographical or local patterns of species, and the prey may determine where the predator occurs.

## 27.4 Parasitism

*Principles*

1 The life cycles of parasites are physiologically integrated with those of their hosts.
2 Some parasites cause immense economic damage.
3 Control of a parasite is more effective if its habits and life cycle are understood.

### STUDY ITEM
### 27.41 The ecology of malaria (*Plasmodium* spp.)

*Principles*

1 The parasite's life cycle is physiologically dependent upon that of its host.
2 Parasites have high reproductive rates.

*Plasmodium* spp. have a complex life cycle. It was selected as an example because of its economic importance, and because so many control measures exist. (For details, see Cox, 1981, and Phillips, 1983, listed in the bibliography to this chapter.)

*Questions and answers*

**a** ***Suggest two advantages to the parasite in promoting bouts of fever in the host which occur every forty-eight hours at night.***

The reasons given here are speculative. Naturally, there may be no 'advantage' at all in the timing of the release of the merozoites from the red blood cells.

**1** The female mosquitoes may recognize victims by sensing the heat radiated from a human. A feverish temperature might increase the likelihood that infected humans might be bitten.

**2** The mosquitoes are active at night.

**3** Humans are inactive at night and especially when feverish.

**4** If the feverish symptoms occur with a periodicity which is a multiple of twenty-four hours, a feverish human may be bitten more frequently than if the symptoms occurred at random times in the day and night.

**b** ***Name another parasite which has two alternative host species and which reproduces within each. In what ways does the possession of two alternative hosts benefit* Plasmodium*?***

The liver fluke *Fasciola hepatica*, for example, has two hosts: the sheep and the snail *Limnaea truncatula*.

*Plasmodium* spp. multiplies in both hosts. The mosquitoes are mobile and a single infected mosquito might infect many humans over a wide area.

**c** ***Explain the advantages to* Plasmodium *of producing short-lived blood phases rather than circulating in the blood plasma.***

Within red blood cells or liver cells the *Plasmodia* are 'invisible' to the immune system of the host. If they circulated continuously in the blood plasma, they would stimulate the formation of anti-*Plasmodium* antibodies and might be ingested or otherwise eliminated by white blood cells.

**d** ***Suggest some methods of eliminating mosquitoes.***

Breeding sites have been considerably reduced by draining swamps and filling in water holes. Paraffin oil, Paris green (copper arsenite), and synthetic insecticides have been sprayed onto water surfaces all over the tropics and subtropics. Fish such as guppies, which eat mosquito larvae, have been successfully introduced.

When organochlorine insecticides such as DDT became availabl control concentrated upon house-spraying with residual insecticide This has been most successful. Two sprayings a year with DDT dieldrin are sufficient. The insecticide remains on the walls and th female mosquitoes, gorged with blood, are killed when they alight to re and digest their food. House-spraying in this manner eradicated malar from Italy, Greece, and parts of South America, and greatly reduced th death rate in Sri Lanka and Mauritius.

Insecticides such as malathion, propoxin, and the pyrethroids ha been used on a large scale to replace the cheap but toxic DDT (see Study item 29.53). Nevertheless, they are relatively expensive and the mosquitoes have become resistant to them too.

e ***Why is a vaccine against merozoites likely to be more effective than a vaccine against sporozoites?***

The sporozoites circulate in the blood for only thirty minutes. It seems unlikely that antibodies would combine with and inactivate all thousand or so sporozoites within that time. Once within the liver cells, sporozoites could not be detected by the immune system. The merozoites, on the other hand, circulate in the blood for several hours each time they are released. They are released at regular intervals not once, but during the whole period of the disease. □

## 27.5 Commensalism

Commensalism is only dealt with briefly. It is often difficult to establish whether an organism 'benefits' from an association.

## 27.6 Mutualism

*Principles*

1 Mutualisms are frequent in nature. Both partners benefit from the association and they are physiologically interdependent.
2 Many of the partners in mutualistic associations are single-celled prokaryotes or single-celled eukaryotes.

### STUDY ITEM

### 27.61 Equilibrated physiological interdependence in *Riftia pachyptila*, the Galápagos vent-worm

This is rather an anomalous example of mutualism, but it is discussed here because
**1** it illustrates well the complexity of physiological interdependence involved in mutualism, and
**2** its unfamiliarity to students will enable them to concentrate upon the principles.

*Assumptions*

1 A knowledge of some of the more important enzymes in photosynthesis.
2 An understanding of the role of cytochromes in aerobic respiration.
3 An understanding of the Bohr effect (*Study guide I*, Chapter 4).

*Principle*

1 The metabolism of both partners in a mutualism may depend on complex interactions between them.

a **_Consider in turn each of the points below. State the inferences which you would draw from each piece of information. At the end, state a hypothesis to explain how Riftia gains its energy._**

**1** *Riftia* lives in permanent darkness and this colour is unlikely to be warning coloration, camouflage, or to attract a mate! It suggests that the worm contains haemoglobin which is exposed at the body surface or seen through transparent tissue.

**2** They cannot eat insoluble food. They must absorb small molecules of low relative molecular mass.

**3** Large surface areas in organisms are characteristic of absorbing surfaces. The obturacular plume presents to the water a considerable surface area over which compounds are taken up and transferred to the blood (see Laubier and Desbruyères 1985).

**4** It seems unlikely that there are enough organic compounds floating around in the sea in the worm's natural habitat to feed the worm. The absorption of soluble compounds by the obturacular plume seems inadequate to feed the worm if this is its only food source.

**5** The bacteria in such numbers are probably mutualistic not parasitic. The presence of the blood capillaries suggests that the blood is supplying them with chemical compounds. If they are not parasitic, saprotrophic (too few organic compounds absorbed by the worm) or photosynthetic (too deep), they must be chemo-autotrophic.

**6** Sulphur is produced by the oxidation of hydrogen sulphide by the bacteria. This is an energy-yielding process, which might supply the ATP and NADPH needed to reduce carbon dioxide to carbohydrates in the Calvin cycle. In that case the blood of the worm must supply the bacteria with hydrogen sulphide and carbon dioxide.

**7** These enzymes are present in the leaves of green plants in the stroma of chloroplasts, where they are involved in the Calvin cycle. Ribulose-5-phosphate kinase converts ribulose monophosphate to ribulose bisphosphate. This, under the influence of ribulose bisphosphate carboxylase, absorbs a carbon dioxide molecule and splits into two molecules of PGA (phosphoglyceric acid). The presence of these enzymes indicates a Calvin cycle. Muscle is not autotrophic and lacks these enzymes.

**8** The blood absorbs hydrogen sulphide from the sea and supplies it to the bacteria. Oxygen is required by the bacteria to reduce the hydrogen sulphide. It is also needed by the cells of the worm and the bacteria to respire. This sulphide-binding protein is probably haemoglobin itself, but if so it must have separate binding sites for oxygen and hydrogen sulphide since they would otherwise react together.

**9** The cytochrome c oxidase system is probably protected in *Riftia* b a sulphide-resistant protein. This compound could be inside the cells, or in the bloodstream. Experiments indicate that this protein is in *Riftia's* blood, and is probably haemoglobin itself. It has such affinity for sulphide that it not only concentrates the sulphide from the worm' environment, but it protects the cytochrome c oxidase in the worm cells from sulphide poisoning.

b ***The blood exhibits a small Bohr effect (*****Study guide I**, ***chapter 4) (a shift of the oxygen equilibrium curve to the right with increased $CO_2$ levels) compared with that of other worm species which inhabit permanent, well-ventilated burrows or tubes. What might be the significance of this?***

In most worms which respire aerobically, haemoglobin gives up its oxygen in the acid conditions created by the high $CO_2$ concentrations in the capillaries. Around the bacteria in the trophosome, however, the $CO_2$ concentration is low since the Calvin cycle in the bacteria is removing it from the blood. The difference in pH between the blood in the capillaries and that in the obturacular plume is minor. Carbon dioxide does not need to be excreted by the worm, since it is required by the autotrophic bacteria inside it.

c ***Compare the problems of gas transport by the blood in*** **Riftia** ***and a mammal.***

The mammal needs to provide its cells with oxygen and to transport carbon dioxide back to the lungs. *Riftia* needs to take up three gases, hydrogen sulphide for the bacteria, carbon dioxide for the bacteria, and oxygen for the bacteria and the worm tissues. The bacteria will not produce oxygen because they use hydrogen sulphide, not water, as their hydrogen donor in photosynthesis. Hydrogen sulphide itself may be toxic to the cells of the worm. The sulphur is deposited in the trophosome. No gas needs to be excreted.

d ***Is it justifiable to call*** **Riftia** ***the first example of an autotrophic animal at the base of a food chain?***

*Riftia* itself is not autotrophic. It contains autotrophic bacteria essential for its survival. Nor is it known to be at the base of a food chain. If free-living chemo-autotrophic bacteria exist near the vents, or other species, such as clams, also harbour similar bacteria, *Riftia's* contribution to a grazing food web is probably minor. Nevertheless, dead *Riftias* probably contribute food to the detritus food web. □

## STUDY ITEM
## 27.62 Crops without chemicals

*Assumption*

1 An understanding of the fact that plants require large quantities of nitrogen for healthy growth, mainly because it is an essential part of proteins.

*Principles*

1 Some nitrogen-fixing prokaryotes live in mutualism with green plants, for example, plants of the legume family.
2 Nitrogen fixation has considerable economic potential.

Aspects of the economic potential of nitrogen fixation are explored by Postgate, 1979, Brill, 1979, and Brill, 1981. (See the bibliography.)

*Questions and answers*

a ***Give the meaning of the following terms:***

***1 eutrophication***

***2 facultative anaerobes***

***3 obligate mutualist***

**1** Eutrophication is the enrichment of water with plant nutrients, particularly nitrates and phosphates. The nitrates come mainly from fertilizers and the phosphates from detergents and sewage. Eutrophication causes increased growth of water weed, population explosions of algae, and stagnation, as bacteria absorb oxygen when they decompose dead algae.

**2** Facultative anaerobes are organisms which can live without oxygen if the need arises, but normally live in microhabitats which contain oxygen.

**3** Obligate mutualists are organisms which must be in partnership with another species in order to survive and reproduce, a partnership from which both partners benefit.

b ***Describe two ways in which nitrogen fixed by* Rhizobium *bacteria in leguminous plants may enter the soil.***

When the plant dies.
Diffusion of organic nitrogen compounds from the roots.
Seed dispersal.
Plant eaten by animal, and egested.

c ***1 Suggest a function of the leghaemoglobin produced in the root nodules of legumes.***

***2 Explain how this benefits the* Rhizobium *bacteria.***

The leghaemoglobin may prevent large *concentrations* of oxygen from building up in the inner cortical cells. This may shield the nitrogenase enzyme in the *Rhizobium* bacteria from attack by oxygen. In fact the leghaemoglobin probably ensures that *rate* of supply of oxygen to the bacterial cells is quite high. After all, it is required by the host cells and by the bacteria for aerobic respiration.

d ***State the sequence of events in* Rhizobium *which result in the synthesis of nitrogenase enzyme.***

The two complementary strands of the *Rhizobium* DNA double helix separate and messenger RNA for the three nitrogenase genes is made along the sense strand of the DNA double helix. The messenger RNA molecules move to the ribosomes, where multiple copies of the three proteins are synthesized.

e ***Give a reason for the comment that 'the presence of nitrogenase in many cereals could turn out to be less than the economic miracle anticipated'.*** *(J.M.B.–modified)*

Nitrogen fixation is an energy-demanding process. The constant need to supply glucose to the bacteria may reduce crop production more than the extra supply of amino acids increases it. It would also be essential to transfer the leghaemoglobin genes to the cereal so that *Rhizobium* in the cereal root was protected against oxygen. □

## 27.7 Dispersal and other historical factors

*Principles*

1. Many species might be able to occupy a much wider geographical area if they could overcome a dispersal barrier, such as the sea, a river, or a mountain range.
2. Humans have altered the distribution limits of many species, either intentionally or unintentionally.

### STUDY ITEM
### 27.71 Dispersal onto Surtsey

*Questions and answers*

a ***Plot the data in table* [S]*64 on a graph.***

See *figure 54*.

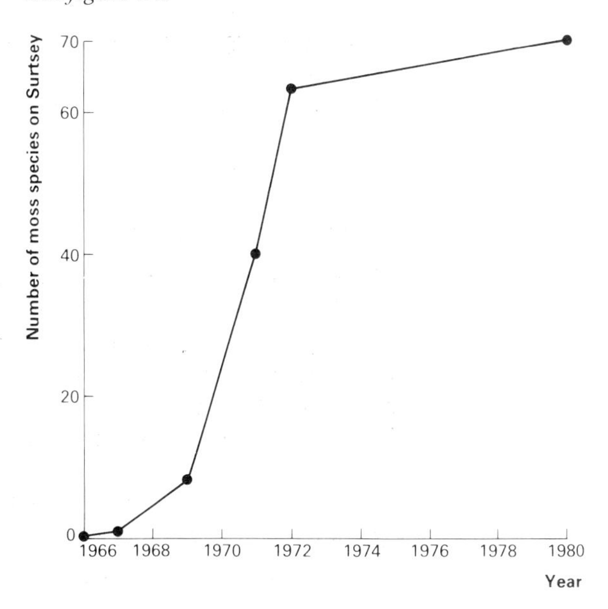


**Figure 54**
The number of moss species present on Surtsey from 1966 to 1980.

b ***Account for the shape of the curve.***

The well-dispersed mainland moss species, especially those which reach bare soil, reached the island within two or three years, presumably by wind-borne spores. Most suitable species had colonized by 1972. When much of the surface of the island had been colonized by moss there was less space for other moss species to invade. Some mosses might have altered their microhabitats so that some colonists of moist soil or moss tussocks could invade.

c ***Why did micro-organisms and birds occur only in very small numbers for the first seven years?***

The micro-organisms required plant detritus, but plants were sparse at first. The birds require seeds or insect food, but higher plants and insects were slow to invade. Herbivorous insects could not become abundant until their food plants appeared on the island.

d ***For each of these groups, suggest two ways in which they might have reached the islands: 1 spiders 2 mosses 3 protozoa 4 higher plants.***

**1** convection currents, bird's feathers
**2** spores by wind or on bird's feet
**3** spores on bird's feet, spores by wind
**4** seeds on birds' feet, accidental dispersal by humans, parachute seeds or fruits.

e ***Explain how the proximity of other islands might affect the invasion of species onto Surtsey.***

Poorly-dispersed species could have reached Surtsey from Iceland by migrating a short distance at a time, via a series of 'stepping-stone islands', eventually reaching Surtsey. If stepping-stone islands are present, the flora and fauna of Surtsey should be richer in species than if they are absent.

f ***Suggest three ways in which sea birds might affect the composition of the flora.***

Disperse seeds or spores on feet. Disperse seeds in their crop or gizzards and regurgitate them. Eat seeds and egest them. Egest phosphate-rich compounds, altering soil properties for invasion of new species. Grazing. Trampling and nesting, killing vegetation.

g ***Speculate on the changes which will take place in the flora and fauna of Surtsey over the next fifty years.***

The island is eroding rapidly and may not last long. Weedy, colonizing species of plants and animals will probably be replaced by those characteristic of more permanent communities, such as grasses, which might eliminate the mosses. This would support a diversity of small insects and bird life. Shrub and tree invasion would occur in inland habitats at low altitudes. On Surtsey, however, the climate may be too windy to support species with exposed buds, and tree invasion might be confined to the leeward cliffs. □

**Practical investigation. *Practical guide* 7, investigation 27F, 'A grassland survey using sampling techniques'.**

## 27.8 Successional changes

*Principles*

1 Successions are predictable changes in plant and animal species at a site.
2 Primary successions occur on bare rock and create soil.
3 Secondary successions occur on soil from which the previous vegetation has been removed.
4 Successions tend to end in a 'climax' state in which the species composition of the community changes less rapidly.
5 Zonations of species often indicate in space that a succession is occurring in time.
6 In many successions, the species alter the environment so that it is more, or less, suitable for the invasion or growth of other species.
7 As succession proceeds, the ratio of productivity to biomass declines until it becomes constant or nearly constant. This is because the dead wood of trees (which could be called 'necromass') is included in 'biomass'.

### Zonation and succession

*Principle*

1 Succession around the edges of a nutrient-rich lake can cause it to silt up. The vegetation may change until a forest or a *Sphagnum* bog becomes established.

*Question and answer*

a ***If a cylinder of peat was taken out of the 'soil' beneath this 'climax' forest, what fragments of vegetation would you expect to obtain?***

We should expect to obtain, from the bottom to the top, (a) a layer of silt, (b) a layer of the remains of aquatic plants, such as pond-weed (*Potamogeton* spp.), (c) a band of sedge-reed peat in which the rhizomes of the reeds and the fruits of the sedges were clearly visible, (d) fen peat containing sedge and moss remains, (e) fen peat with alder cones, and finally (f) drier soil with tree leaf litter, acorns etc.

In some samples the plant remains towards the base are old enough to be dated by radiocarbon dating. This indicates that the whole process, from open water to forest, may take several thousand years.

### STUDY ITEM
### 27.81 Changes in production and biomass during succession

*Principles*

1 With time, the productivity of a maturing forest stand reaches a peak and then declines.
2 During forest succession, the ratio of productivity to biomass steadily decreases.

*Questions and answers*

**a** ***Plot the data in table 65 on a graph.***

See *figure 55*.

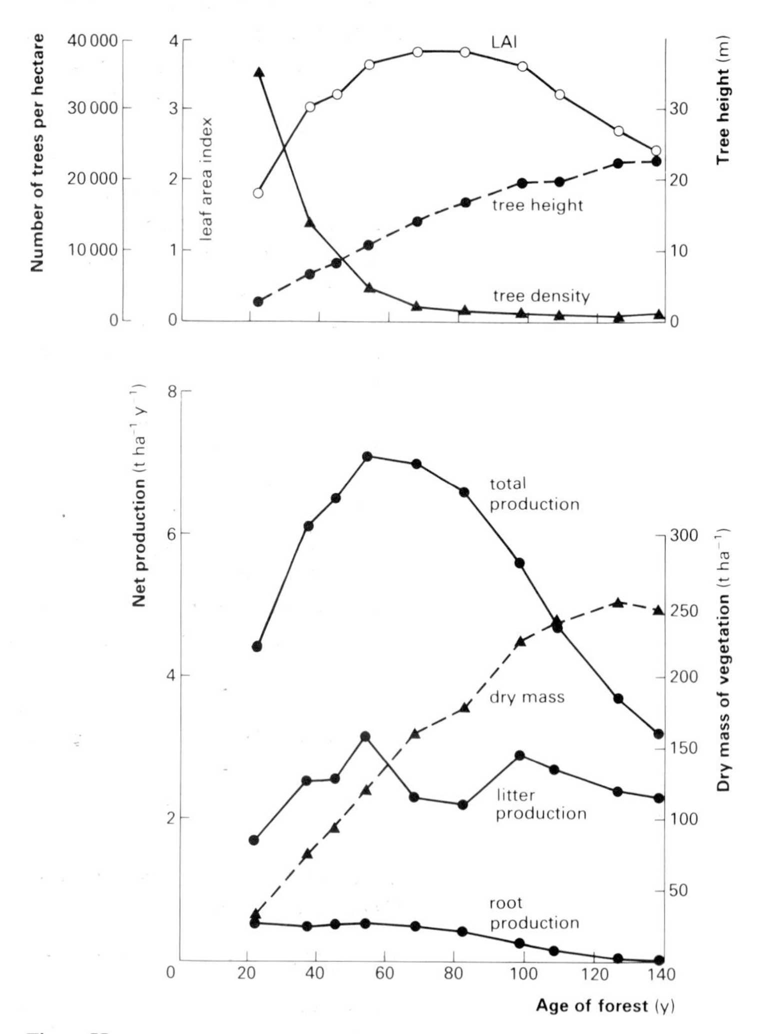


**Figure 55**
Production and biomass in stands of the Norway Spruce (*Picea abies*) of different ages.

**b** ***Examine these results and state the main changes which are occurring with time during the development of a mature stand of Norway Spruce. Pay some attention to the ratios of photosynthesizing to respiring to dead tissue.***

As the trees age, intra-specific competition thins them. Tree numbers decline exponentially with age, suggesting that the chance that a tree will die is constant throughout the life span.
Tree height increases very rapidly at first and then much more slowly. At first (22 years) the leaf area index is low because the plants are small. It reaches a peak, and then declines as some trees die and clearings appear.

The dry mass of the vegetation increases rapidly at first but begins to decline in the oldest stands, presumably because of tree death. During the growth of the stand the net annual dry mass production increases to a peak and then declines. Over the first fifty years net production increases year by year. As the trees grow, a higher proportion of the incident light energy is trapped, as suggested by the increase in the leaf area index. At this stage, the ratio of photosynthesizing leaf area to respiratory volume is high.

As the stand matures the volume of respiring tissue increases. The tissues around the cambium and the phloem, and the whole of the non-woody part of the root system, are a respiratory burden on the plant. Once the leaf area index reaches a maximum, root production declines because of the increasing volume of respiring tissue.

Eventually, whole trees are blown down and branches fall, reducing the leaf area index. Net primary production decreases still further.

Net production of litter follows total net production, of which it is a component. Litter includes branches, twigs, leaves and cones. At first, small trees produce little litter. At the peak of production (50–68 years) litter-fall declines, but this may be an artefact of sampling. An estimate of tree dry mass production is obtained by subtracting litter production from total production for each year.

The production of roots is fairly constant for the first 68 years but then declines, reflecting perhaps the loss of tree vigour, imminent death, high respiration rates, and the complete exploitation of soil in the root zone.

**c** ***At what age is the spruce growing fastest?***

Sixty-eight years (after subtracting litter production from net primary production).

**d** ***Calculate the ratio of production to biomass for each of the stands. Explain the trend.***

Total production to dry mass ratio:

| Stand age (years): | 22 | 37 | 45 | 54 | 68 | 82 | 98 | 109 | 126 | 138 |
|---|---|---|---|---|---|---|---|---|---|---|
| Ratio: | 14 | 8 | 6.9 | 5.9 | 4.3 | 3.7 | 2.9 | 2.0 | 1.5 | 1.3 |

The ratio declines markedly with stand age. This is characteristic of ecosystems in succession.

The figure for the net production of living biomass includes the energy deposited in trunks and branches as dead biomass. Dead biomass increases as the trees age. This accounts for much of the decline in the ratio of productivity to biomass as the stands mature.

When new leaves are produced early in the life of a stand they are exposed to high light intensity and on average their photosynthetic rate is high. Leaves produced in older stands 'cost' the plant just as much metabolic energy to produce, but yield lower returns because they are shaded.

In young stands the ratio of chloroplasts to mitochondria is high. This ratio declines as the stands mature. The respiring volume increases with time.

e ***Calculate the ratio of litter fall production to total production for forests of each age. What do these data indicate about the trees as they mature?***

Total production ratio:

| Stand age (years): | 22 | 37 | 45 | 54 | 68 | 82 | 98 | 109 | 126 | 138 |
|---|---|---|---|---|---|---|---|---|---|---|
| Ratio of litter production to total production: | 3.9 | 4.1 | 3.9 | 4.4 | 3.3 | 3.3 | 5.2 | 5.7 | 6.5 | 7.2 |

After 100 years, the trees divert less of their energy to permanent structures and more to branches and leaves with a shorter expectation of life. They may also shed large dead branches which were produced early in the succession, thus increasing litter fall without an increase in net production.

f ***How would the mass of detritus at the soil surface change during succession?***

It would increase rapidly at first, but in a mature stand the rates of addition and decomposition may equalize. □

## Succession and the climax

*Principles*

1. Successions often end in 'climax' vegetation, in which the species composition of the vegetation changes only slowly.
2. The type of 'climax' vegetation which develops depends considerably on human activity.
3. Most communities may not reach the climax stage because catastrophic disturbances, like fires and hurricanes, and local small-scale disturbances, like the fall or death of a tree, trigger secondary successions.

*Questions and answers*

a ***Suggest some reasons why saplings of species A might be able to grow better under the canopy of trees of other species than of trees of species A. Remember that biotic factors such as those outlined in this chapter, not just physical and chemical factors, may be involved.***

Trees of different species tend to cast different degrees of shade. In some cases trees which cast a dense shade can be 'nurse plants', providing suitable conditions, perhaps protected from frost or evaporation, for saplings of other species to become established beneath them. The leaf litter produced by a tree might be toxic to its own seedlings or saplings. Species-specific parasitic fungi may kill all the seedlings near a parent tree. Only those seedlings which appear at some distance from the parent may become established. Similarly, seed predators such as beetles, squirrels, and jays may eat all the seeds or fruits which fall near the parent tree.

**b** ***How could you test your hypotheses?***

By controlled experiments in the field, properly replicated and randomized.

# PART II The *Practical guide*

The investigations related to this chapter appear in *Practical guide 7, Ecology*.

## INVESTIGATION
## 27A Intra-specific competition in *Drosophila*

### ITEMS NEEDED

Two fruit fly cultures, one of pure-breeding wild-type (+ +), and the other of pure-breeding vestigial winged (vg vg) 1/class
Dried yeast

Ethoxyethane (diethyl ether)
Specimen tubes of *Drosophila* culture medium (similar in size to empty tube); access to further tubes of medium at weekly intervals 3/group

Balance
Binocular microscope or dissection lens 1/group
Card, piece of 1/group
Fly etherizer (*figure (P)*7) 1/group
Foam bungs 3/group
Forceps
Glass jar population cages (*figure 56*) 3/group
Marking pen, spirit-based 1/group
Paint brush, soft 1/group
Pins 2/group
Specimen tubes, empty, about 80 mm long × 25 mm internal diameter
Wall tile, white, glazed, about 150 mm square 1/group

(*Study guide* 27.1 'Competition'; Study item 27.12 'Shepherd's purse (experiment design)'.)

**Population cages.** The population cages are conveniently made from jars in which honey is sold. To insert the bung in the metal top, a hole of standard size is conveniently made as follows. Sharpen the end of a steel pipe and punch out the hole in a metal lid with a heavy hammer, using a block of wood as a firm base.

*Assumption*

1 A knowledge of monohybrid Mendelian inheritance.

*Principles*

1 Competition for resources can result in selection.
2 Pure cultures must always be compared with mixed cultures when interactions are suspected.

If only a few cages are to be set up and no other fruit fly experiments are in progress, a supply of dried 'instant' fly medium is useful. It is more expensive than 'home made' medium. It has the advantages, however, that a standard mass can be placed in each tube, and exact replicates of the food tubes prepared in small numbers each week.

Desiccation of the medium is likely to be a major problem. Push plenty of cottonwool into the medium in each tube when the yeast is added. This provides a good site for pupation.

It is important to discuss external factors such as temperature and humidity which could influence the outcome of the experiment. Such factors must be controlled as far as possible.

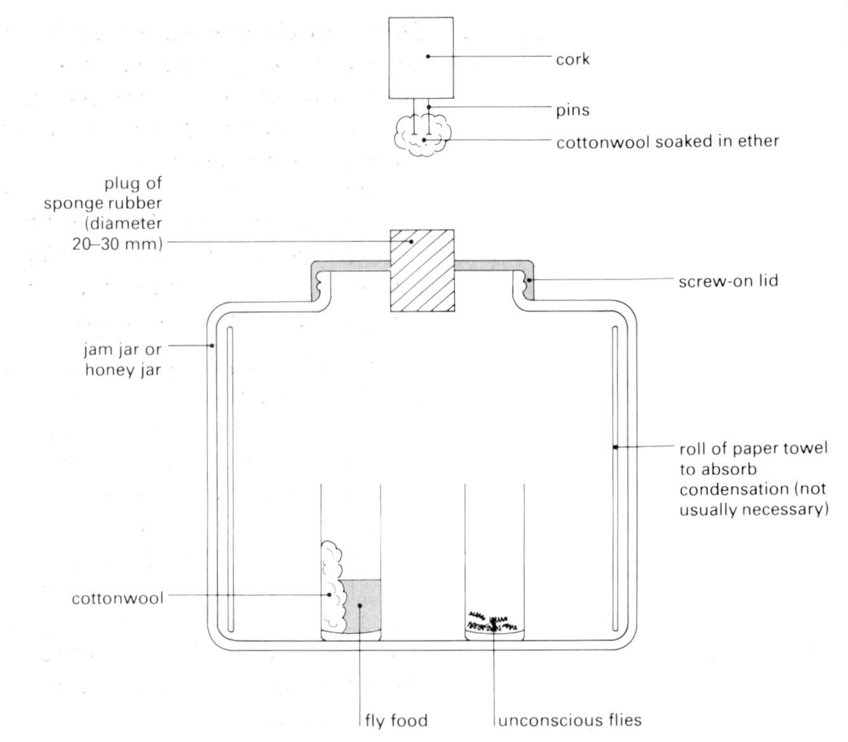


**Figure 56**
A glass jar *Drosophila* population cage suitable for use in the experiment.

*Questions and answers*

a ***What was the purpose of setting up the two populations of pure cultures?***

As controls. If either genotype is unable to reproduce well under the conditions then the number of flies produced in the two cultures may be consistently different.

b ***The gene for vestigial wings is recessive to the wild-type gene. How may this influence the relative numbers of vestigial-winged and normal-winged individuals in the population in jar B, the mixed population? Do the results support your hypothesis?***

If virgin flies can be provided, in the mixed culture their progeny ($F_1$) are likely to be all nearly wild type. The number of vestigial flies will therefore fall in the first generation and rise again in the second as the $F_1$ heterozygotes produce vestigial-winged offspring.
Then the number of vestigial-winged flies should fall more and more slowly as they become rarer and rarer, as a higher and higher proportion of the vg genes will be protected from selection in heterozygotes. If no selection was operating, the proportions of the three genotypes should remain constant from one generation to the next, according to the Hardy–Weinberg law (see *Study guide II*, Chapter 20.)

c ***Three possible hypotheses to explain the decline in the frequency of the vestigial-winged flies are listed below. Suggest some experiments which would distinguish between these hypotheses.***
***1 Vestigial-winged flies cannot efficiently compete for certain resources with wild-type flies.***
***2 Vestigial-winged flies are less able to survive some environmental stresses, such as unsuitable humidity, than the wild-type flies.***
***3 Fruit flies use their wings during courtship. A female need only mate once in her life. A male may mate several times. Vestigial-winged males might be at a disadvantage in courtship.***

A wide variety of responses to this question are possible. All three hypotheses could be correct. To test hypothesis **1**, mix large numbers of very young larvae of the two phenotypes and allow them to develop in competition. To test hypothesis **2**, set up several glass culture jars containing pure cultures of the two phenotypes. Place in each jar a solution of propane-1,2,3-triol in water, to keep the relative humidity of the air at a certain level, and a petri-dish containing agar blackened with carbon black. Count the eggs laid and hatched by each phenotype in each relative humidity. To test hypothesis **3**, set up several glass culture jars. To each jar, add one vestigial-winged virgin female and two males, one wild-type and the other vg vg. Rear the progeny of each female separately.

d ***It has been reported that vestigial-winged fruit-flies are frequent in the wild on very small islands in the Pacific Ocean. Suggest a hypothesis to explain this finding.***

Vestigial-winged flies are less likely to be blown out to sea by the continual strong winds. This advantage outweighs other disadvantages. One would expect a student to suggest that something in the environment of a small island favours vestigial-winged flies. Very small islands, for example, may be too small to support predators such as spiders. Selection in fruit-flies for wings as a means of avoiding predators might be relaxed.

## INVESTIGATION
## 27B Inter-specific competition between two *Lemna* species

(*Study guide* 27.1 'Competition'; Study item 27.13 'Competition between two flour beetle species in a limited environment'.)

*Assumptions*

1 A knowledge of the biology of water plants such as *Lemna* and their interaction with the environment.
2 An ability to handle data and plot relevant graphs, including the use of logarithmic scales.

*Principles*

1 The composition of a community may be influenced by competition between the species within it.

### TEMS NEEDED

*emna minor* and *Lemna trisulca* other species can be used, or *zolla filiculoides* according to ocal availability). The leaves hould be washed to remove other lants and animals before use.

ontainers (such as ice-cream) pproximately 400 $cm^3$ apacity 3/group
offman clips for control of air d water flow through the pparatus 4/group *(continued)*

**ITEMS NEEDED** *(continued)*

Pump or air line (optional) 1/group
Syphons (glass), rubber and glass tubing, beaker 4/group
Tank of pond water (large) 1/group

2 The isolation of individual environmental factors causing competition is often very difficult to achieve.
3 Inter-specific competition does not necessarily lead to the extinction of one species and the domination of another.

*Lemna* is a 'modular' organism (*Study guide II*, Chapter 28). It must be borne in mind that *Lemna* plants of the same species collected from the same place may be parts of the same genetic individual. Thus, although this experiment involves competition between two different *Lemna species*, only two *genotypes* may be competing.

*Questions and answers*

a ***Explain briefly how 1 variables were controlled and 2 the number of living leaves were counted in your investigation.***

**1** The use of siphons, aeration, and equal illumination should be mentioned.
**2** There are difficulties in deciding what constitutes a leaf. Do many small leaves equal one large one? What is the most acceptable measurement of growth?

b ***Plot a graph on log-normal graph paper to show the pattern of growth of each of the* Lemna *species when grown in isolation. What evidence did you find of intra-specific interaction during the experiment?***

The log graphs of each species growing in isolation should level off as the rate of production of new leaves declines. The interpretation of log-normal graphs should be discussed.

c ***Plot two graphs, on linear graph paper, one for each species of* Lemna*, comparing the number of leaves produced in the single species cultures with the number of leaves produced in the mixed culture. In the conditions of this experiment, what evidence do you have that inter-specific interaction is a significant factor in controlling the growth of either species of* Lemna*? Explain your answer.***

Growth rate and final populations may be lower in mixed culture than in single species culture.

d ***Interaction between plants growing together in a habitat may involve competition for environmental resources. What resources might restrict the growth of* Lemna *in your experiment? Select the resource which you consider to be the main limiting factor and suggest an experimental procedure which you could use to investigate its effect in more detail.***

Light, nutrient ions, pH, temperature, carbon dioxide, oxygen. There is the possibility that *L. triscula* produces a soluble carbohydrate capable of stimulating the growth of *L. minor*. Competition for oxygen may limit the uptake of nutrient ions.

If possible the collection of *Lemna* should be carried out by the pupils themselves so that the two species can be observed in their natural habitat.

## INVESTIGATION
## 27C Inter-specific interaction between clover and grass

### ITEMS NEEDED

Lawn or pasture in which white clover (*Trifolium repens*) grows in patches amongst grass. The established pattern of mowing or grazing should continue during the duration of the experiment

Ammonium sulphate (20 g dissolved in 4 $dm^3$ water is enough for a 1 $m^2$ quadrat)
Potassium nitrate or potassium sulphate for control quadrats
Tap water

Camera, capable of taking close-ups 1/class
Clip boards, graph paper, writing implements 1/group
Hammer 1/class
Measuring tape, 30 m 1/group
Quadrats, 1 × 1 m, preferably gridded into 10 × 10 cm squares with string or copper wire 1/group
Watering can 1/class
Wooden or metal pegs 1/group

(*Study guide* 27.1 'Competition'; Study item 27.62 'Crops without chemicals'.)

*Principle*

1 Many lawns and pastures have *Trifolium repens* growing in patches and this mixture is persistent. Clover does not normally die out on a neglected lawn. Indeed, it may become even more abundant when the lawn is neglected and soil nitrogen is scarce. Patches of clover may travel across a lawn, as may be confirmed by mapping clover patches each year for a few years.

One explanation for the co-existence of lawn grasses and white clover is as follows. Grass grows well when supplied with soil nitrogen, but many soils, especially compacted ones, lack nitrogen. When the grass is mown with a box on the mower and the cuttings are removed, the loss of nitrogen is continuous and debilitating to the grass. Clovers help to re-introduce soil nitrogen and thus aid grass growth. Immediately next to the clovers, the grass may grow rapidly. Grass leaves which penetrate clover patches are often long and bright yellow-green.

As a result of its increased growth, the grass may eliminate the clover. The clover, however, may spread into adjacent areas which are deficient in nitrogen. A mixture of species thus persists.

A simple way to investigate this dynamic situation is to watch what happens when one of the species is mechanically removed at the boundary. When one member is persistently removed, the other usually spreads to colonize the bare ground. Situations like this are known as 'reciprocal replacements'. They are well-documented, for example, amongst forest trees in North America.

In this experiment we can confirm that the two species act in this way, but the descriptive results cannot explain *why* one species replaces the other. Adding a nitrogenous fertilizer makes the grass grow well, and clover dies out. This indicates that low levels of soil nitrogen previously limited grass growth. Nitrogen improves the vigour of the grass. It increases the ability of its leaves to intercept light, and its roots to mine the soil for water and nutrient ions. Grass tillers may be able to grow up more rapidly through the stolons of white clover, and overtop them, decreasing their vigour. Then a reduction in the supply of light or water to the clover might decrease the ability of its mutualistic bacteria to fix nitrogen, thus slowing its rate of production of new stolons. Even if clover grew faster under the influence of a nitrogenous fertilizer, its prostrate form would ensure that it was overtopped by grass. All this assumes that the addition of fertilizer did not cause an appreciable change in the pH of the lawn soil, which might favour grasses at the expense of clover.

The results will be more dramatic if the quadrats are replicated. One quadrat, or one fertilized and one unfertilized quadrat, can be delegated to each group of students.

It will probably be best to let the experiment run for some weeks and then arrange a class discussion when the evidence has been collected.

If lawn space is limited, run the experiment as a demonstration. If no lawn is available, use turves in seed boxes in the laboratory or greenhouse. If your lawn is waterlogged or too rich in nitrogen, turves with clover can be lifted and placed in seed boxes or kipper boxes on cinders or gravel. The turves dry out quickly, or rain leaches the nitrates quickly, and it is then possible within a few days to start the experiment.

If no turves are available, in March sow grass and clover seed in loam. Allow the mat to form – this will take about five weeks – before starting the experiment.

If *Trifolium repens* is absent, other legumes will probably behave in similar fashion, but perhaps less dramatically.

Mapping should be kept to the absolute minimum. It is of course desirable to record the results in more detail and to make some quantitative measurements of growth, but the changes which occur are usually so striking that mapping and photographic record are convincing enough.

Ammonium sulphate is a convenient nitrogenous fertilizer. If enough replicates are available, potassium nitrate and potassium sulphate in tap water can be added to some of them. Each of these compounds may affect the soil pH. Changes in soil pH can be documented by testing soil samples from the experimental plots before and soon after fertilizer application with an electrical pH meter or Sudbury soil test kit.

*Questions and answers*

**a** ***Does the evidence suggest that in the absence of any one species the other might eventually colonize the lawn?***

When one species is removed mechanically this normally allows the other to invade its territory. This suggests that either could colonize that whole lawn. The soil and climate are suitable for both 'species'. However, the grasses in the lawn may be a mixture of species with different micro-habitat requirements. One of these component species, on its own, may be incapable of colonizing the whole lawn. Remember also that other grass and herbaceous species would colonize patches of lawn vacated by dead grass or created by animal artefacts such as worm casts or ant mounds. Only intense human interference could maintain a lawn of a single species.

**b** ***What sort of change might prevent this from happening?***

The invasion of a vigorous new species of herbaceous plant, such as the daisy (*Bellis perennis*) or a marked change in the mowing regime. A drastic change in the micro-environment of the lawn might make the conditions in part of it unsuitable for one of the lawn species. For instance, the growth of a tree over the site might prevent ryegrass

(*Lolium perenne*) from growing in the shaded area. Bare patches under trees are often colonized by annual meadow grass (*Poa annua*) which is more tolerant than ryegrass of shade.

c ***In what ways does the disturbance change the micro-habitat? How might this affect the interaction between clover and grass?***

If the aerial parts of one species are removed this will increase the evaporation of water from the soil, increase the light intensity at the surface, and increase the amplitude of temperature fluctuations in the local area. More rain will reach the soil surface. If buds nearby perceive the removal of the adjacent shoots, perhaps because they receive more light, they will be stimulated to produce tillers or stolons which rapidly colonize the space.

d ***Does grass grow more vigorously when extra nitrogen is supplied? If so, what advantage could the increased growth confer on the grass?***

Yes; it is lusher and greener. Nitrogen permits better growth of root and leaf. Fertilized grass should produce greater wet and dry masses than untreated grass. This allows the plant to intercept a greater proportion of the incident visible light. The increased rate of root growth enables the plant to intercept more water and nutrient ions.

e ***Does clover grow better with additional nitrogen? How do you explain the results which you have obtained?***

Not usually. The root nodule bacteria seem to provide adequate nitrogen for the clover (Chapter 27). Incidentally, many plant species capable of growing in grasslands of low nutrient status cannot respond to nitrogen application by increased growth. Such species are often slow-growing, and do not produce the inducible enzyme nitrate reductase in large quantities even when nitrate fertilizers are applied.

f ***What evidence is there that grass responds in the same way to added nitrogen and to being near clover? If there is a similarity, can you put forward an hypothesis to account for it?***

Same appearance. Clover perhaps supplies extra nitrogen.

g ***Is there any evidence of competition between clover and grass? If so, for what might they be competing?***

When clover is removed, grass invades, but it might have done so even if grass had not been removed. When grass is cut away, clover invades, but without fertilization it might have invaded the grass area anyway. The results of the experiment do not show that the two species are competing for a limited resource or that the growth of at least one of them is retarded by the interference.

**h** ***Do the results of your experiments justify the conclusion that clover and grass are competing for a limited amount of nitrogen in the soil?***

No. This is a sensible suggestion, but not proved. This question is intended to emphasize the problem of defining competition. Nitrogen is a resource which both species require. There is always enough nitrogen for the clover, in the form of nitrogen gas in the soil atmosphere, provided that the clover is vigorous enough to continue to supply its nodule bacteria with glucose. There is unlikely to be enough nitrogen for the grass in unfertilized pastures, in the form of nitrate. If the clover roots take up nitrate or ammonium ions which would otherwise be used by the grass, then clover is competing with the grass for nitrogen. In fertilized plots the grass roots are probably taking up nitrate ions at maximum rate, and nitrogen supply is not the resource which is limiting the growth of either grass or clover.

## INVESTIGATION
## 27D The holly leaf miner (*Phytomyza ilicis*) and its parasitic insects

### ITEMS NEEDED

*Ilex aquifolium* (holly) leaves, preferably collected by students

Container for holly leaves 1/group
Dissecting instruments
Microscope, binocular 1/group

(*Study guide* 27.2 'Interactions apart from competition' and 27.4 'Parasitism'.)

*Assumptions*

1 A knowledge of the life-cycle of insects.
2 Outline knowledge of the life history of an insect parasitoid such as the ichneumon fly.

*Principles*

1 Parasitic organisms are widespread and often co-exist with their hosts, causing minimal damage.
2 The population of one parasite may be largely regulated by a second parasite (hyperparasite).

This investigation was first suggested by Lewis and Taylor, 196 from which further details can be obtained.

*Questions and answers*

**a** ***What percentage of the holly leaves were affected by the holly leaf miner? Compare your result with those of other groups. How can the differences in the size of infestation be explained?***

Considerable variation will be found, even when neighbouring trees are examined. This could be explained by the amount of sunlight falling on different trees, by local air currents influencing flight and oviposition by the adult flies, or by small differences in the texture of the leaves. There are several possible project topics here.

**b** ***What effect does the miner have on the holly tree? Were heavily infested trees less vigorous than trees with more healthy leaves?***

The proportion of the total leaf area affected is usually very small, an heavily infested trees show no sign of loss of vigour. Affected leaves may be replaced earlier than healthy leaves.

c ***If the original population of the leaf miner is taken to be the number of leaves with mines, what percentage of them has survived? What percentage has been parasitized?***

Samples taken up to February will contain mainly healthy larvae, but the percentage of dead and parasitized larvae and pupae will be found to increase rapidly in spring. Pupils may point out the risk of collecting mined leaves from the previous year, since holly is an evergreen.

d ***In one study only 2 per cent of pupae examined in late May contained healthy leaf miners. What impact do the parasites have on the population of the miner? What would happen if the parasites were eliminated?***

Such a high level of parasitism must have a large influence on the population. Class discussions could follow up various aspects of this study, such as the parasitic food web (note that the parasites are all smaller than their hosts) or biological control of economically harmful predators and parasites.

## INVESTIGATION
## 27E A study of inter-specific associations

### ITEMS NEEDED

Aphids on a green leaf 1/group
Flower of white dead-nettle (*Lamium album*) or similar 1/group
Freshwater shrimp (*Gammarus* sp.) or freshwater louse (*Asellus aquaticus*) with *Vorticella* attached 1/group
Slide of mosquito/similar insect 1/group
Slide of honey bee legs 1/class
Slide of T.S. lichen 1/class
Specimen of common lichen on twig 1/group
Specimens of a plant gall such as oak apple, Robin's pinchushion 1/group

0.7 % saline

Microscope, binocular 1/group
Microscope, monocular 1/group
Microscope slides and cover-slips
Watch-glasses

(*Study guide* 27.2 'Interactions apart from competition', 27.4 'Parasitism', 27.5 'Commensalism', and 27.6 'Mutualism'.)

*Principles*

1 The classification of relationships between organisms is not always simple and straightforward.
2 Careful examination of individuals is necessary before the nature of the relationship becomes clear.
3 There is a spectrum of relationships which blend into one another.

*Assumptions*

1 The ability to use microscopes and to observe.
2 A background of information concerning the types of relationship to be seen (Chapter 27).
3 Some knowledge of simple dissection techniques.

*Questions and answers*

*A Aphid and green plant*

a ***What type of association exists between the plant and the aphid?***

Parasitism.

b ***Explain your reasons for your answer to part* a.**

The aphid could be either a predator or a parasite. The word predator, however, is usually used to refer to an animal which eats animal food. A predator often seeks out its prey, kills it, and eats it. Predators are usually larger than their prey. An aphid eats plant food and is smaller than the plant on which it feeds, on which it depends for its life cycle.

*B Mosquito and mammal*

**a** ***What evidence do you see to suggest that the mosquito may have the same relationship to the mammal as the aphid has to the flowering plant?***

Superficially, the mouthparts look similar. Both sets of mouthparts are sharp and elongated and can penetrate the tissues of the host to extract fluid. Mosquitoes and aphids both act as vectors of disease.

**b** ***What is the major difference between the aphid/flowering plant and the mosquito/mammal relationships?***

Mosquitoes only feed briefly on their host from time to time. Many aphids are permanently 'plugged in' to the phloem of their host plants.

**c** ***What is the relationship between the mosquito and the malarial parasite?***

The female mosquito is the secondary host for the malarial parasites (*Plasmodium* spp.). The malarial parasite does not harm the mosquito but it may harm the human, its primary host.

*C Gall wasp (or fly) and flowering plant*

**a** ***What is the relationship of the insect to the gall?***

It lives within the gall and feeds on the tissues within the gall. It does not seem to markedly harm the host plant.

**b** ***How has the plant responded to the activities of the gall-former?***

Very specifically. The galls grow to a characteristic size, shape, and colour, they grow on particular parts of the plant and are often confined to very localized areas.

**c** ***Does the plant appear to have been harmed?***

The relationship appears to be very finely balanced, and even densely galled plants are able to grow well. There must be some damage, however, except in the case of the nitrogen-fixing bacteria in the leaves of some plant species, such as Douglas fir (*Pseudotsuga menziesii*) where the relationship is positively beneficial to the 'host'.

*D Honey bee and flowering plant*

**a** ***How would you describe the relationship of the bee to the flower?***

Since there is benefit to both organisms, the relationship is superficially mutualistic.

**b** ***From your examination of the pollen and the stigma, do you think that the flower is specialized for insect pollination in the structure of these two parts?***

Large sticky or spiny pollen grains are typically found. The stigmas may be small, specially shaped, sticky and rough.

*E Freshwater shrimp and* Vorticella

a ***What similarities are there in the feeding mechanisms of the two organisms?***

Both are filter feeders.

b ***What are the consequences of the type of behaviour observed?***

The current created by the vibration of the limbs of *Gammarus* may waft food particles towards the *Vorticella* living on it. On the other hand, the currents set up by *Vorticella* are probably so weak that *Vorticella* neither helps nor harms *Gammarus*. This is therefore a case of commensalism – one partner benefits, but the other seems unaffected.

*F Lichens*

a ***What are the consequences of the arrangement of the fungal and algal components?***

The algae, towards the top of the thallus, can intercept much light. The fungal hyphae around them reduce water loss, and can absorb nutrient ions from both air and soil.

b ***How does the association between the two mutualists affect the types of habitat in which the lichens grow?***

Fungi can normally grow in the dark. Lichens cannot. Lichens often colonize bare rock surfaces. Fungi could not colonize bare rock because it lacks a food source. The alga could not on its own withstand the occasional high temperature and the resulting desiccation. Yet lichens can withstand these extreme conditions.

## INVESTIGATION
## 27F A grassland survey using sampling techniques

TEMS NEEDED

ɅIeasuring tape 1/group
ʾegs for marking out the
reas 40/class
Quadrat frame (wire or wood),
0 × 50 cm 1/group
Random number tables
–99 1/class

(*Study guide* 27.7 'Dispersal and other historical factors'; Study item 27.71 'Dispersal onto Surtsey'.)

*Assumptions*

1 The ability to identify the commoner dicotyledonous species found in grassland.
2 A knowledge of the use of co-ordinates to identify positions on a grid.
3 The ability to use a Wilcoxon signed-ranks test for matched pairs and to interpret the results.

*Principles*

1 Discontinuities within a habitat can be investigated by taking random samples using a quadrat frame.
2 Simple estimation of the percentage covers of a few common species can reveal or confirm differences in the vegetation without damaging the community. Thus the method is ideal for long-term investigations.

Advance preparation is essential. Careful selection of a suitable site and the availability of keys or guides to identification are required. If 'natural' sites are not available, it may be possible to arrange that an area of playing field is left unmowed or untreated with fertilizers or weed-killer. Interesting results can be obtained with minimal preparation if a 20 m length of string is tied to a tree surrounded by grass, and quadrat samples are studied every metre along the radial transect so obtained. Alternatively, the vegetation on animal artefacts such as ant-hills and mole-hills can be compared with the surrounding vegetation. Results obtained by groups in previous years should be retained and produced for comparison.

*Questions and answers*

**a** ***Describe the difference in the distribution pattern and numbers of plants found in the two areas.***

The answer to this question will differ, according to local circumstances.

**b** ***Relate the differences in the two communities to differences in the environment. Suggest as many factors as you can which might account for the distribution pattern of the plants. Which may be the most important? Give reasons for your answer.***

The answer to this question will differ, according to local circumstances.

**c** ***Which species seem to be 1 most tolerant and 2 least tolerant of the environmental factors which you have listed above?***

The answers to **b** and **c** will depend on the area selected, but grasses are tolerant of mowing, as are 'rosette' plants and clover. The least tolerant species only do well in long meadow grass and are usually not rosette-formers.

**d** ***Which features of these tolerant species might be responsible for their ability to survive?***

Plasticity, and the ability to tolerate shoot and root competition.

**e** ***Some of the plant species may exist in similar abundance in each habitat, but show considerable variation in growth form. Such species are said to exhibit plasticity if the variation is environmentally-induced rather than the result of genetic differences between members of the same species. An example is shown in* figure [P]12. *Describe or draw any examples of plasticity which you observe in your areas.***

A useful discussion of the origins of plasticity could be based on students' observations. Students might be able to suggest simple experiments to distinguish environmentally-induced plasticity from genetic differences.

f ***Weeds can establish themselves after grass has been sown. How would seed size, seed numbers, speed and rate of germination and vegetative reproduction affect the establishment of weeds in grassland?***

Plants establishing themselves produce large numbers of efficiently-dispersed seeds, rapid and often irregular germination, rapid growth and flowering. If they are persistent then there is often some means of vegetative reproduction. Rapid germination in the light allows establishment in open habitats and irregular germination preserves a reservoir of seeds in the soil so that exposure to the light will cause rapid germination. These are some of the features of opportunist plants. A local weed survey, using the techniques employed in this experiment, together with growth experiments in the laboratory can provide the basis for interesting project work.

## PART III Bibliography


BENNETT, D. P. and HUMPHREYS, D. A. *Introduction to field biology*. 3rd edn. Edward Arnold, 1980.

BRILL, W. J. 'Nitrogen fixation: basic to applied'. *American Scientist*, **67**(4), p. 458, 1979. (A general introduction to the mutualism of *Rhizobium* with legumes.)

BRILL, W. J. 'Agricultural microbiology'. *Scientific American*, **245**(3), p. 146, 1981. (Thorough and interesting discussion of the prospects and techniques for producing cereals which can fix their own nitrogen.)

CARLQUIST, S. 'Chance dispersal'. *American Scientist* **69**(5), p. 509, 1981. (An analysis, with plenty of examples, of the modes of dispersal of plants in particular onto islands.)

COX, F. E. G. 'The malarial parasites, *Plasmodium* spp.' *Biologist*, **28**(1), p. 9, 1981. (Concise introduction to the malarias of humans and other animals.)

DARLINGTON, A. and HIRONS, M. J. D. *The pocket encyclopaedia of plant galls*. Blandford, 1968.

DIXON, A. F. G. Studies in Biology 44. *The biology of aphids*. Edward Arnold, 1973.

DOBSON, M. J. 'When malaria was an English disease'. *Geographical Magazine*, **54**(2), p. 94, 1982. (Fascinating account of the effects and eradication of malaria in south-east England.)

DOWDESWELL, W. H. *Ecology: principles and practice*. Heinemann Educational, 1984.

FREDERIKSSON, S. 'Life develops on Surtsey'. *Endeavour, N.S.*, **6**(3), p. 100, 1982. (Well-illustrated description of the development of the fauna and flora on this new volcanic island.)

HARPER, J. L. *The population biology of plants*. Academic Press, 1977.

HILL, T. A. Studies in Biology No. 79, *The biology of weeds*. Edward Arnold, 1977. (The chapter on 'What weeds do' summarizes the various ways in which weeds affect crops.)

HUTCHINGS, M. J. 'Ecology's law in search of a theory'. *New Scientist* **98**, p. 765, 1983. (Discusses why the −3/2 power law applies intra-specific competition in plant populations.)
JONES, M. L. '*Riftia pachyptila* Jones: observations on the vestimentiferan worm from the Galápagos rift'. *Science*, **213**, p. 333, 1981. (Describes the structure of the worm.)
KING, T. J. and WOODELL, S. R. J. The use of the mounds of *Lasius flavus* in teaching some elementary principles of ecological investigation. *Journal of biological education* **9**(3/4), pp. 109–13, 1975.
LAUBIER, L. and DESBRUYÈRES, D. 'Oases at the bottom of the ocean'. *Endeavour*, N.S. **9**(2), p. 67, 1985. (Fascinating, well-illustrated description of the hot water springs beneath the ocean, in which *Riftia* lives.)
LEWIS, T. and TAYLOR, L. R. *Introduction to experimental ecology.* Academic Press, 1967.
PHILLIPS, R. S. Studies in Biology No. 152, *Malaria.* Edward Arnold, 1983. (Excellent summary of the life cycles and control of these important parasites.)
PONTIN, A. J. *Competition and coexistence of species.* Pitman, 1982. (Clear, detailed analysis of the evidence for competition and its effects in both laboratory and wild populations.)
POSTGATE, J. 'Nitrogen fixation and the future of the world's food supply'. *Biologist*, **26**(4), p. 165, 1979. (Very brief summary of the applied potential of nitrogen fixation.)
SOUTHWARD, A. J. et al. 'Bacterial symbionts and low $^{13}C/^{12}C$ ratios in tissues of Pogonophora indicate unusual nutrition and metabolism'. *Nature* **293**, p. 616, 1981. (Deals with some of the evidence for mutualistic bacteria inside vestimentiferan worms, like those of the Galápagos Rift communities.)
TAYLOR, J. C. 'The introduction of exotic plant and animal species into Britain'. *Biologist*, **26**(5), p. 229, 1979. (Lists and discusses many of the more dramatic cases of the introduction of species.)

# CHAPTER 28 POPULATION DYNAMICS

*Review of the chapter's aims and content*

1. The fluctuations in population size from year to year in various species of animals and plants, including humans, are described and accounted for.
2. The fundamental distinction between unitary ȧnd modular organisms is made. In modular organisms, the 'modules' to be counted might be rosettes, leaves, buds, branches, or polyps.
3. Methods of counting organisms are briefly discussed. Methods of sampling plant communities, and the capture–mark–recapture technique for small mobile animals, are described.
4. Fluctuations in population size must be seen as the consequences of birth and immigration on the one hand, and death and emigration on the other.
5. The S-shaped ('logistic') curve of population increase with time, characteristic of new species entering a favourable environment, is illustrated with reference to bacteria and yeast.
6. Population sizes remain constant when the birth-rate and death-rate are equal. Different species, however, have different expectations of life and different patterns of death with age.
7. Research on the great tit (*Parus major*) in Wytham Wood, near Oxford, has pinpointed many of the factors which help to keep its numbers relatively stable from year to year, such as territories, competition for food, and predation.
8. Population sizes are affected by factors which act either in a density-dependent or a density-independent manner.
9. The populations of many species are regulated by parasites or predators. Agricultural pests can sometimes be controlled by introducing a suitable predator or parasite.
10. Some species, such as lemmings, exhibit cyclic fluctuations in population size.
11. The human population is rapidly increasing. Population increase is faster in less developed countries, where birth rates remain high and death rates have fallen. The balance between population and food supply, and the prospects, is discussed.

# PART I The *Study guide*

## 28.1 How can population size be assessed?

*Principles*

1. Sampling techniques must be used to estimate the population sizes of species in nature.
2. The capture–mark–recapture technique is suitable for estimating the population size of some of the species of small, mobile animals.

**Practical investigations.** ***Practical guide*** **7, investigation 28A, 'Sampling methods for small invertebrates', and investigation 28B, 'Capture–mark–recapture technique'.**

### STUDY ITEM

**28.11** The capture–mark–recapture method (short answer question) (*J.M.B.*)

*Questions and answers*

**a** ***1 Calculate the estimated population size for the grasshoppers in the meadow.***

***2 Suggest a reason why some of the original marked grasshoppers were kept enclosed in a cage and not released.***

**1** Thirty marked grasshoppers were recaptured (M) on Day three. In order to capture all the 180 marked individuals (C) released into the population on Day one, a sample of six times greater than the 150 individuals recaptured (R) would have been required. Thus the total population size was $6 \times 150 = 900$. Alternatively, the population size = $CR/M = (180 \times 150)/30 = 900$.

**2** The twenty marked grasshoppers were kept to find out how many of them died of natural causes in the next two days. The marking may have lowered their life expectancy.

**b** ***Explain how the marking–recapture method as practised here can introduce large errors in the estimation of population size.***

If marked individuals found beneath stones were preferentially chased and recaptured, the population size would be overestimated. If the marked individuals were more sluggish than unmarked ones they would escape more slowly when stones were raised and stand more chance of being recaptured.

**c** ***Suggest two further circumstances in which population estimates using marking–recapture methods might be inaccurate.***

Marking may kill individuals. Paint marks on one individual may repel others. If a long time elapsed between capturing and recapturing, many unmarked individuals are born, and many marked individuals die. If larger, older individuals are preferentially captured and marked, they may die first. If individuals migrate into or out of the area, this will upset the estimate. □

## 28.2 Population growth

*Principles*

1. The increase in numbers with time, in a population reaching a new environment, may follow a S-shaped ('logistic') curve. A 'lag' phase of slow population growth is followed by a 'log' phase of rapid increase in numbers. This, in turn, is followed by population stability.
2. Limiting factors act on dense populations to make birth and death rates approximately equal.

**STUDY ITEM**

28.21 The growth of a population of yeast

*Assumption*

1 An understanding of the differences between logarithmic and linear scales on graphs.

*Questions and answers*

a ***What phase in population growth does A in figure [S] 325 represent? What is happening to the cells at this stage?***

A is the 'lag' phase. The organisms are possibly adjusting to a new environment. Population increase is slow since there are few individuals in the population to reproduce.

b ***Describe the manner of increase in the number of cells at stage B.***

Yeast cells divide by budding and the daughter cells separate. The increase in numbers can be described as logarithmic, exponential, or geometric. When the first half of the graph is plotted with a logarithmic vertical scale, a straight line is obtained (see Study item 28.22).

c ***What difference would there be in the graph if the population growth had not been limited after time X?***

The graph would continue to rise logarithmically.

d ***What is happening in the culture to give the shape of the curve at C?***

The birth rate is declining, the death rate is increasing, or both these processes are happening at once.

e ***Name two factors which could be causing the shape of the curve at D, assuming that the temperature remains constant.***

Lack of oxygen, exhaustion of nutrients, accumulation of toxic waste products.

f ***What would be the shape of the graph if we used a sample of the culture, in its stationary phase, to inoculate a fresh flask of nutrient broth?***

Repetition of the growth curve in figure (S)325.

g ***Would you describe the possible interactions between the yeast cells at C as competition? For what resources could they be competing?***

Yes, except for the accumulation of toxic waste products; these are not a resource in short supply. Yeast cells could compete for oxygen, organic nutrients or inorganic nutrients. □

**Practical investigation. *Practical guide* 7, investigation 28C, 'Population growth in *Chlorella*'.**

## STUDY ITEM

## 28.22 Population growth of bacteria

*Assumption*

1 A knowledge of how to calculate the $\text{logarithm}_{10}$ of a number.

*Principle*

1 Mathematical models can be used to interpret experimental data.

*Questions and answers*

a ***Plot a graph, on your own graph paper, of the equation $x = 2^n$ for values of $n = 0$ to $n = 5$. The symbol n represents the number of generations which have elapsed.***

See *figure 57.*

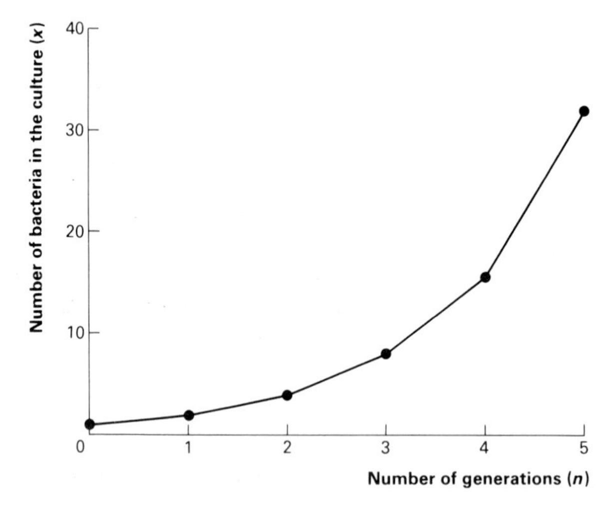


**Figure 57**
Graph of the equation $x = 2^n$, for values of $n = 0$ to $n = 5$.

b ***Compare your graph with that in* figure [S] 327 *which shows the observed growth rate of a population of the bacterium* Lactobacillus bulgaricus. *What conclusions do you draw from this comparison?***

The similarity between the observed results and the model is striking. Thus the hypothesis on which the model was constructed probably explains the way in which the growth of the population occurs. The assumptions are that each bacterium divides into two when it reproduces, and that all the bacteria do so simultaneously after a constant time interval.

c ***For* Lactobacillus bulgaricus, *plot a graph of the $\log_{10}$ numbers of the bacterium against the time in minutes. Use for the horizontal axis the same linear scale as in* figure [S]327. *This graph is most simply constructed on log-linear graph paper. If the bacteria divide by binary fission 1 do they divide at regular intervals? 2 what is the generation time?***

See *figure 58.* The cells divide at regular intervals. The generation time is about 17 minutes.

☐

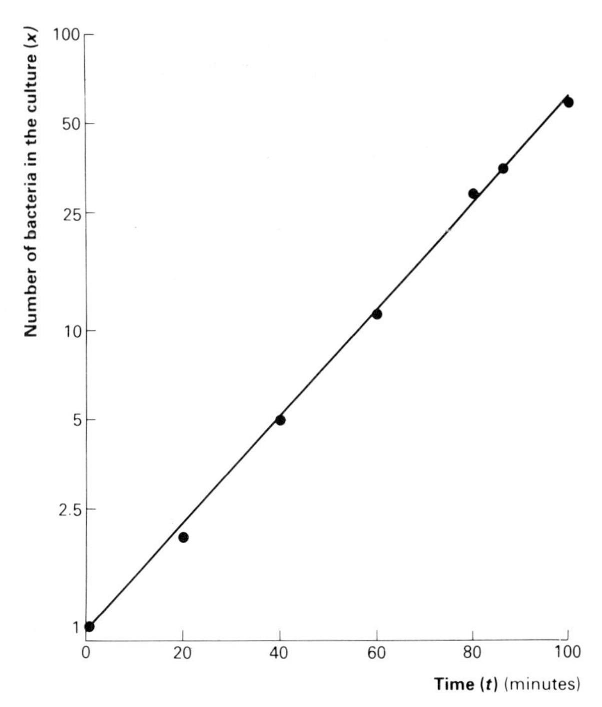


**Figure 58**
Graph of numbers of the bacteria (logarithmic scale) against time in minutes. The time taken for the numbers to double is constant at 16–17 minutes.

## STUDY ITEM
## 28.23 Age structure of a fish population

*Principle*

1 A population can remain a constant size indefinitely with the same relative proportions of individuals in each age class.

*Questions and answers*

a ***How can the ages of fish be determined?***

Fish scales show growth rings similar to those found in cross-sections through woody stems. Each ring corresponds to one year's growth. The numbers of rings in the scales can be related to the length of the fish. The frequency distribution of ages in a fish population can therefore be approximately determined from a frequency distribution of fish lengths.

b ***How did the total catch differ from year to year?***

1950–51 6017
1951–52 6951
1952–53 7650
It increased.

c ***Is it justifiable to conclude from these data that the plaice were more abundant in some years than others?***

No. The results do not show how efficient the trawlers were at finding high numbers of plaice locally, nor how efficient they were at catching them.

**d** ***Concentrating on the age-classes 4–11 only, calculate for each catch figure the percentage of the year's catch which it represents.***

| *Age of plaice* | *Per cent of the year's catch in each age class* | | |
|---|---|---|---|
| | 1950–51 | 1951–52 | 1952–53 |
| 4 | 38.6 | 37.0 | 18.6 |
| 5 | 28.6 | 29.6 | 37.0 |
| 6 | 6.4 | 14.1 | 17.1 |
| 7 | 3.3 | 3.8 | 6.8 |
| 8 | 1.5 | 2.2 | 1.6 |
| 9 | 1.5 | 1.0 | 1.4 |
| 10 | 1.3 | 0.8 | 0.8 |
| 11 | 0.9 | 0.8 | 0.5 |

**e** ***On the basis of these percentages, calculate the proportion of plaice which survived until the next year, for each of the age-classes surrounded by the rectangle.***

| *Age of plaice in year 1* | *Per cent survival to following year* | |
|---|---|---|
| | 1950/1 to 1951/2 | 1951/2 to 1952/3 |
| 4 | 77 | 100 |
| 5 | 49 | 58 |
| 6 | 59 | 48 |
| 7 | 67 | 42 |
| 8 | 67 | 64 |
| 9 | 53 | 80 |
| 10 | 61 | 62 |

**f** ***Examine the percentages which you have calculated in part* e. *How do the chances of survival of the fish differ 1 between one year and the next and 2 between ages?***

**1** On the whole the chances of survival are similar in both years.
**2** Perhaps the 4-year-old fish stand more chance than others of survival. Survival, however, does not seem to depend much on age.

**g** ***Does the age-distribution change significantly with time? What will be the age-distribution in ten years time?***

No, it does not change. In 1962–3 the frequency distribution of ages would probably be the same as in 1952–3.

**h** ***Discuss the changes which might occur in the age-distribution if***
***1 fishing intensity increased so that twice as many fish were caught per unit time.***
***2 fishing ceased.***
***3 the mesh size of the nets was greatly reduced.***

The effects depend on whether or not the fish caught are an important proportion of the whole population.

**1** If catches are increasing, the catch will probably have more impact on the older than the younger age-classes, because the results suggest that the fish begin to be caught when they reach four years old. This will produce a rapidly-growing fish population with a higher percentage of fish in the younger age-groups.

**2** If fishing ceased a higher proportion of older individuals would survive, since these were preferentially caught. A new stable age-distribution would result with more fish in the higher age-classes.

**3** A smaller mesh size would kill many small fish, thus reducing 'recruitment' into the adult breeding population. The density of young fish, and ultimately the whole population, would decline. □

## STUDY ITEM

## 28.24 Expectations of life of different animal species

*Principle*

1 Species differ in the pattern of their mortality with age. This pattern is partly genetic and partly environmental.

*Questions and answers*

**a** ***State what each of the curves in* figure [S] 328 *shows. Discuss the reasons for the differences between the curves.***

From top to bottom, the three curves in figure (S)328a are known as Deevey type I, II, and III curves. Dall mountain sheep produce a small number of large offspring which are long-lived. They live in a cold environment with no large predators and little disease. The shape of their curve reflects the fact that many of them die of natural causes at the ends of their normal life-span. In lapwing, the probability of death is constant with age. The sardine, however, produces vast numbers of fertilized eggs but most of the small fry are eaten or are hostages to fortune.

Figure (S)328b illustrates that the environment affects survival pattern. In the wild the night herons are exposed to predation, competition for food, and disease. Cosseted in a zoo, infant mortality is reduced and many individuals live out their natural life span. Similarly (figure (S)328c) infant mortality in humans was high in London in the 17th century because of disease and possibly competition for food. When these constraints were largely removed the cosseted human population of the USA in 1930 suffered fewer early deaths.

**b** ***Comment on* figures [S]328b *and* [S]328c *in the light of the differences in the curves in* figure [S]328a.**

Figures (S)328b and 328c show that extreme environmental influences can have a considerable impact on the survival curve. In nature, however, the survival curves in figure (S)328a would be little influenced by normal environmental fluctuations, for two reasons. Firstly, the scale of infant mortality depends on whether a species invests its energy in a few large offspring (dall mountain sheep), each with a high probability of survival, or a large number of tiny offspring (sardine). Secondly, populations are exposed to a variety of density-dependent factors which regulate their population size. If, for example the mortality of lapwing chicks was slightly higher than usual, the remaining birds might have more chance of survival in the competition for food. If young sardines survived for longer than usual in a dense population, their numbers would soon be reduced by predation or disease.

In the habitat of the dall mountain sheep, the zoo and the USA in 1930 all the animals were living in environments virtually free from disease and predation. Amongst lapwings and night herons in the wild, and humans in London in the 17th century, disease and other mechanisms of population regulation were operating. □

## 28.3 Population limitation

*Principles*

1 Many species, despite high potential reproductive rates, have population sizes which fluctuate between narrow limits from one year to the next.

2 The population sizes of animals and plants in nature are regulated by a complex of interacting factors, such as competition for food, territories, predation and parasitism.

### STUDY ITEM

### 28.31 The population dynamics of the great tit (*Parus major*) in Marley Wood, Wytham Woods, near Oxford

*Questions and answers*

a ***Study the nine graphs shown in* (figures [S]333–[S]341).**

***1 Summarize what each graph shows.***

***2 Discuss the reasons for the shape of each graph.***

***3 Define the conditions in which a nest might produce most offspring and those in which it might produce fewest offspring.***

**1, 2** Figure (S)333. The higher the density of breeding pairs, the fewer young were fledged per pair.

Competition for food? Predation?

Figure (S)334. The warmer the weather in spring, the earlier the great tit breeds.

Improved food supply?

Figure (S)335. The earlier the eggs hatch, the higher the proportion of young which leave the nest.

Improved food supply? Later seasons produce less food.

Figure (S)336. In ten years when the winter moth pupated early, great tits hatched early.

Not cause and effect! Some years were more suitable than others for early reproduction in great tits and early hatch of winter moth caterpillars in leaf buds of oak.

Figure (S)337. The higher the density of breeding pairs, the fewer eggs are laid per nest.

Competition for food?

Figure (S)338. The more eggs laid per nest, the greater the chance that an egg will not hatch.
Predation?
Figure (S)339. The higher the density of breeding pairs, the greater the proportion of the nests which are predated by weasels. Weasels form a 'search image' for great tit nests when nests are locally abundant.
Figure (S)340. The more young birds per nest, the less they weigh on average.
Lighter eggs laid? Less food per bird provided by parents.
Figure (S)341. The heavier a nestling on the fifteenth day after hatching, the greater its chances of subsequent survival. More food reserves? More active?
**3** A nest will produce the most offspring if it is isolated and the winter is mild, and least if it is surrounded by other great tit nests and the winter is cold. For when population density is low, more young are fledged per nest (figure (S)333). When March is warm, the earlier the great tit breeds (figure (S)334) and the higher the proportion of young which leaves the nest (figure (S)335).
The contradictions in these graphs will be pointed out by good students. Young great tits cannot avoid death from one cause or another. For example, if a winter is cold the density of nests is unlikely to be high next spring because of winter death. The lower the density of breeding pairs, the more eggs are laid per nest (figure (S)337). The more young birds per nest, the less they weigh on average (figure (S)340) and the less their chance of subsequent survival (figure (S)341). This conflicts with the results in figure (S)333.

**b** ***Explain why great tits require more energy in summer than in winter.***

In winter the great tits spend most of the daylight hours searching for food, and since they are homoiotherms, an increased food intake is necessary to maintain their body temperatures under cold conditions. About 75 per cent of the great tits alive in the autumn of 1962 disappeared in the severe winter of 1962–3.

**c** ***List the various factors which can 'regulate' the population size of great tits in Marley Wood, that is, factors which tend to bring the population size down when it is high, but which cause little mortality when the population is small.***

Territories
Competition for food between birds in adjacent nests
The higher the density of breeding pairs, the fewer eggs are laid per nest
The more eggs laid per nest, the greater the chance that an egg will not hatch
Predation from weasels
Predation by sparrowhawks
Competition between young birds for food
Factors such as death from winter cold should not be mentioned because they are not density-dependent.

**d** ***Why are butterflies, aphids, and locusts common in some years and rare in others?***

When the climate is favourable a high proportion of eggs form individuals, but when the climate is unfavourable there is massive mortality. The death rate is influenced mainly by the climate. The death rate does not depend upon predation or upon competition between individuals for food, mates, or shelter. □

**Practical investigation. *Practical guide* 7, investigation 28D, 'Population regulation in the water flea, *Daphnia*'.**

## STUDY ITEM

## 28.32 Density-dependent, density-independent, and inverse-dependent factors

*Principle*

**1** All three types of factor affect population size. Populations in the stable phase, however, are regulated by density-dependent factors.

*Questions and answers*

**a** ***State for each of the lines A–H whether it represents a density-dependent, density-independent or an inverse density-dependent relationship. (Inversely density-dependent factors have the opposite effect to density-dependent factors – they kill a larger proportion of the population when the population size is small than when it is large.)***

**A** Inverse density-dependent
**B** Inverse density-dependent
**C** Density-dependent
**D** Density-dependent
**E** Inverse density-dependent
**F** Density-dependent
**G** Density-independent
**H** Inverse density-dependent

**b** ***Explain why the shooting of a wood-pigeon population is likely to have only a short-term impact on the population size.***

Next year the few remaining females might be better nourished, because of reduced competition for food. They might lay more eggs and produce better-fed chicks. When the young birds leave the nest, competition between them for food might be less intense than before, and the risk of predation (for example, shooting again by the farmer) reduced. Each individual would have a greater life expectancy than previously, and the population size would soon reach its previous hig[h] level.

c ***Discuss for each of the following factors whether it is likely to be density-dependent, density-independent, or inversely density-dependent.***

**1** ***a large population of seedlings being grazed by a single slug;***

**2** ***death of birds from hypothermia in winter;***

**3** ***death of birds from hypothermia in winter if there is only a limited number of warm places where they can shelter;***

**4** ***death of birds from hypothermia in winter if their low body temperature depends on how much food they have collected in competition with other birds in the population;***

**5** ***the mortality of locust eggs in dry sand;***

**6** ***a pest population being attacked by an insect parasitoid capable of locating high concentrations of its prey.***

**1** Inversely density-dependent. The slug can only eat a certain number of seedlings. This comprises a smaller proportion of a large seedling population than a small one.

**2** Density-independent. The same proportion of birds should be susceptible to low body temperatures in large and small populations.

**3** Density-dependent. In a small population most birds will survive in warm places, but in a large population, only a few will find these places and the rest will die.

**4** Density-dependent (probably).

**5** Density-independent.

**6** Density-dependent.

d ***The graph in* figure[S] 343 *illustrates cyclical fluctuations in population size of a predator and its prey, of a type which can often be achieved in laboratory cultures. Explain these cycles of abundance in terms of the density-dependent factors acting on the populations.***

Prey numbers increase and the predator increases as food becomes more abundant. Increased predation of prey reduces prey numbers (more competition for food amongst prey might also be implicated). Lowered food supply reduces predator numbers. When predator numbers are low, prey can increase again. Oscillations in both species are out of phase.

e ***Discuss the significance of density-dependent factors in the regulation of the population sizes of animal and plant species in nature.***

All three types of factor can operate on populations of animals and plants. If density-independent factors are important (certain insects) numbers may fluctuate wildly from year to year (for example, populations of the desert locust *Schistocerca gregaria*). Density-dependent factors damp these potential oscillations.

Well-known examples of other factors which operate are: Territorial regulation of population size in the tawny owl (*Strix aluco*)

(Southern, 1970; this and the references that follow are listed in the bibliography to this chapter);
Competition for food in deer on the Kaibab plateau and moose on Isle Royale (Lack, 1954);
The population dynamics of three buttercup species (Harper, 1977);
The role of the *Encarsia* parasitoid in controlling greenhouse whiteflies (Study item 28.33, Samways, 1981).

f ***Imagine a population of plants, such as that of creeping buttercup (*****Ranunculus repens*****) in a meadow. The plants grow by producing stolons, overground runners which produce new stolons at their tips. Discuss the various factors which might keep its population size relatively constant from year to year.***

In dense populations, the probability that a mature plant will produce a daughter is lower than in sparse populations. Furthermore, the life expectancy of a daughter decreases as the population density increases.
□ Reproduction by seed is of little relevance.

**Practical investigation. *Practical guide 7*, investigation 28E, 'The responses of a predator to changes in the number of its prey.'**

**STUDY ITEM**

28.33 Pest control in glasshouses: biological control or pesticides?

*Principles*

1 Insect parasitoids are capable of reducing the population sizes of their hosts to low and economically-acceptable levels.
2 They are often used for this purpose in 'biological control', an alternative strategy to the use of pesticides for the destruction of agricultural pests.

*Questions and answers*

a ***What are the difficulties of integrating biological control with pesticide application in the same control programme?***

The parasite or predator may be susceptible to the same pesticide as its prey.

b ***The graphs in*** **figure [S] 345** ***show an attempt to control a spider mite under experimental conditions with*** **a** ***insecticide and*** **b** ***a predatory mite. Comment on the graphs.***

With insecticide, repeated applications were necessary to keep the mite population size below the level at which it might cause economic damage. Each application killed the majority, but not all, of the mites, whose populations then underwent rapid increases. Control by the predatory mite was successful and continuous, probably because it was density-dependent.

**c** ***Compare the advantages and disadvantages of the biological control of pests (adding a parasite or a predator to control the pest) with the use of pesticides to control the pest, under the following headings:***

*1 speed of action*
*2 specificity to pest*
*3 toxic effects*
*4 cost of programme*
*5 host resistance to pesticide, parasite, or predator*
*6 effect on beneficial insects*
*7 ability to kill a range of potential pests*

**1** Pesticides achieve more rapid control, essential if the pest is already causing economic damage.
**2** Insecticides may kill useful parasites and predators as well as pests. If parasites and predators reproduce more slowly than pests, this may stimulate a population explosion of the pest. Parasites are usually host specific. Predators often eat a range of prey species, on which they can maintain themselves when pest population densities are low.
**3** Some pesticides have long-term toxic effects. They accumulate in soil and only slowly decompose. Their residues may alter the relative proportions of species in the soil fauna. Parasites and predators have no toxic effects.
**4** Pesticides are expensive. They must first be researched and developed, and tested for side-effects. They must then be manufactured or extracted. Spray equipment on aircraft, on vehicles or on the ground is expensive to operate. Biological control, apart from the initial research expenditure, costs nothing.
**5** When pest resistance appears to a pesticide, the only remedy is to introduce a different pesticide. If a pest becomes resistant to a parasite or predator, there will be selection in the parasite or predator population for those individuals which can still attack the pest. This will overcome the resistance. Natural selection will maintain the balance between parasite and host, predator and prey.
**6** Beneficial insects should not be affected by a biological control agent, but might be killed by a pesticide.
**7** Pesticides usually kill a wider range of species than biological control agents. Pesticide application may kill a wide range of potential pests. A parasite or predator might reduce the population of its target species, but a similar species, unaffected by the predator or parasite, might undergo a population explosion.

**d** ***Imagine that there was a widespread outbreak of the sugar-cane borer (*****Diatraea saccharalis,** ***Lepidoptera), on sugar cane (*****Saccharum** ***spp.). Draw a flow diagram to represent the measures which you might take to assess the scale of the problem and to control the pest.***

☐ See *figure 59.*

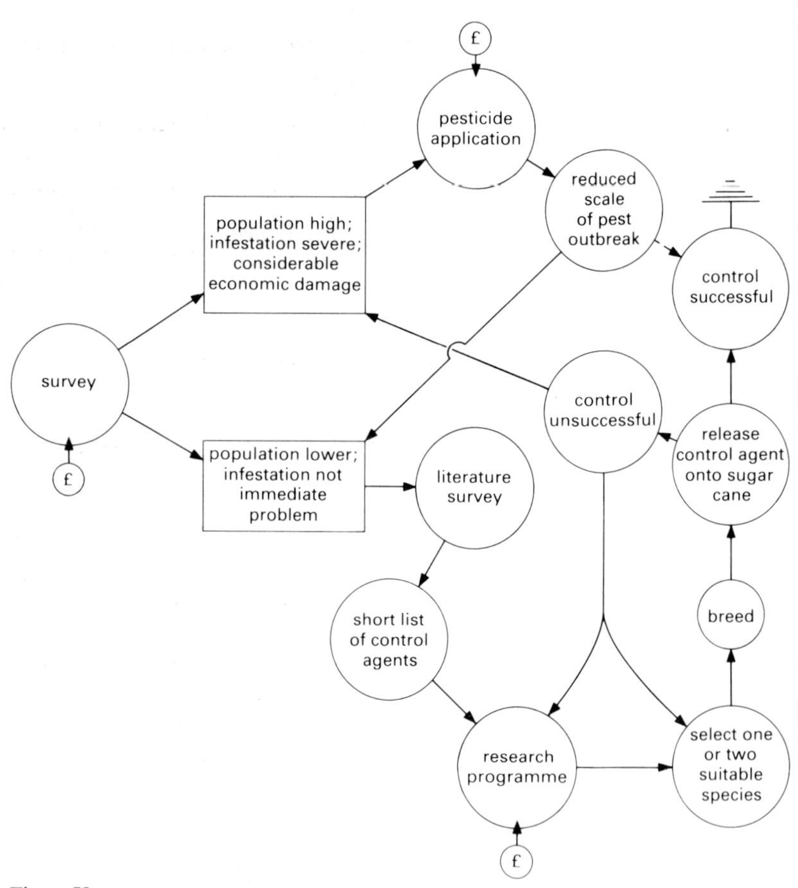


**Figure 59**
Hypothetical scheme for counteracting an infestation of the sugar cane borer.

## 28.4 Population cycles

*Principle*

1 Some species have regular cycles of abundance and scarcity. In most cases the reasons for these rhythmical cycles are obscure.

*Cyclical fluctuations in populations of lemmings* (Lemmus *spp.*)

*Principles*

1 Lemmings in the Arctic exhibit population cycles with peaks three and a half years apart. Shortage of food when the population density is high probably causes the rapid decline in numbers.

*Questions and answers*

a ***How could you investigate the hypothesis that lemming migrations are necessary to regulate the lemming population?***

Select several areas which will be subjected to intense trapping for lemmings over several months. Fence half of the areas, in a way which ensures that lemmings cannot leave them, and leave the other half unfenced. Compare the population fluctuations in the two areas.

**b** ***How might hunting have affected these fluctuations?***

Hunting could have produced the fluctuations in the snowshoe hare population if the Hudson Bay Company shot a significant proportion of the animals. There is also the possibility that Cree and Oijuba Indians in the area might have contributed to the fluctuations by hunting the hares. The peaks in lynx numbers, however, follow those in hare numbers. It seems most likely that their numbers are influenced more by the abundance of hares than by hunting.

## 28.5 Human populations

*Principles*

1 The human population is increasing rapidly.
2 The rate of population increase is greater in the 'less developed' than the 'developed' nations.
3 The developed nations have undergone the 'demographic transition'. Initially a slowly increasing population, with high birth and death rates, underwent rapid increase. This was because the death rate was lowered by improvements in public health, sanitation, medicine and nutrition. Later, the birth rate declined, and populations stabilized.
4 Those 'less-developed' nations with vigorous birth control programmes are beginning to exhibit reduced birth rates.
5 The human population will ultimately stabilize because there are limits to agricultural productivity.

### STUDY ITEM
### 28.51 The current World distribution of population increase

*Principles*

1 The 'developed' nations have achieved population stability, but numbers are still rising steeply in most of the less wealthy nations.
2 This is because of a much higher birth rate in the less wealthy nations.

*Questions and answers*

**a** ***Discuss the influence of geographical position and economic development on the population characteristics of the countries listed in table*[S]*69.***

In many of the 'less-developed' countries (LDCs), birth rates are high, family size is high, the population is increasing rapidly, and a high proportion are less than fifteen years old. In the 'developed' nations (DCs) birth rates are much lower, family size is small, the population increases slowly, and only 20–24 per cent are under fifteen years old. The countries near the equator, in general, have the highest death rates, especially if the small proportion of elderly people is taken into account. This is partly because of the incidence of diseases such as malaria, sleeping sickness, bilharzia (schistosomiasis) and filariasis, all of which are far from being conquered.
The greater the degree of economic development and the higher the income per head, the lower the birth rate and the lower the infant

mortality. West Germany and Sweden, in fact, have declining populations.

b ***Suggest why couples in economically-developed countries decide to have fewer children.***

In pastoral countries, children help to produce food from an early age, and eventually do work to support the adults. In developed countries children are consumers, not producers. They cost a lot to feed, clothe, and educate. Many parents would rather buy consumer goods than bring up an additional child. A family size of two or three in a developed country is usually enough, because infant mortality is so low, to ensure that two or three children will reach reproductive age. Until recently in the tropics and subtropics, infant mortality was so high that a family would need six or seven children to make it likely that two or three would reach maturity. □

**STUDY ITEM**

**28.52 The population–food equation: the Indonesian experience**

*Principle*

1 The current increase in the human population of the world can only be sustained by a comparable improvement in world crop yields.

*Questions and answers*

a ***Explain the italicized terms 'marginal land', 'subsidized inputs', 'non-photosensitive', 'integrated', 'productivity'.***

Marginal land is not ideal for crop growth because of difficulty of access, nutrient deficiency, aridity, or waterlogging, but it can be cultivated in times of economic necessity.
Subsidized inputs are subsidized fertilizers and machinery.
Non-photosensitive means that flowering time in these crop strains does not depend upon day or night length.
Integrated pest control does not rely merely on pesticides, but on crop sowing time, biological control, etc., to reduce the chances of pest infestation.
Productivity is the net production of the crop, which is the rate at which it accumulates energy or dry mass per unit area per unit time.

b ***Discuss the possible advantage of short-stalked (paragraph 2) varieties of rice over long-stalked varieties.***

Short-stalked varieties may be more resistant to 'lodging', being blown down by storms. Short stalks are made more rapidly than long stalks, allowing the crop to flower earlier.

c ***Suggest three advantages of growing groundnuts or soybeans (paragraph 3) in the annual cycle instead of merely another rice crop.***

Groundnuts and soybeans are legumes. They harbour nitrogen-fixing *Rhizobium* bacteria in root nodules, and their protein contents are higher than non-legumes. These species add nitrogen compounds to the soil. Pests of rice do not accumulate so rapidly in a crop rotation as they would in continuous rice culture. The legumes remove other nutrient elements in different relative proportions to rice, and at different levels in the soil. In continuous rice culture, however, the same soil nutrients are depleted time after time.

d ***Explain how continuous and overlapping cultivation and increased application of nitrogenous fertilizers (paragraph 4) can encourage the brown plant-hopper.***

Continuous and overlapping cultivation allowed the brown planthopper to disperse from the remains of the old crop onto the new one, and from one crop to another in the same region. The longer the period between cultivations, the more planthoppers die. A higher nitrogen content in the foliage would allow the insect to accumulate amino acids and proteins more rapidly and shorten the generation time.

e ***Suggest three ways in which the application of pesticides might have reduced the harvest of fish (paragraph 4).***

Invertebrates normally eaten by fish may be killed by pesticides. Fish accumulate high concentrations of persistent pesticides from food web.

f ***Discuss methods of diversifying Indonesian agriculture so that the recent increases in crop production can be maintained. Bear in mind that Indonesia's recent oil and gas boom seems to be over.***

Energy subsidies (see figure (S)365) are now more costly since fuels must be imported. Traditional methods are valuable, though labour-intensive. Possibilities include:
Increase crop rotations involving legumes.
Grow legumes simultaneously in several fields in the same area.
Bring marginal land into cultivation for a range of suitable crops.
Grow perennial crops, with overseas markets, to attract foreign exchange with which to buy fertilizer.
Grow a wider variety of grass crops apart from rice.
Adopt fish-farming in paddy fields or in sewage farms.
Search native flora for plants yielding drugs or foods which can be exported as delicacies. □

# PART II The *Practical guide*

The investigations related to this chapter appear in *Practical guide* 7, *Ecology*.

## INVESTIGATION

## 28A Sampling methods for small invertebrates

### ITEMS NEEDED

Ethoxyethane (diethyl ether)
Methanal (formaldehyde), 2 %

Cover traps (pieces of wooden tile, or house bricks soaked overnight in water) 1/group
Etherizer, large 1/group
Identification keys
Petri dishes
Pitfall traps (see, for instance, Dowdeswell, 1984; see the Bibliography)
Pooter or sieve 1/group
Water traps (washing-up bowls, preferably painted bright yellow 1/group)

(*Study guide* 28.1 'How can population size be assessed?'.)

*Assumptions*

1 Some knowledge of the types of animals found on the soil surface.
2 An ability to use simple keys to identify organisms.
3 An ability to sort and group data and to classify it according to the type of treatment it requires.

*Principles*

1 The use of several different sampling methods may reveal some of the underlying patterns of animal life.
2 All sampling methods have certain advantages and disadvantages which must be taken into account when assessing the significance of the results.

These methods of catching organisms are described in Dowdeswell, 1984.

*Questions and answers*

a ***Compare the types and number of invertebrates caught by the different kinds of traps. Which categories of animal only appeared in one type of trap? Suggest reasons for the differences and describe briefly how you could test one of your explanations experimentally.***

Typical results are shown in table 27. Water traps tend to catch more flying animals, pitfalls catch larger carnivores, and cover traps catch less active cryptozoic organisms. The results can be related to feeding methods, sensory perception, ability to escape, and so on.

b ***Which organism appeared in the traps most frequently? What information have you obtained about 1 its spatial distribution (where it is found), 2 its temporal distribution (when it is captured), 3 its population size (total numbers in the habitat)?***
***What further experiments could be performed to confirm your impressions?***

Students should be encouraged to speculate on the basis of their results, but also to realize the limitations imposed by the number of traps, the trapping methods, and the frequency of their visits. Further experiments should concentrate on reducing these limitations. Capture–mark–recapture (investigation 28B) can also be attempted on suitable species.

| *Pitfall traps* | | *Cover traps* | | *Water traps* | |
|---|---|---|---|---|---|
| *Animal* | *Number* | *Animal* | *Number* | *Animal* | *Number* |
| Springtails | 52 | Earthworms | 15 | Greenflies | 85 |
| Ground beetles | 40 | Woodlice | 15 | Flies | 19 |
| Greenflies | 4 | Slugs | 11 | Parasitic wasps | 9 |
| Spiders | 3 | Snails | 4 | Rove beetles | 4 |
| Centipedes | 3 | Centipedes | 2 | Bark lice | 2 |
| Flies | 2 | Ground beetles | 2 | Spiders | 2 |
| Harvestmen | 2 | Ants | 2 | Sawflies | 2 |
| Mite | 1 | Millipedes | 2 | Plant bugs | 2 |
| Rove beetle larva | 1 | Rove beetle | 1 | Other beetles | 1 |
| Woodlouse | 1 | Spider | 1 | Thrip | 1 |
| Ground beetle larva | 1 | Rove beetle larva | 1 | Bee | 1 |
| Parasitic wasp | 1 | Other beetle | 1 | | |
| Rove beetle | 1 | | | | |
| Ant | 1 | | | | |

**Table 27**
Typical invertebrate catches using different trapping methods.

**c** ***What proportions of carnivores, omnivores, and herbivores were captured in the traps? Does this reflect their proportions in the community?***

The answers will depend on the particular results obtained, but it is unlikely that any of these trapping techniques collect a random sample.

**d** ***What differences appeared between habitats? Concentrate on the species which were found in several traps. Compare the number of each species in the two habitats by a Mann–Whitney U-test (see*** **Mathematics for biologists*****). For those species which were significantly more abundant in one habitat than another, try to explain their distribution patterns. You may be able to suggest what they eat, or how they select their habitat.***

Again, the answers to this question will depend on the outcome of the sampling.

## INVESTIGATION
## 28B Capture–mark–recapture technique

TEMS NEEDED

ꞏeads, poppet, of two ꞏolours 200/group of one colour ꞏnd 40/group of another
ꞏarrier bag 1/group
ꞏollecting apparatus, such as nets
ꞏaint, cellulose, quick-drying
ꞏaintbrush, fine 1/group

(*Study guide* 28.1 'How can population size be assessed?'; Study item 28.11 'The capture–mark–recapture technique'.)

*Principle*

1 It is difficult to estimate the population sizes of mobile, unitary organisms.

*Assumptions*

1 Students must be familiar with the biology of the species being investigated.

**2** Unless the criteria concerning boundaries, abundances, and so on are fulfilled, the results of the investigation will not provide an estimate of the population size. In grassland, for instance, a vertical migration in response to changes in humidity and temperature takes place in several different species. This behaviour may alter as a result of disturbance, yielding different population results.

**3** Students must understand that physical disturbance of the habitat may change the distribution of the animals. This can easily happen, for example, when an area of grass is flattened by sweeping for insects.

**4** Populations in a breeding state may be altered, or even prevented from breeding at all, by a disturbance of their breeding patterns.

**5** An understanding of the principles behind the use of the Lincoln Index.

The method described was devised by Lincoln in the United States in 1930 to estimate water-fowl populations, using coloured rings for marking. Several related methods exist. They are described in Dowdeswell, 1984, and in Southwood, 1978 (see the bibliography).

*Questions and answers*

**a** ***What was the mean size of the population? Explain how you have arrived at your answer and give an indication of the confidence with which you express your answer.***

Students should use the Lincoln index to calculate the answer, and show their working.
If several estimates of population size have been made they can be averaged. The 95 % confidence limits of the mean can then be worked out (see *Mathematics for biologists*).

**b** ***What are the principal errors encountered in your investigation? How did they affect the result, and how could you take account of them in estimating the true size of the population?***

Small samples, failure to disperse, emigration, immigration, hatchings, predation, climatic changes, and deaths from paint toxicity may all be encountered in such investigations.

**c** ***What do you consider to be the principal practical use of an investigation of this sort? Consult the available literature and find an example where such an investigation has had important implications for humans.***

Control of insect pests such as locusts, flour beetles, and cockroaches will serve as examples.

## INVESTIGATION
## 28C Population growth in *Chlorella*

### ITEMS NEEDED

*Chlorella* culture

Growth medium (recipe below)

Colorimeter with blue (preferably 410 nm) filter 1/group
Cottonwool, non-absorbent
Ehrlenmeyer flasks, 250 $cm^3$ 3/group
Gro-lux tubes – artificial light is not essential, but gives more consistent results
Haemocytometer (Neubauer pattern) 1/group
Syringes for sampling, 10 $cm^3$ and 1 $cm^3$

(*Study guide* 28.2 'Population growth'; Study item 28.21 'The growth of a population of yeast'. *Study guide* 28.3 'Population limitation'; Study items 28.31 'The population dynamics of the great tit (*Parus major*) in Marley Wood, Wytham Woods, near Oxford' (investigation 28Ca) and 28.32 'Density-dependent, density-independent, and inverse density-dependent factors' (investigation 28Cb).)

**Growth medium.** potassium nitrate $KNO_3$, 0.02 g; dipotassium hydrogen phosphate $K_2HPO_4$, 1.0 g; magnesium sulphate $MgSO_4 \cdot 7H_2O$, 0.2 g; calcium chloride $CaCl_2 \cdot 6H_2O$, 0.02 g; iron(III) chloride $FeCl_3 \cdot 6H_2O$, 0.01 g; $Na_2EDTA \cdot 2H_2O$, 0.01 g; boracic acid $H_3BO_3$, 2.0 mg; cobalt sulphate $CoSO_4 \cdot 7H_2O$, 1.0 mg; manganese (II) sulphate $MnSO_4 \cdot 5H_2O$, 1.5 mg; ammonium para-molybdate $(NH_4)_6Mo_7O_{24} \cdot 4H_2O$, 0.15 mg; sodium vanadate $NaVO_3$, 0.01 mg; zinc sulphate $ZnSO_4 \cdot 7H_2O$, 1.0 mg.

The cobalt and vanadium salts may be omitted if desired. From these, made up as concentrated stock solutions, make up to 1 $dm^3$ with distilled water. The final medium is adjusted to pH 9.0 with dilute NaOH solution.

Aeration of the cultures is not essential, but does have a stimulating effect on growth rate, if not the final density of cells in the culture.

*Principles*

1 An understanding of the S-shaped logistic growth curve of population increase is an essential pre-requisite for any study of population regulation.
2 Logarithmic (exponential) growth is characteristic of population expansion.

*Assumptions*

1 Ability to use a colorimeter.
2 Knowledge of the factors influencing growth in algae.
3 Ability to plot graphs on logarithmic and linear scales and to assess the significance of the data obtained.

*Questions and answers*

a ***Does the evidence suggest that the growth of the population was logarithmic, arithmetic, or of some other kind? Name the period during which the growth was most nearly logarithmic.***

The graph should show evidence of logarithmic growth in the mid period of the experiment and possibly from the very start.

b ***What factors appear to be controlling growth in the conditions of your experiment?***

Light, temperature, concentration of nutrients, absence of some essential nutrient.

c ***In what way does the supply of energy to the* Chlorella *population differ from the supply of energy to most populations in the wild?***

In the experiment, light energy is supplied continuously.

d ***What factors could be altered to examine their effects on the pattern of growth of this population?***

Light, concentration of nutrient ions in the medium, continuous supply/removal of waste, temperature, competition, aeration, excess or lack of carbon dioxide, pH, effects of metal ions such as copper or lead.

e ***In what ways was it useful to use a colorimeter? What factors may cause inaccuracies in the method?***

*Advantageous* – smoother curve, smoothes out the inaccuracies, quick and easy method.
*Disadvantageous* – may disguise some aspect of *Chlorella* biology which is important. In this case the *Chlorella* forms autospores which accumulate in large old cells and are then released when the cell breaks. This process sometimes becomes synchronized after sub-culturing so that the number of cells increases in a step-wise fashion.

## INVESTIGATION
## 28D Population regulation in the water flea, *Daphnia*

### ITEMS NEEDED

Culture of *Daphnia* 1/class
Dried yeast

Sucrose

McCartney bottles 3/group
Minimum five groups, to provide at least five replicates a class, of each of the three treatments
Petri-dishes
Pipette, wide mouth, dropping

(*Study guide* 28.3 'Population limitation'; Study item 28.32 'Density-dependent, density-independent, and inverse density-dependent factors.')

**Water.** Use a proprietary artificial pond water, or filtered pond water from a pond containing *Daphnia*. One drop of diluted Indian ink allows the *Daphnia* to ingest particles of carbon and to survive in some tap waters.
**Yeast.** Use a concentration of 0.01 g yeast to 100 $cm^3$ water. Add 0.01 g sucrose to the water. Keep the culture warm for about 30 minutes. Then add 0.1 $cm^3$ to each of the McCartney bottles.

The yeast culture can be kept to provide the yeast suspension which will be added after a week to the new *Daphnia* culture. Note, however, that the addition of too much yeast to the *Daphnia* culture is fatal.

**Transferring the *Daphnia*** Wide-mouthed pipettes must be used to avoid damage to the *Daphnia*. Care must be taken to ensure that air does not become trapped in the carapace of the water fleas during transfer. The use of ovigerous females is desirable, but not essential to the success of the experiment.

The size of the initial population, feeding rate, and the size of the containers can be varied according to the availability of apparatus, space, and convenience.

*Assumptions*
1 Some familiarity with the biology of *Daphnia*.
2 An understanding of the need for observation as well as counting of the individuals.

3 A recognition of the need for careful experimental technique. Feeding rates should be changed if the numbers of *Daphnia* increase sufficiently.

4 An understanding of the need for care in the maintenance of the organism.

*Principles*

1 Population regulation mechanisms act in different ways. Regulation of a population has to be density-dependent if violent fluctuations or uncontrolled changes are to be avoided.

2 An investigation of a laboratory population helps to isolate the mechanisms of population regulation, and can help the student to understand the mechanisms of population regulation which operate in the wild.

3 These principles are also relevant to an understanding of physiological homeostasis.

*Questions and answers*

a ***Which treatment regulated the* Daphnia *population to a mean value?***

Treatment C.

b ***Which treatment was the least successful in regulating the* Daphnia *population to a mean value?***

Treatment B. When predation is relaxed the population expands in the same way as in control experiment A. On the other hand, heavy predation may occasionally cause the *Daphnia* to become extinct. The population size will show wild and uncontrolled swings.

c ***Describe the kind of fluctuations in size which might occur in a population subject to density-dependent control.***

Population will probably show fluctuation about a mean value. The amplitude of these fluctuations should diminish the longer the experiment lasts.

d ***What does the variation in population density have in common with homeostatic mechanisms?***

Both apply a regulating action above and below a mean value.

e ***What biological factors have been ignored or avoided in the experimental procedures suggested?***

Age, reproductive habits, generation time, reproductive rate, availability of food and oxygen, and interaction between individuals which may change the rate of reproduction.

f ***What procedure could you use to eliminate or to make allowance for these biological factors in your experimental design?***

At each stage only return one type of *Daphnia*, that is, reproducing (ovigerous) female organisms, making sure that only young and immature individuals are culled. This ensures continuity of the culture and its ability to recover from predation. It illustrates one of the factors which may come into play in natural populations in which a whole generation of organisms may be removed by differential predation.

## INVESTIGATION
## 28E The responses of a predator to changes in the number of its prey

### ITEMS NEEDED

Box for collecting discs 1/group
Discs, approximately 1 cm diameter and preferably textured on one surface (plastic tokens, bottle tops, sandpaper discs, metal rings or rubber cap liners for McCartney bottles are suitable) 225/group
Graph paper
Paper, 60 × 60 cm ruled with 2 cm squares (commercially available)
Stop-clock 1/group

(*Study guide* 28.3 'Population limitation'; Study item 28.3a 'Density-dependent, and inverse density-dependent factors.)

*Assumptions*

1 An ability to interpret graphs.
2 An outline knowledge of population dynamics – reproduction, death rate, immigration, and emigration.

*Principles*

1 Construction of a model can be a useful tool in the investigation of a biological process.
2 A model does not necessarily demonstrate all aspects of the biological process under investigation.

This investigation can be followed up by a real predator/prey experiment, such as predation by dragonfly nymphs on *Daphnia* at different densities (Roberts 1974), and the results compared with the results of the simulation.

*Questions and answers*

a ***What happens to the number of prey killed as the population increases in size? Explain the shape of the graph at higher population densities.***

The number of prey killed increases rapidly as the population increases, but the rate of increase slows down at the higher population densities. Catching, eating, and digesting the prey take up a greater proportion of the time than finding them.

b ***In what ways does the graph showing the percentage of prey killed against density differ from the previous graph? What additional information does it provide?***

The graph should be a straight line with a negative slope. It shows that the predator becomes less effective at killing the prey population as numbers of the prey increase.

c ***What would happen to a population which was increasing in numbers if this model was correct?***

If the prey population was not made extinct by the predator when the prey population was small, the population size of the prey would increase exponentially.

d ***Would this model fulfil the conditions given in the introduction, and give a density-dependent regulation?***

No. The point could be further illustrated by selecting an imaginary birth rate (such as 100 discs per survivor) and calculating the number in the next generation!

e ***Describe ways in which you could alter the model to achieve a better density-dependent regulation. Explain how the predators would achieve regulation at all densities.***

Increase the number of predators at high densities of prey and decrease them at low densities. Increase the rate of emigration at high densities and the rate of immigration at low densities. At high densities of prey there might be increased efficiency of predator, learning, co-operation between two predators, or the speeding up of digestive processes.
Combinations of one or all of the above methods.

f ***What factors, other than predation, can exert density-dependent control on a population?***

Availability of food, territories, disease, infertility, death rate and birth rate, plus combinations of these.

# PART III Bibliography


BEDDINGTON, J. R. and MAY, R. M. 'The harvesting of interacting species in a natural ecosystem'. *Scientific American* **247**(5) 1982. (Explores the food web based on krill in the South Atlantic and how krill and its predators should be harvested to maintain the population sizes of all the species in the long term.)
BENNETT, D. P. and HUMPHRIES, D. A. *Introduction to field biology*. 3rd edn. Edward Arnold, 1980.
CHINERY, M. *A field guide to the insects of Britain and Northern Europe*. 2nd edn. Collins, 1976.
CLOUDSLEY-THOMPSON, J. L. and SANKEY, J. *Land invertebrates*. Methuen, 1961.
DALEY, M. and HILLIER, D. Computer simulation of the population growth (*Schizosaccharomyces pombe*) experiment. *Journal of biological education*, **15**(4), p. 266, 1981.
DOWDESWELL, W. H. *Practical animal ecology*. Methuen, 1959.
DOWDESWELL, W. H. *Ecology: principles and practice*. Heinemann Educational, 1984.
GWATKIN, D. R. and BRANDEL, S. K. 'Life expectancy and population growth in the third world'. *Scientific American*, **246**(5), 1982, p. 33.

(Discussion of the influence of changes in birth and death rates on world population increase.)
HARPER, J. L. *The population biology of plants.* Academic Press, 1977. (A modern classic. Contains a discussion of buttercup dynamics.)
KERNEY, M. D. and CAMERON, R. A. D. *Field guide to the land snails of Britain and Northern Europe.* Collins, 1979.
KREBS, C. J. *Ecology. The experimental analysis of distribution and abundance.* Harper and Row, 1978. (This standard university text includes a critical discussion of the theory of population regulation and describes plenty of investigations.)
LACK, D. *The natural regulation of animal numbers.* Oxford University Press, 1954. (Still a pleasure to read and a valuable source of case-histories.)
LINCOLN, R. J. and SHEALS, J. G. *Invertebrate animals. Collection and preservation.* British Museum and Cambridge University Press, 1979.
MYERS, J. H. and KREBS, C. J. 'Population cycles in rodents'. *Scientific American,* **230**(6), 1974, p. 38. (Discusses the fluctuations and the genetic differences between individuals at different stages in the cycle.)
PAVIOUR-SMITH, K. and WHITTAKER, J. B. *A key to the major groups of British free-living terrestrial invertebrates.* Blackwell Scientific Publications, 1968.
PERRINS, C. M. 'The great tit, *Parus major*'. *Biologist,* **27**(2), 73, 1980. (An elegant summary of over thirty years work on population regulation in this bird in woodlands near Oxford and in Holland.)
ROBERTS, M. B. V. *Biology, a functional approach, Students' manual.* 4th edn. Nelson, 1986.
SAMWAYS, M. J. Studies in Biology No. 132. *Biological control of pests and weeds.* Edward Arnold, 1981. (A clear, lively, and well-illustrated introduction, containing an account of the interaction between whiteflies and *Encarsia.*)
SOUTHERN, H. N. 'The natural control of a population of tawny owls (*Strix aluco*)'. *Journal of Zoology,* **162**, 197, 1970. (Describes a long-term study of the population regulation of a territorial predator.)
SOUTHWOOD, T. R. E. *Ecological methods.* 2nd edn. Chapman and Hall 1978. *World population data sheet* of the Population Reference Bureau Inc. (1337 Connecticut Avenue, N.W., Washington DC 20036, USA), 1982. (Vital population statistics for every country in the world, updated each year.)
WRATTEN, S. D. and FRY, G. L. A. *Field and laboratory exercises in ecology.* Edward Arnold, 1980.

# CHAPTER 29 COMMUNITIES AND ECOSYSTEMS

*A review of the chapter's aims and contents*

1 This chapter concentrates upon the quantitative study of whole communities. Particular attention is paid to energy flow and nutrient cycling in natural and artificial ecosystems. Economic consequences are emphasized throughout.

2 After a brief consideration of food chains, food webs, and trophic levels, ecological pyramids are discussed.

3 Energy flow and nutrient cycling are introduced. Some ecosystems are closed, but many are open, that is, they exchange considerable quantities of organic matter and nutrients with other ecosystems.

4 The factors which influence the rates at which plants (primary production) and animals (secondary production) accumulate dry mass are discussed with reference to crop plants and bullocks. Results from a variety of ecosystems are used to analyse the factors which determine the rate at which energy passes from one trophic level to another.

5 The roles of detritivores and decomposers in the breakdown of detritus are briefly mentioned.

6 Nutrient cycles are discussed in detail. Amongst the topics dealt with are the carbon cycle of the whole biosphere, the nitrogen cycle in a temperate forest, nutrient limitation of algal growth in seas and lakes, and nutrient cycling in the Amazonian rain forest.

6 The impacts of selected pollutants on ecosystems and individual species are considered.

7 The brief synopsis of nature conservation is designed to make students think critically about its value.

# PART I The *Study guide*

## 29.1 Introduction

*Assumptions*

1 A familiarity with the major types of biome on Earth.

2 An understanding of the concept of energy, and its measurement in kilojoules.

3 An understanding that energy cannot be destroyed.

4 Familiarity with the nutrient elements required by plants and animals, and their roles in the organism.

*Principles*

1 An ecosystem is a community of interacting organisms, coupled with their environment. In the system energy flows and nutrients cycle.

2 Ecosystems can be investigated quantitatively. Rates of nutrient cycling and energy flow can be measured.

3 Quantitative analysis of ecosystems is simplified if each species population is placed in one or more feeding levels, known as trophic levels. Each species can be categorized as a primary producer, a primary, secondary, or tertiary consumer, a detritivore, a decomposer or a parasite. Energy flow or nutrient cycling between trophic levels can then be investigated.

4 This allows ecosystems to be validly compared, on the same basis.

## STUDY ITEM

## 29.11 Trophic levels and food webs

*Principles*

1 It is often difficult to determine what an organism eats.

2 Many species occupy two or more trophic levels.

3 Feeding interactions in ecosystems are so complex that some species cannot be placed easily in a trophic level.

*Questions and answers*

a ***How could the diet of the following organisms be determined?***

***1 fish***

***2 predatory tiger beetle living on the forest floor***

***3 caterpillar grazing***

***4 zebra grazing on the Serengeti plains (do not kill the zebra!)***

***5 vulture***

***6 earthworm***

1 Dissect it and examine the gut contents.

2 Students will find this difficult, but the question provides the opportunity to mention the following technique. Collect possible prey species by pitfall trapping. Inject each possible prey in a ground-up state into a different rabbit or horse. Extract blood and prepare serum which should contain antibodies against the potential tiger beetle prey species. Extract gut contents of tiger beetle, separate into small portions, test against each serum in turn. Intensity of coagulation shows prey species eaten and relative quantities.

3 Watch it.

4 First catch your zebra. Take faecal sample from anus, clear, stain, examine with high power monocular microscope. Epidermal cells of grasses still visible and identifiable to species. Species other than grasses, however, may have been completely digested. Digestibility can be investigated by feeding tame zebra with diet of single species for a time under standard conditions and examining faecal samples.

5 Watch them, and examine gut contents.

6 Examine gut contents. The origin of detritus, however, is difficult to determine.

Gut contents soon after ingestion may provide a better indication of diet than faecal samples. Experiments in which organisms are presented with a range of possible prey items in the laboratory are less reliable because the prey species are not encountered in the same relative proportions, or under the same conditions, as in the wild.

**b** ***Discuss the trophic levels occupied by the following organisms:***
***1 bug which sucks phloem sap***
***2 earthworm***
***3 blackbird which eats an earthworm***
***4 fungus associated with an alga in a lichen***
***5 'carnivorous' plant such as the sundew.***

**1** Parasite. It does not kill and eat the plant, but lives attached to its surface, taking organic compounds from it and probably contributing nothing in return.
**2** Detritivore.
**3** Does not fit into a particular feeding category – nor would, for example, a cat which eats the blackbird. It is most convenient when drawing food webs to place the detritivores at the same level on the diagram as primary consumers, because some organisms eat both.
**4** Since the fungus obtains all or most of its organic carbon atoms from the alga it is a primary consumer.
**5** Simultaneously a primary producer and a secondary or tertiary consumer.

□

## STUDY ITEM
## 29.12 Ecological pyramids

*Principle*

1 Pyramids of energy flow are more valuable than either pyramids of number or pyramids of biomass.

*Questions and answers*

**a** ***Why are pyramids of numbers not a valid basis for comparing ecosystems?***

Because every organism, whatever its mass, life span, or significance in the community, is accorded equal status. In the sea, for example, a whale weighing 130 tonnes would be regarded as one unit. So would a bacterium on its surface, although its volume is $10^{-5}$ that of a single whale *cell*!

**b** ***How might the pyramids of numbers and biomass in* figures [S]353 *and* 354 *differ at different times of year?***

In the grassland and the forest there would be about the same numbers of primary producers in winter and summer, but far fewer primary consumers in winter. In the sea the numbers of primary producers would be drastically reduced in winter, the primary consumers would decline and the secondary consumers would be at about the same numbers; but this partly depends on the timing of fish reproduction.

**c** ***The mass of primary consumers in the English Channel is more than five times that of the primary producers which support them. How can the organisms survive for some time at these biomasses?***

Because the turnover of algal cells is so rapid. At any one time the biomass of phytoplankton is small. Seventy-five per cent of the cells are eaten. They are rapidly replaced by division of the remaining cells, so providing a continuous food supply for the longer-lived primary consumers. The total mass of algae eaten by a primary consumer during its lifetime is of course much greater than its own mass, but the algae consumed represent several generations.

d ***Why is the pyramid of energy flow for an ecosystem always pyramidal in shape* (figure [S]355)*?***

The organisms at each trophic level 'waste' energy through respiration and egestion. The compounds in their bodies therefore contain less potential energy than the compounds in the food which entered their mouths. In addition, many organisms, or parts of organisms, die before they can be eaten.

e ***Should one use the energy eaten by the organisms or the energy assimilated by the organisms to construct the pyramid? What is the difference?***

The distinction is not worthwhile for plants. The proportion of the energy assimilated by an animal will depend on whether it is eating plants or animals. Assimilation of plant food is much less efficient. If total energy intake is plotted this dietary difference is not confused with the efficiency of energy transfer between trophic levels.

f ***Suggest how you might estimate a pyramid of energy flow for 1 a forest 2 a meadow and 3 the sea.***

For a forest, net primary production can be estimated in two main ways. Firstly, the litter falling in a year can be caught in litter traps, and the net production of the trunks, branches, and roots estimated and added to it. Secondly, the carbon dioxide concentration can be continuously monitored at different levels in the air above the forest over several representative days in the year. The greater the photosynthetic rate, the steeper the gradient in carbon dioxide concentration from the atmosphere to the forest. Results can be integrated over the whole year. The harvesting method neglects losses to herbivores, and the carbon dioxide method includes carbon dioxide produced in the respiration of all the other trophic levels.
Net primary production in a meadow can be estimated by harvesting samples at intervals and determining their dry mass. Samples must be taken at frequent intervals, otherwise leaves will be formed and die between samples. A correction must be made for the biomass eaten by herbivores. Alternatively, the carbon dioxide uptake and release by certain areas of vegetation can be monitored with a field infra-red gas analyser.
In seas, net primary production is conveniently estimated by the 'light and dark bottle method'. Identical bottles are filled with identical water samples and suspended at the same depth. In the dark bottle, all organisms respire. In the light bottle, there will also be some photosynthesis. The oxygen concentration in the water must be

measured at the beginning and end of the twenty-four hour period. The difference at the end corresponds to the oxygen produced by photosynthesis.

Energy flow through primary and secondary consumers is difficult to estimate. The food intake of individuals can be measured under laboratory conditions (see *Practical guide* 7, investigation 29B. 'The energetics of a stick insect (*Carausius morosus*)'). If their numbers in the wild are known, energy flow through a whole trophic level can be estimated. This is only practicable for the largest and most abundant animals in the ecosystem.

The unit used to draw a pyramid of energy flow is kilojoules. In order to determine the energy content per dry mass, dry tissues are burnt in a bomb calorimeter. Carbon dioxide intake is converted to energy equivalents by assuming a balanced equation for photosynthesis, and that each glucose molecule would release 2 880 $kJ\ mol^{-1}$ on hydrolysis.

In the following investigation, students construct a food web for a series of laboratory aquaria, and estimate a pyramid of biomass.

**Practical investigation. *Practical guide* 7, investigation 29A, 'A quantitative study of an ecosystem'.**

## 29.2 Energy flow and nutrient cycling in ecosystems

*Principles*

1 The biosphere can be viewed as a mosaic of different types of ecosystems.
2 Ecosystems exchange nutrients and energy with each other.
3 Energy flow and nutrient cycling in an ecosystem can be measured, and modelled on computers.

## 29.3 Energy flow

*Principles*

1 The net primary production of the plants in the ecosystem is the rate at which they accumulate dry mass.
2 The net primary productivity of an ecosystem is influenced by environmental factors such as water availability and nutrient supply.
3 A crop which covers the ground area may trap a maximum of six per cent of the total incident solar radiation in the energy of organic compounds over periods of a few weeks.
4 The rate at which mass is accumulated by the animals in an ecosystem is known as its secondary production.
5 The secondary production of a particular species is influenced by its diet, whether it is homoiothermic or poikilothermic, and the ages and sizes of its individuals.
6 Energy flow through an ecosystem can be measured.
7 Different ecosystems differ in their patterns of energy flow.

**8** Detritivores and decomposers attack detritus. The energy in their diet is ultimately radiated to outer space but the nutrient ions they release become available for uptake by the primary producers.

**9** The rate of decomposition is influenced by temperature and the ratio of carbon to nitrogen in the detritus. Acid anaerobic conditions, low in nitrogen, slow down decomposition and may lead to peat formation.

## STUDY ITEM

## 29.31 The productivity of plants

*Principle*

1 The efficiency with which a crop converts photosynthetically active useful radiation into dry mass differs at different stages of its growth.

*Questions and answers*

a ***Give brief explanations of the following terms:***
***1 light-saturated***
***2 net primary production.***

**1** Light saturation occurs when an increase in light intensity has no effect on photosynthetic rate.
**2** Net primary production is the rate at which the plant accumulates dry mass or energy. It is expressed in kJ $m^{-2}$ $year^{-1}$. It represents the excess of energy gain in photosynthesis over energy loss in respiration and other metabolic reactions.

b ***Explain the difference between 'photosynthetically-active useful radiation' and 'total solar radiation'.***

Approximately fifty per cent of the total solar radiation is in the ultra-violet or the infra-red and cannot be trapped by pigment molecules for its energy to be used in photosynthesis.

c ***1 Which one of the three methods of measurement of net primary production given above would be best for testing a model of the effect of the environment on production in a natural plant community?***
***2 Give reasons for your choice.***
***3 State two disadvantages of using the method.***

**1** Method **1**, the disappearance of carbon dioxide from an area of vegetation enclosed in a plastic container.
**2** This method would allow the environment within the container to be changed.
**3** The apparatus reduces wind movement. It may trap solar energy, thus becoming warmer than its surroundings. Control measurements, those meant to be taken without any alteration in microclimate, will therefore be unrepresentative of natural conditions.

d ***The leaves and plants in a community are continually dying. What effect will this have on the harvesting technique for the direct measurement of net primary production?***

Single harvests or infrequent harvests underestimate net primary production. In the intervals between harvests, leaves or whole plants will die or rot. Their contribution to the net primary production will not be measured when the living vegetation is harvested. To compensate for this, harvests should be taken as frequently as possible throughout the growing season. Dead leaves and plants should be collected, separated, and weighed separately at each harvest.

e ***Account for the low efficiency of production of agricultural crops compared with algal cultures in the laboratory.***

In laboratory cultures the algal cells reproduce rapidly to produce a maximum density of absorbing cells. All the cells photosynthesize. Optimal levels of carbon dioxide, temperature and nutrients are maintained in the culture. The high carbon dioxide concentration reduces the rate of photorespiration. The agitation of the culture solution enables all the cells to be exposed to light for short periods in turn.

The soil is bare between successive crops and a crop leaf cover takes some time to become established. Many of the cells do not photosynthesize. Carbon dioxide at 0.03 per cent may limit crop growth. At least in $C_3$ plants, photorespiration may be rapid.

f ***Suggest two reasons why plants with the $C_4$ mechanism of photosynthesis are often more efficient at converting solar energy to dry mass than $C_3$ plants.***

$C_4$ plants tend to use water more efficiently than $C_3$ plants, so that $C_4$ plants may still photosynthesize in situations in which $C_3$ plants have closed their stomata. Photorespiration occurs in $C_3$ plants, and increases rapidly with temperature, whereas photorespiration hardly occurs in $C_4$ plants. Maximum photosynthesis in $C_4$ plants occurs at higher temperatures than in $C_3$ plants. Thus warmer climates may favour $C_4$ plants. □

## STUDY ITEM

## 29.32 Energy flow and agriculture

*Principle*

1 Many homoiothermic animals are poor converters of net primary production into their own mass.

*Questions and answers*

a ***How much energy is trapped by the plants in photosynthesis?***

$21\,436 + 1968 = 23\,404 \text{ kJ m}^{-2} \text{ year}^{-1}$.

b ***What is the efficiency of net primary production over the year, expressed as per cent of total visible light energy?***

21 436/1 046 700 = 0.0205 = 2.05 per cent.

c ***What proportion of the net primary production of the ley was eaten by the bullocks? Why is this proportion so low?***

3 056/21 436 = 0.142 = 14.2 per cent.
About forty per cent of the plant mass is underground and inaccessible to grazers. Much of the rest either dies, or is eaten by other herbivores, before the bullocks reach it. The proportion of the vegetation eaten by bullocks depends on their numbers per hectare. The vegetation near cowpats is avoided by bullocks several weeks after a fresh cowpat has been deposited.

d ***What proportion of the energy in the plants eaten by the bullocks was devoted to increasing the dry mass of bullocks (secondary production)?***

125/3 = 41.67 = 0.41 per cent.

e ***Would you advise that farmers should concentrate more on improving the net primary productivity of grasses than on improving the conversion efficiencies of the bullocks?***

Yes, at least from the point of view of energy content. More humans can survive on a field of crop grass than by eating the herbivores which graze it. Bullocks waste a great deal of energy, despite the mutualistic protozoa and bacteria in the rumen.

f ***Why does our diet probably contain more meat from herbivores than from carnivores? Which carnivores, if any, do you eat?***

Herbivores are so wasteful of energy that they can support only a lower density of carnivores. The biomass of herbivores in ecosystems is greater than that of carnivores, except perhaps in aquatic ecosystems, which provide fish, some of which are the only carnivores which many of us eat.

g ***In some tropical regions, native herbivores such as zebra, wildebeest, and antelope yield more energy per unit area than introduced species of cattle. Suggest some reasons why this is so.***

Introduced cattle are more susceptible to native diseases than introduced mammals. Introduced cattle do not select the most nutritious grasses, as native herbivores do. Introduced cattle often overeat the vegetation. They may specialize on palatable species, thus increasing the proportion of unpalatable species and lowering potential cattle production in the future. □

The following experiment allows the components of energy flow through stick insects to be estimated in a laboratory setting. The differences in energetics between young and old animals may also be studied.

**Practical investigation. *Practical guide* 7, investigation 29B, 'The energetics of the stick insect (*Carausius morosus*)'.**

## STUDY ITEM

### 29.33 The efficiency of energy transfer between trophic levels

*Principles*

1 Energy transfer from one trophic level to another has been quantified for many ecosystems.

2 Transfer of energy from body mass at one trophic level to body mass at the next depends on the proportion of the bodies eaten, the proportion assimilated and the proportion devoted to respiration.

*Questions and answers*

a ***What patterns emerge?***

In aquatic ecosystems based on phytoplankton as the primary producers, a higher proportion of energy enters the grazing food web than in other ecosystems.

b ***State the main factor which seems to affect the proportion of food energy assimilated (absorbed across the gut wall into the bloodstream and absorbed by the cells). Explain in biochemical terms the principal differences which emerge from table* [S]72.**

Carnivores assimilate a higher proportion of the energy they ingest than herbivores. The animal food eaten by carnivores is similar in composition to their own bodies and lacks cellulose, hemicelluloses and pectins. Herbivores eat cellulose cell walls as well as the plant cells inside them. Even ruminants cannot fully digest cell walls. Much of the energy ingested by herbivores is therefore egested.

c ***Discuss, giving reasons where appropriate:***
***1 why, in both non-social insects and other invertebrates, herbivores have the lowest production efficiencies; is this what you would expect?***
***2 the inefficiency of fish and social insects;***
***3 the difference in production efficiency between poikilotherms and homoiotherms;***
***4 the possible influence of size on the production efficiencies of mammals.***

**1** This is unexpected, and difficult to explain. Perhaps carnivores rest between meals, whereas the inefficient assimilation of herbivores requires them to expend energy continually in finding and ingesting food. The food of detritivores is similar to that of herbivores, yet their conversion efficiencies are greater.

**2** Fish and social insects are active and mobile all the time.

**3** Poikilotherms are much more efficient energy converters than homoiotherms, presumably because mammals and birds have to respire rapidly just to maintain their body temperatures.
**4** Large mammals possibly devote more of their assimilated energy than small mammals to net production. This is because their bodies have less surface area per unit volume (or mass). Each kilogram of tissue has to respire less to maintain body temperature in a large mammal than in a small one.

**d** ***Using the data in tables*** **[S]** ***70 to 72, make some speculative calculations of the per cent of net plant production which ends up in the herbivore production in 1 a pond, with phytoplankton being grazed by invertebrate animal plankton and 2 an oak forest, being grazed by non-social insects. Show your working clearly.***

Pond:
Assume that 40 per cent of phytoplankton production is eaten by zooplankton.
Zooplankton, as herbivores, might assimilate 40 per cent of this energy.
Zooplankton are non-insect herbivores in which only 21 per cent of assimilated energy is devoted to production.
Thus $0.4 \times 0.4 \times 0.21 = 0.033$ of plant production ends up in zooplankton production = 3.4 per cent.
(For the Silver Springs data (figure (S)364), however, we have a ratio of $(1609 + 4599)/(14\,146 + 22\,953) = 0.27 = 27$ per cent.)

Oak forest:
Assume that 10 per cent of oak tree production is eaten by invertebrates.
Invertebrate herbivores assimilate 40 per cent of this.
Non-social insect herbivores devote about about 40 per cent of assimilated energy to production.
Thus $0.1 \times 0.4 \times 0.4 = 0.016$ of plant production ends up in insect herbivores = 1.6 per cent.
(For the oak forest data in figure (S)363, $1500/(2920 + 5840 + 15\,330) = 6.2$ per cent of plant production ended up in herbivore production.)

**e** ***Do you think that the '10 per cent rule' is likely to be valid for many ecosystems? Is it more likely to be valid for secondary consumers eating herbivores than primary consumers grazing plants?***

The rule is unlikely to be valid for the transfer of energy from primary producers to herbivores. The results in table (S)70 show that in most ecosystems only a small proportion of the net primary production is eaten by herbivores. This alone is only ten per cent in many ecosystems.
Carnivores, on the other hand, usually eat the whole of their herbivorous prey. They are efficient at assimilating herbivorous food, and devote a higher proportion of the assimilated energy to dry mass than herbivores. The ten per cent rule may well apply to them. The data in this Study item do not state, however, how much of the herbivore dry mass is ingested by carnivores in a variety of ecosystems. □

## STUDY ITEM

## 29.34 Energy flow in a grassland ecosystem

This question provides a simple introduction to energy flow through a whole ecosystem.

*Questions and answers*

**a** ***Which of the labelled boxes represent producers, primary consumers, and secondary consumers?***

'Grasses and herbs' are producers.
'Seed-eating birds', 'Common green grasshoppers' and '*Apodemus sylvaticus* (field mice) are primary consumers.
'Spiders' are secondary consumers.

**b** ***How much energy is trapped in photosynthesis by the grasses and herbs?***

$(20.4 \times 10^6) + (36 \times 10^5) = 24 \times 10^6$ kJ m$^{-2}$ year$^{-1}$.

**c** ***Why are the grasses and herbs so inefficient at trapping the incident light energy?***

See *Study guide II*, figure (S)358. The photosynthetic rate of the grasses may be limited through much of the year by cold temperatures or shortages of nutrient ions.

**d** ***What is the net production of each of the following groups of animals 1 seed-eating birds 2 common green grasshoppers 3 spiders?***

**1** $(60\,000) - (59\,200) = 800$ kJ m$^{-2}$ year$^{-1}$.
**2** $(444\,000) - (374\,300) = (69\,700)$ kJ m$^{-2}$ year$^{-1}$.
**3** $700 - 500 = 200$ kJ m$^{-2}$ year$^{-1}$.

**e** ***How much energy is lost through respiration and faeces by field mice?***

$116\,000 - 2000 = 114\,000 = 114 \times 10^3$ kJ m$^{-2}$ year$^{-1}$.

**f** ***State in detail what happens to the energy which enters 'death and other pathways'.***

Some of this energy enters insects, snails, and slugs which graze the vegetation. Some enters fungal parasites. The rest, in dead stems, leaves and roots, is shredded and eaten by detritivores. Decomposers such as bacteria and fungi attack both the detritus and the faeces of detritivores. The energy is ultimately lost to outer space.

## STUDY ITEM

## 29.35 Energy flow in different ecosystems

*Principle*

**1** Comparisons between ecosystems are valuable because they focus attention on those factors limiting primary and secondary production.

*Questions and answers*

a **Compare the efficiency with which the incident light energy is incorporated in dry mass by the plants in each ecosystem.**

Salt marsh: $(33\,475 + 1278)/2.5 \times 10^6 = 0.0139 = 1.4$ per cent
Forest: $(5840 + 2920 + 15\,330)/4.6 \times 10^6 = 0.0052 = 0.52$ per cent
Silver Springs: $(14\,146 + 22\,953)/17.1 \times 10^6 = 0.0216 = 2.16$ per cent

b **Compare the proportion of plant production which enters the grazing and the detritus food webs in all three ecosystems. Account for the differences.**

| *Figures in kJ $m^{-2}$ $year^{-1}$* | *Net production* | *Grazing web* | *Detritus web* |
|---|---|---|---|
| Salt marsh: | 34 753 | 1278 (4 %) | 33 475 (96 %) |
| Forest: | 24 090 (not 18 250) | 2920 (12 %) | 15 330 (64 %) |
| Silver Springs: | 37 099 | 14 146 (38 %) | 22 953 (62 %) |

Speculatively, it may have been difficult for herbivorous invertebrates living on angiosperms to adapt to the salt marsh because of the rapid environmental changes between exposure and submergence. In forests, plant defences against herbivores are well-developed. In Silver Springs, however, the herbivores are much larger than the primary producers and can filter them rapidly from the water.

c **Which of these ecosystems would yield the largest annual crop, and how would you harvest it?**

Silver Springs. It might be difficult to harvest the crop before the herbivores eat it. Water might be passed through a filtration system, such as a rotating drum with very small holes in it, to filter off the algae.

d **Which of these ecosystems, if any, appear to be in energy balance?**

The forest is a storage ecosystem, with 5840 kJ $m^{-2}$ $year^{-1}$ devoted to trunks, branches, roots, and litter storage (see Study item 27.81). It is accumulating energy. Silver Springs and the salt marsh are rapidly losing detritus downstream. None of these ecosystems is in energy balance. □

### Decay and decomposition

**Practical investigation. *Practical guide* 7, investigation 29C, 'A study of decomposer organisms in the soil'.**

## 29.4 Nutrient cycling

*Principles*

1 All elements cycle in ecosystems at different rates.

**2** The cycles of carbon, oxygen, hydrogen, water, nitrogen, and sulphur are 'open'. Their gaseous phases permit considerable exchange between ecosystems. The cycles of the other elements, for example, calcium, potassium, sodium, magnesium, and iron, are 'closed'.

### STUDY ITEM
### 29.41 The carbon cycle

*Assumption*

**1** An understanding of gaseous exchanges in photosynthesis and respiration.

*Principle*

**1** The carbon dioxide concentration in the atmosphere, which is strongly affected by the balance between photosynthesis and respiration, is slowly increasing.

*Questions and answers*

**a** ***Describe how you might estimate the annual flux of carbon dioxide from the atmosphere into the primary producers in the sea. Discuss the difficulties which you might encounter and the ways in which they might be overcome.***

The method employed must distinguish between carbon dioxide which diffuses from the air and dissolves in the sea, and that which is absorbed by plant plankton. The 'light and dark bottle' method can be employed at various depths and at various times of year, and the results integrated. In this technique, identical bottles containing water samples are suspended next to one another for 24 hours. The oxygen concentrations in the water in both bottles are measured at the beginning and the end of the experiment. In the dark bottle, all the organisms respire; in the light bottle, all the organisms respire and the plants photosynthesize during the day. The light bottle should contain more oxygen at the end of the 24 hours. The difference in oxygen levels in the light and dark bottles estimates oxygen production in photosynthesis. This should be equivalent in volume to the carbon dioxide absorbed in photosynthesis.
Difficulties, which cannot satisfactorily be overcome, are: zooplankton eat phytoplankton. Phytoplankton photorespires as well as respires (see *Study guide I*, Chapter 7, Section 7.7). Zooplankton egest faeces, which are decomposed by respiring bacteria. Carbon dioxide concentration in the bottles changes in a different manner from that in the water outside. The glass of the bottle absorbs some light. Geographical variation in primary production of phytoplankton is immense.
Recent studies suggest that this technique may have underestimated primary production by the smallest cells, the nannoplankton. Primary productivity in the sea may be five or six times higher than previously suspected.

**b** ***Many bacteria which live in watery habitats are chemo-autotrophic. Where do they fit into the diagram of the carbon cycle (*figure [S]368*)?***

Chemo-autotrophic bacteria use energy from chemical oxidations to synthesize high energy organic compounds from carbon dioxide and a hydrogen donor, which could be water or hydrogen sulphide. Thus they occupy the same position in the carbon cycle as green plants, which also take up carbon dioxide from the air or water.

c ***Atmospheric methane levels have risen at 1.7 per cent a year since at least 1975. Suggest some reasons for this increase.***

Methanogenic bacteria occur in anaerobic swamps and in the guts of ruminants and termites. Possibly as tropical forests are cleared, termites spread and increase in numbers. Their methane output has been estimated at 150 million tonnes a year. Many tropical forests are being converted to ranches on which ruminants graze. Free-living methane-producing bacteria may also have contributed to this increase.

d ***Suggest two possible reasons why the extra carbon dioxide is not removed by plants in photosynthesis.***

The primary productivity of most plants is limited by lack of nutrient ions, particularly nitrates and phosphates (see Study item 29.43). In many areas lack of water or low temperatures are the limiting factors. Carbon dioxide probably only limits plant growth in $C_3$ plants at high light intensities and at reasonably high temperatures when nutrients are abundant. Such situations are rare. The number of chloroplasts in the world might be declining, because of the rapid rate of forest felling.

e ***Speculate on the effect of phytoplankton in the oceans on the carbon cycle. If the phytoplankton were killed, how might the atmospheric levels of carbon dioxide change?***

The uptake of carbon dioxide by the phytoplankton during photosynthesis in daylight may create the diffusion gradient down which carbon dioxide travels into the sea. If the phytoplankton were killed less carbon dioxide might dissolve. The extra carbon dioxide produced each year might then all remain in the atmosphere, hastening the onset of any climatic problems.

f ***Climatologists are waiting for a clear signal, above the random fluctuations of climate, of global warming. What types of signal might indicate that the climate was warming up?***

At present some glaciers are growing and others are melting. If glaciers all over the world began to melt this would signal a climatic warming.

g ***What would be the effect of a temperature increase on 1 the global rate of photosynthesis and 2 the global rate of respiration of decomposers?***

**1** Most plant production is at present probably limited by nutrient deficiency. If the climate warmed, there would be an increase in photosynthesis in those ecosystems where low temperature limits it (for example, Alpine regions and tundras). Plant respiration rates, however, would also increase. Any increase in net photosynthesis in these areas would be balanced by a decrease in photosynthesis in other biomes, since the arid zones would spread if the evaporation rate increased.

**2** Decomposer respiration rates would increase. The carbon dioxide released would warm the climate further. The warmer climate might increase still further the respiration rate of decomposers, and so on. In this situation climatic warming would occur rapidly.

**h** ***When a clear signal of climatic warming is received, what steps could nations take to limit the increase in global carbon dioxide levels? Discuss the problems.***

Reduce the burning of fossil fuels. Concentrate on the production of energy by hydroelectric power, tidal power, nuclear fission, and nuclear fusion. Remove carbon dioxide by chemical means. Dependence of industry and domestic users on fossil fuels creates political problems, both national and international. □

## STUDY ITEM

### 29.42 The nitrogen cycle in a temperate forest ecosystem

*Assumptions*

**1** A basic understanding of the chemical composition of the various inorganic ions containing nitrogen, such as ammonium, nitrate(III), and nitrate(V).

**2** An understanding of the terms oxidation and reduction.

*Principle*

**1** Felling of temperate forests causes nutrient losses.

*Questions and answers*

**a** ***Suggest some of the inputs and outputs of nitrogen which have been omitted from the simplified nitrogen budget for Hubbard Brook shown in* figure [S] 370.**

Organisms produce some ammonia and oxides of nitrogen, which are then dissolved in rainfall. Oxides of nitrogen, mainly from industrial sources, contribute to the acid rain which is deposited in these northern hardwood forests. The trees support a grazing food web through which about 10 per cent of their primary production passes. Ten per cent of the nitrogen atoms absorbed each year probably take the same pathway. Legumes in the community will fix nitrogen gas because of the *Rhizobium* bacteria in their root nodules. Many ecosystems, but not this one, receive nitrate(V) or ammonium compounds in fertilizers, derived from the fixation of nitrogen by the industrial Haber–Bosch process.

b ***If this ecosystem were allowed to mature into a climax system, what differences would you see between its nitrogen budget and the nitrogen budget in* figure [S] 370*?***

The fifty-five year old system illustrated in figure (S)370 is still accumulating nitrogen. In a mature system, there would be no net accumulation of nitrogen compounds in tree branches, stems and roots, nor in litter. The extra litterfall which occurs in a mature system adds more nitrogen atoms each year to the soil surface. This is compensated for in a mature system by extra denitrification and leaching of nitrate(V) to groundwater. The leaching of nitrate, for example, might occur in gaps created by tree falls. There, higher temperatures at the soil surface would increase the rate of nitrification, but there are no plants to absorb the extra nitrate(V) produced.

c ***Most estimates suggest that nitrogen atoms are recycled much more rapidly in tropical than in temperate ecosystems. Suggest three reasons why this is the case.***

Decomposition of leaf litter is more rapid in the moist tropics than in temperate regions, since temperatures at the soil surfaces are higher and the decomposers respire faster. In the tropics, root systems are extensive and associated with mycorrhiza. They efficiently exploit a higher proportion of the soil volume than in temperate systems. Roots metabolize and grow faster in the tropics. In the tropics there is no cold season in which decomposition comes to a halt.

d ***Discuss the reasons why the output of nitrate increased so dramatically when the forest was felled.***

More dead branches and leaves than usual, containing organic nitrogen compounds, reached the soil surface and were decomposed. High soil temperatures (less shading) increased the rate of bacterial activity in nitrification, producing more nitrate(V) per unit time. The volume of water leaching through the soil increased, since soil water was no longer transpired. Some rainfall no longer hit the canopy and was evaporated without ever reaching the soil. In the felled area, there were no living roots at first to take up this extra nitrate(V), which was therefore leached to the outflow streams.

e ***On the basis of the numbers in* figure [S] 370*, how long might it take before the nitrogen lost was replaced?***

Nitrate losses exceeded 340 kg $ha^{-1}$ of nitrogen atoms in three years. In the intact ecosystem illustrated, the gain of nitrogen atoms from nitrogen fixation and precipitation exceeds the losses from leaching and denitrification by 16.7 kg $ha^{-1}$ $year^{-1}$. The total nitrogen loss of 340 kg $ha^{-1}$ would take on this basis about twenty years to replace. If felling causes the permanent death of the nitrogen fixers, however, the net nitrogen income from precipitation (precipitation minus leaching) would be 2.5 kg $ha^{-1}$ $year^{-1}$. The 340 kg $ha^{-1}$ nitrogen lost over three years would take 340/2.5 = 136 years to replace.

f ***Would the results have been different if the felled vegetation had been removed?***

Yes. The felled vegetation was removed in Hubbard Brook Watershed W101, clear cut in the autumn of 1970. Losses of nitrogen in the first two years after cutting amounted to 95 kg ha$^{-1}$ in stream water, and 144 kg ha$^{-1}$ in tree trunks. The total of 239 kg ha$^{-1}$ in the first two years resembles the losses of nitrogen from Watershed Two (236 kg ha$^{-1}$) in the first two years.

g ***How might the method of felling and/or after-treatment be modified so as to reduce the loss of nitrate from the ecosystem after felling?***

By strip cutting. Cut 25 m wide strips at 75 m intervals across the forest in one year, and then repeat in years two and three until all the forest is felled. It is to be hoped that the remaining plants will take up some of the nitrogen compounds released from the felled strips. Existing plants might also reduce temperatures at the soil surface and intercept some of the precipitation. Regrowth should be allowed on the cut strips since the invading plants will begin to conserve soil nitrogen.

h ***Discuss the possible effects on rivers and streams of the increase in sediment load and nutrient ion concentration after forest felling.***

The increased sediment load of clay and silt particles eroded into streams might cause silting downstream and reduce photosynthesis. The nitrates might contribute to 'eutrophication', the enrichment of water with nitrates and phosphates. This is frequent in streams, ponds and lakes in areas of dense human population and industrialized agriculture. Algae in the water course undergo a population explosion. As the algae die, they are decomposed by bacteria, which in their aerobic respiration begin to deoxygenate the water. This may kill fish and cause stagnation. The gases and other compounds produced (see Study item 26.41) may taint drinking water.

i ***Discuss the following statement made by the investigators. 'These facts strongly suggest to us that the forest... is almost totally dependent for its existence on this thin (2–20 cm), rather fragile, organic layer.'***

According to the nitrogen contents in circles in figure (S)370, this organic layer of litter plus humus does contain $(3600 + 1100)/(26 + 351 + 181 + 1100 + 3600) = 0.89 = 89$ per cent of the nitrogen in the ecosystem. It is in constant flux. It holds water, and is intensively exploited by plant roots. If this thin layer of soil were removed, plant growth would be retarded, because inorganic nitrogen compounds would not be so rapidly available to tree roots. Nevertheless, it is equally true that the organic layer depends on the forest for its persistence.

□

**STUDY ITEM**

## 29.43 Nutrient cycling in lakes and oceans

*Principle*

1 In the upper layers of temperate oceans and lakes, the productivity of the phytoplankton may be limited in summer by lack of nutrients such as nitrates and phosphates.

*Questions and answers*

a ***Why does the warmer water float on top of the cooler water?***

It is less dense.

b ***Account for the increase in the numbers of producers during March.***

Increased temperature, increased light intensity.

c ***Suggest two reasons for the decline of the producers in May.***

Overgrazing by herbivores. Competition for nutrient ions – the nutrient ions such as nitrates and phosphates are lost from the epilimnion and cannot be replaced because of thermal stratification.

d ***Account for the shape of the curve for primary consumers.***

The curve rises in spring as their food supply, the primary producers, increases. It declines in summer as primary producers decline. The zooplankton are also eaten by predators in abundance. Primary consumer numbers stabilize in August and September as autumn storms overturn the thermocline and primary producers, freed from nutrient limitation, increase in numbers again. Primary consumers then decrease as their food supply becomes scarce in winter.

e ***Amongst the factors which might alter this idealized picture are, first, some species of algae and cyanobacteria are more palatable and digestible to carnivores than others, and second, some cyanobacteria fix nitrogen gas from the atmosphere. Speculate on the influence of these factors on the curves of relative abundance shown in* figure [S] 371.**

The palatable and digestible cells in the phytoplankton will be eaten preferentially by the primary consumers. The less palatable species, such as the diatoms, will be avoided. The autumn peak in primary producers contains a much higher proportion of the cells of unpalatable species than the spring peak.
The cyanobacteria fix nitrogen gas as well as photosynthesizing. Their reproduction is not limited by lack of nitrate(V) like the other phytoplankton. Cyanobacteria comprises a higher proportion of the cells in the phytoplankton in autumn than spring.
For an account of the fluctuations in phytoplankton numbers in freshwater lakes see Porter, 1977, listed in the bibliography to this chapter. □

**STUDY ITEM**

### 29.44 Nutrient cycling in the Amazonian tropical rain forest

*Principle*

1 Nutrient losses when tropical rain forests are felled are more serious than in their temperate counterparts because in tropical forests more of the nutrients in the ecosystem are in the vegetation.

*Questions and answers*

a ***Calculate for each element at each site the ratio of its mass in the soil to its mass in the vegetation.***

| | *Nitrogen* | *Phosphorus* | *Potassium* |
|---|---|---|---|
| Temperate forest | 11 | 23 | 1.23 |
| Tropical forest | 2.5 | 0.1 | 0.8 |

b ***Is the felling of the temperate or the tropical forest more likely to produce soil suitable for the growth of plants? Which of the three elements in the tropical soil is most likely to be deficient?***

Temperate forest soil would be more fertile, at least on the basis of its nitrogen and phosphorus content. Phosphorus is the element which, in the tropical soil, is most likely to be deficient.

c ***If leaf litter decomposes five times more rapidly in tropical than in temperate forests, when leaves fall one might expect five times the nutrient ions to be suddenly released in tropical soils than in temperate soils. Why, then, is the nutrient concentration in tropical forest soils often*** **lower** ***than in temperate soils?***

In tropical forests the roots and the mycorrhizal fungi exploit a greater proportion of the soil volume than in temperate forests. They will intercept most of the nutrients which are released. Uptake of nutrients by roots exposed to nutrients is also more rapid in tropical than in temperate soils because at the higher temperatures roots grow faster, encountering new nutrient sources all the time. More energy from cellular respiration is also available for ion uptake by cells.
In the tropics there is no cold season. In temperate zones, however, leaf fall often occurs just before the cold season, in which the rates of decomposition, nutrient release, root growth, and nutrient uptake are very slow. □

## 29.5 Pollution

*Principle*

1 Pollution can affect the species composition of communities and the distribution patterns of species in nature.

**Practical investigation. *Practical guide* 7, investigation 29D, 'A comparison of the growth of tolerant and non-tolerant seedlings when exposed to metal ions'.**

## STUDY ITEM

### 29.51 The influence of pollution on river communities

*Assumption*

1 An understanding of nitrification in the nitrogen cycle.

*Questions and answers*

a ***Explain in detail the data in* figure [S] 374 *on the basis of the rates of decomposition of organic matter documented in* figure [S] 373.**

The organic waste decomposes steadily with distance from the point of discharge. Its compounds are oxidized but bacteria need not be involved. The suspended solids settle out quickly, or are rapidly decomposed by bacteria. High bacterial populations near the point of discharge absorb oxygen for aerobic respiration and lower the dissolved oxygen level in the water.
These bacteria will be attacking organic compounds, many of which contain nitrogen and phosphorus. The nitrogenous waste is released as ammonium ions and the phosphorus as phosphate ions. Nitrifying bacteria, such as *Nitrosomonas* and *Nitrobacter* further downstream, act on the ammonium ions converting them first to nitrate(III) and then to nitrate(V) ions.

b ***What are the ecologically important physical and chemical characteristics of water polluted by organic residues?***

The water near the discharge point will be turbid and unlikely to support rapid photosynthesis. Further downstream, the suspended solids disappear and algal photosynthesis can occur in the clear water. Near the discharge point, oxygen for cellular respiration will be scarce. Ammonia is toxic in high concentrations. Further downstream, with the oxygen concentration rising, nitrates and phosphates will become important nutrient ions, promoting the rapid reproduction of algae.

c ***What evidence is there for three fairly well-defined communities downstream from the point where the effluent was discharged? List the chief members of each community.***

The abundances of species along an environmental gradient tend to follow a series of bell-shaped curves. Communities exist when some of these bell-shaped curves coincide, so that species are habitually found together. *Two* 'communities' certainly exist. Near the point of discharge, sewage fungus, bacteria, and tubificid worms predominate. These are characteristic of the most highly-polluted anaerobic rivers in Britain. Further downstream, algae, the alga *Cladophora* spp. and the clear water fauna occur together. Although midge larvae (*Chironomus*) and water lice (*Asellus*) occupy intermediate positions, they do not form a community with distinct boundaries.

d ***Name two relevant physiological characteristics which the members of the polluted water community might possess.***

They must be able to survive with a low concentration of available oxygen and to withstand toxic chemicals.

e ***What is the practical value of the information obtained from these studies of river pollution?***

It is possible to identify polluted waters by the organisms which they contain.

f ***The pollution of a river with organic matter is sometimes measured in terms of the biochemical oxygen demand (B.O.D.) of its waters. This is shown in the curve labelled 'oxygen used by decomposing organic waste' in* figure[S] 373. *From your knowledge of respiration and the rate of diffusion of oxygen from air to water, suggest the reasons underlying the use and value of this term.***

Oxygen is used in the oxidation of the unstable organic wastes, and in aerobic respiration by bacteria. The more polluted the water with organic residues, the faster it absorbs oxygen, and the higher its 'biochemical oxygen demand'. The oxygen to satisfy this demand comes from the photosynthesis of water plants, rather low in many polluted rivers, and from oxygen which dissolves into the river from the atmosphere. The B.O.D. can be determined on water samples in the laboratory and is a convenient measure of the degree of pollution of water with organic wastes. □

## 29.52 The influence of sulphur dioxide on lichen patterns in Britain

*Principles*

1. Some lichen species are more susceptible to aerial pollutants than others.
2. The level of sulphur dioxide in the atmosphere is probably the major pollutant influence on lichen distribution.
3. The lichen species present in an area can be used to indicate the degree of pollution with sulphur dioxide.

*Questions and answers*

a ***Suggest how the $^{14}C$ isotope was incorporated into the fungus in the experiment which yielded the results illustrated in* figure [S]379.**

It was supplied as $^{14}CO_2$, created by treating $^{14}C$-labelled sodium hydrogen carbonate with hydrochloric acid in a closed container containing the lichen. It was absorbed in photosynthesis by the algal component of the lichen.

b ***Relate the results in tables* [S]74 *and 75 to the data in* figure [S]379.**

The results in table (S)74 suggest that the sulphur dioxide concentration in the atmosphere is more likely to influence lichen distribution than the intensity of smoke.
When sulphur dioxide dissolves in water it will form sulphate(IV) ions. The results in figure (S)379 indicate that sulphate(IV) affects lichen species in different ways. Those lichens most susceptible to sulphate(IV) under laboratory conditions are those limited to the least polluted areas in the wild.

**STUDY ITEM**

### 29.53 The effect of DDT on the bald eagle

*Principles*

1 DDT and its derivatives accumulate in the body fat of organisms and disrupt metabolism.

2 The highest levels of DDT are found in predatory species at the tops of food webs.

*Questions and answers*

**a** ***Summarize what the graph in figure [S]381 shows.***

DDE is a toxic metabolic derivative of DDT. The levels of DDE in bald eagle eggs were much higher before DDT was banned (94 p.p.m.) than afterwards (29 p.p.m.). The mean number of young per breeding area declined fairly rapidly from 1966 to 1972–74, and then rose again from 0.4 (1974) to 1.1 (1981).

**b** ***Construct a hypothesis to explain the decline and recovery of the bald eagle population in the area shown.***

DDE residues in the body fat of female adults either interfered with egg production or were deposited in the eggs, disrupting metabolism. It is unlikely that the chicks were fed with DDT-contaminated food in the nesting areas. After DDT was banned, the food of the female eagles began to contain fewer DDE residues and the egg production and/or chick survival increased.
The problem with this hypothesis is that if we assume that DDT is hardly excreted by the birds it is difficult to explain why the DDE residues in their eggs decline. Perhaps the eggs contain a significant proportion of the DDE in a female's body. The most plausible hypothesis is that some old females with bodies rich in DDE die and are replaced by young ones which have not fed on DDT-contaminated food. This decreases the average DDE content per egg and increases the average breeding success.

**c** ***How could you test your hypothesis experimentally?***

This would be difficult unless the birds can be induced to breed in a zoo. Some birds could be fed a DDT-rich diet whilst others were fed on similar uncontaminated food, and the effect on egg contamination and breeding success could be monitored.

**d** ***Suggest three other explanations for the fluctuations in bald eagle numbers.***

Climatic changes. Fluctuations in food supply. Changes in abundance of predators or parasites.

**e** ***Why do birds at the tops of food webs have higher levels of DDE in their body fat than organisms lower down?***

Each carnivore accumulates the DDE from all the numerous prey items which it eats, and concentrates it in body fat. The concentration of DDE becomes higher in its body fat than in the fat of any of its prey items. This process, continued up the food web, results in the highest DDE concentrations in the body fat of the top carnivores.

f ***DDE in food webs does not break down. Suggest two reasons why the average DDE concentration in bald eagle eggs declined from 1974 to 1981.***

Successive eggs produced by the same female perhaps have successively smaller DDE contents. Old birds with high DDE die and young birds with less exposure to DDE replace them. Perhaps birds excrete DDE, or bacteria can break down DDE in detritus. □

## 29.6 Nature conservation

*Principles*

1 Biologists consider nature conservation desirable and it can often be justified on economic grounds as the best use for available land.
2 Nevertheless, nature conservation is often difficult to justify in economic terms. It then becomes one of a variety of conflicting interests to be considered when the optimum use for a piece of the biosphere is discussed.

In discussing nature conservation a critical, objective approach is essential.

### STUDY ITEM

### 29.61 The value of nature conservation (essay)

*Question and answer*

a ***Discuss whether the conservation of wild species of animals or plants is either necessary or desirable, using named species of animals or plants to illustrate your arguments.***

Good answers will mention the following reasons for nature conservation:
Wild ecosystems economically valuable for hunting, grouse moor, game parks, angling, and tourism.
Forestry provides long-term sources of timber.
Some organisms in wild ecosystems represent potential food resource if properly managed (for example, fish, whales, and red deer).
Forests act as watersheds to conserve water and prevent soil erosion.
Ecosystems act as early warning systems for humans for effects of pollutants.
Publicity generated by efforts to conserve certain wild species brings in much money for purchase of nature reserves.
Wild ecosystems must be preserved for psychological and recreational reasons; freedom of the human spirit etc.
Animal and plant species have a right to exist.
Humans are the 'guardians' of nature and must sensitively care for animal and plant species.

Many species are attractive and elegant, with interesting habits; they provide humans with aesthetic pleasure.
Wild ecosystems are needed for ecological research.
The 'balance of nature' can only be preserved by maintaining whole ecosystems intact.
Some insects in nature act as biological control agents, keeping potential pest populations low.
The wild ancestors of crop plants must be preserved for future breeding programmes (genetic conservation).
Most angiosperms have not yet been screened for possible drugs.
Some of the factors to be considered when evaluating sites from the point of view of nature conservation are discussed by Spellerberg, 1983 (see the bibliography). □

# PART II The *Practical guide*

The investigations related to this chapter appear in *Practical guide* 7, *Ecology*.

## INVESTIGATION
## 29A A quantitative study of an ecosystem

(*Study guide* 29.1 'Introduction'; Study items 29.11 'Trophic levels and food webs' and 29.12 'Ecological pyramids'.)

### ITEMS NEEDED

Model pond ecosystems set up in large washing-up bowls, containing pond plants and animals, kept outside or in the laboratory
Meat, piece of, small

Iodine in potassium iodide solution
Light green or fast green stain for cellulose

Balance 1/class
Centrifuge 1/class
Glass rod 1/group
Keys to common pond organisms
Microscopes, stereo, with high power 1/group
Microscopes, binocular 1/group
Paint brushes
Petri dishes, for examining larger organisms in drops of water
Pipettes, dropping
Pond nets 1/group
Specimen tubes
Spoons, plastic
Syringes
Table lamp
Wire netting, if bowls outside, to reduce predation by birds

*Assumptions*

1. An elementary knowledge of the concept of trophic levels, including the role of photosynthesis in plants.
2. The ability to use a stereo microscope, and elementary techniques for examining small organisms under high power monocular microscopes (hanging drop, raised cavity slip and cavity slide methods).
3. The ability to stain and identify cellulose, using light green solution, and starch, using iodine solution.
4. The ability to use keys for identifying pond organisms.

*Principles*

1. Light is captured by plants through photosynthesis.
Subsequently the energy is passed on to carnivores, detritivores, and decomposers.
2. The availability of food inevitably plays an important part in influencing the spatial distribution and population size of a species.

It is possible to waste a great deal of time and effort on this work. It should be limited to a carefully planned series of investigations, perhaps delegated to different groups in the class.

Keep at least one of the tank communities in the dark to demonstrate the need for light.

Microscope slides or Petri dishes should be hung in pond water for several weeks before the class work starts, so that the colonizing organisms can be seen in abundance. This will not be successful unless the tanks contain a variety of species.

It is useful to make sure that *Daphnia*, *Cyclops*, and some of their relatives are present.

Gut contents of herbivores provide good corroborative evidence for the feeding habits which have been observed. This is one way to decide whether an organism feeds by scraping organisms off the leaf surface, or whether it eats the leaves themselves. In this case, examine the leaves also with a stereo microscope. Unfortunately, food organisms are rapidly broken up and degraded in the gut. For this reason, the foregut is the best part to examine.

Despite the limited range of species in small aquaria, it will often be possible to show that some species occupy different feeding niches. Students sometimes find it difficult to understand that an animal's ecological niche refers to all the factors which influence its role in nature. By contrast, the place where it lives, considered on a small scale, is its microhabitat.

As far as detritivores and decomposers are concerned, plenty of detritus is seen in the aquaria. Some organisms, such as *Asellus*, planarians, and the annelid worm *Lumbriculus*, will be observed eating it. Gut contents are unlikely to be instructive; if the class has time to examine gut contents, examine those of herbivores.

The diets of different species of snails can be compared by an analysis of faecal pellets. Students are often fascinated by the experiments in which carnivores are placed in water in specimen tubes with a range of potential prey, and watched closely. Both these investigations are suitable for project work.

It is valuable to round off the investigation with a discussion of applied aspects, such as the setting up of aquaria with the optimum proportions of organisms, the maintenance of fish ponds, fish farming, and the stocking of rivers with fish.

*Questions*

**a** ***Does each species of herbivore feed on a specific range of plant species? Are different herbivores competing for the same plant food? If so, can you devise an experiment to test for competition? How do you think that each herbivore recognizes its food plants? How could you test experimentally your ideas about food recognition?***

–

**b** ***Do the carnivores have specific food preferences? Are there species which carnivorous predators do not catch and eat? Suggest some reasons why some herbivores are avoided as food.***

–

**c** ***Do the herbivores tend to feed for a greater proportion of the time than the carnivores? Can you explain your observations in terms of the problems in feeding faced by herbivores and carnivores in general?***

–

d ***Comment on the shape of your pyramid of biomass.***

e ***Is there evidence that the carnivores might control the population size of herbivores? Would the relationship between carnivores and herbivores be any different under natural conditions?***

## INVESTIGATION
## 29B The energetics of the stick insect (*Carausius morosus*)

### ITEMS NEEDED

*Carausius morosus* (stick insect) 1/group
*Ligustrum ovalifolium* (privet) shoots

Balance, weighing to three decimal places 1/class
Cavity microscope slides 2/group
Cylinder cage 1/group
McCartney bottles 2/group
Parafilm, one piece 1/group
Petri dishes (glass) or crystallizing basins (250 $cm^3$)
Respirometer (see Mackean, 1976, or *Practical guide* 2, investigation 5E, p. 17)

(*Study guide* 29.1 'Introduction'; Study item 29.12 'Ecological pyramids'; and *Study guide* 29.3 'Energy flow'; Study item 29.32 'Energy flow and agriculture'.)

*Assumptions*

1 A knowledge and understanding of the use of a respirometer.
2 Some understanding of the principles of energetics.
3 An ability to handle simple data conversions.
4 An appreciation of the differences in the energy budgets of different types of organisms.

*Principles*

1 It is possible to make a fairly ac urate estimate of the energy budget of an organism in laboratory cond..ions.
2 The physiology of an organism has an important bearing on its distribution and way of life.
3 The limits of accuracy imposed by different techniques can be seen to have an effect on the results of this experimental investigation.

Further information about stick insects appears in Clark, 1974, and Slatter, 1980 (see the bibliography).

*Questions and answers*

a ***How closely do your results fit the equation? What are the main sources of inaccuracy?***

There should be fairly close agreement, but values for the energy content of fresh privet leaves, live insects and their eggs will vary. Other sources of inaccuracy include the differences in the insect's behaviour in the cylinder cage and the respirometer, variations in physical factors such as temperature and light, and changes in water content. If time or equipment precludes the use of a respirometer, the energy used in respiration should be estimated by subtraction before proceeding to question **b**.

**b** ***Living organisms differ in the efficiency with which they convert their food material into biomass. The ratio of production (P) to respiration (R) provides a measure of food conversion efficiency. Calculate the P/R ratio for your stick insect and relate it to the values for other species given in table [P]5.***

The ratio should be in the range of the other poikilothermic herbivores. It is likely to depend on the temperature and the age of the insect. In adult insects the ratio of production to respiration is likely to be very low.

**c** ***What effect would a change in the mean temperature have on the conversion efficiency as estimated by the P/R ratio? Explain your answer.***

Changes in temperature would be expected to affect the rate of respiration more than production. Thus, as the temperature rises, the ratio will fall, indicating a fall in efficiency. This would make an interesting project.

**d** ***How would you expect the energy budgets of young and old stick insects to differ?***

There is no easy or straightforward answer to this question, but it has been included to make students think about the factors involved.

Slatter, 1980 (see the bibliography), provides plenty of data. Young insects devote more of their assimilated energy to production than old ones. In old insects, with a fresh mass above a gram, production has become almost zero and the insect devotes most of its assimilated energy to respiration. In fact old insects often decline in mass during the experiment, especially if they have produced eggs.

The respiration rate will depend not only on the age of the insect but also the temperature and its activity during the week.

## INVESTIGATION
## 29C A study of decomposer organisms in the soil

(*Study guide* 29.3 'Energy flow'; see 'Decay and decomposition'.)

### ITEMS NEEDED

Soil sample containing a high level of humus

Agar, cellulose
Agar, malt
Agar, nutrient
Agar, soil extract

Insulating tape
Keys to soil organisms
McCartney bottles 4/group
Micro-ecology tubes (Philip Harris Biological), sterilized before use
Microscope – preferably with phase contrast 1/group
Microscope slides
Plant pot, 10 cm diameter 1/group

**Cellulose agar, malt agar, nutrient agar, and soil extract agar** should be available in McCartney bottles kept at 50 °C in water bath.

Cellulose agar is made from 1 g cellulose powder (fine), 2.0 g agar, 0.05 g yeast extract and made up to 100 $cm^3$ with distilled water.

Malt and nutrient agars are commercially available.

Soil extract agar is made by shaking soil with warm water, settling and filtering, and making the filtrate up with 2 % plain agar. It is then sterilized.

*Assumptions*

1 Ability to use a microscope with phase contrast.
2 Knowledge of the use of sterile technique.
3 Some knowledge of the processes of decay and the organisms involved.

*Principles*

1. Many different organisms, belonging to a variety of taxonomic groups, are involved in the important processes of decay.
2. The use of enrichment media can encourage the growth of these organisms and enables them to be concentrated in a restricted habitat.
3. A complex food web exists, entirely based on energy released by decomposers.

*Questions and answers*

a ***What other environmental factors, apart from a high soil humus content, would create conditions suitable for these soil organisms?***

A high temperature; aerobic conditions (although very rapid decomposition can take place anaerobically); presence of certain mineral ions (such as accelerators used in compost making, which provide a high ratio of nitrogen to carbon); plenty of moisture.

b ***Devise an experiment, using this technique, to investigate the effect of one of these factors on the variety of organisms found in the micro-tubes.***

Careful consideration will need to be given to the problems of controlling environmental factors in the soil.

c ***In what ways is the investigation likely to provide a restricted idea of the decomposition processes in the soil?***

Larger organisms are excluded; colonization will be inhibited; early stages in decomposition are bypassed.

d ***Compare the types of organisms you have found in the different media. Which of the media contains 1 the largest variety of organisms, 2 only fungi and the herbivores which feed on them, 3 mainly bacteria, 4 soil amoebae, and 5 eelworms?***

Results will depend on the media used and precise conditions. The largest variety of organisms is to be expected where the widest range of nutrients is available – in soil extract or nutrient agar. Bacteria are favoured by nutrient agar and fungi grow best on malt agar. The viscosity of agar may affect the numbers of motile animals such as eelworms and amoebae.

e ***Explain how the size of the tubes affected the variety and numbers of organisms found in the tubes.***

The size will affect the results in two ways: larger tubes will admit larger organisms but they also contain more food material so that more complex food webs can develop within them.

f ***Construct a food chain or a food web from your observations on the tubes. Look for interactions between the organisms involved.***

## INVESTIGATION

## 29D A comparison of the growth of tolerant and non-tolerant seedlings when exposed to metal ions

### ITEMS NEEDED

Seeds of *Festuca rubra* (creeping red fescue) varieties Merlin and Dawson. Less than 1 g of each variety is required. The seeds can be purchased from W. W. Johnson & Son, London Road, Boston, Lincs.

Rooting solutions

Plastic or polystyrene beakers (small yogurt pots would be suitable, but glass absorbs and releases ions and should be avoided) 10/group
Nylon stocking material, or alkathene beads ref 9388 Black 9025 to form a layer two beads deep on the surface of the culture solution. Ruler or other scale graduated in mm.

(*Study guide* 29.5 'Pollution'.)

**Rooting solutions.**
**a** 1/10th strength Hoagland's or Sachs solution, minus phosphate, adjusted to pH 7.
**b** 1/10th strength Hoagland's or Sachs solution, minus phosphate and containing a solution of one of the ions shown in table 28.

| *Salt* | *Salt concentration in stock solution* (g $dm^{-3}$) | *Metal* | *Metal level in test solution* (mg $dm^{-3}$) |
|---|---|---|---|
| $CuSO_4 \cdot 5H_2O$ | 0.393 | Copper | 0.5 |
| $ZnSO_4 \cdot 7H_2O$ | 0.660 | Zinc | 7.3 |
| $Pb(NO_3)_2$ | 3.837 | Lead | 12.0 |
| $NiSO_4 \cdot 7H_2O$ | 1.913 | Nickel | 2.0 |

**Table 28**
Recommended concentrations of metals for use in tolerance tests (Bradshaw and McNeilly, 1981)

In each case 5 $cm^3$ of stock solution is made up to 1 $dm^3$ with the Hoagland's solution to give the appropriate level of the metal in the test solution.

*Assumptions*

1. An understanding of the need to replicate to allow for natural variation.
2. An appreciation of the factors needed for the normal growth of seedlings.
3. An understanding of the Mann–Whitney U-test.

*Principles*

1. Heavy metal pollution is a significant environmental factor.
2. The development of plant varieties resistant to toxic pollutants is of increasing economic importance.

*Questions and answers*

**a** ***Do the two varieties differ in their root growth in the normal culture solution? Do you think that the mean values are sufficiently similar to enable you to make a comparison between the effects of the metal ions on the two varieties?***

The differences between the two varieties are of little importance. It is the differences in their responses to the metal ions which is significant.

**b** ***Which of the two varieties shows the greatest tolerance to the presence of metal ions?***

Variety Merlin has been bred for resistance to toxic ions, particularly lead and zinc. Its resistance to nickel and copper is not reported.

c ***Does either of the varieties show tolerance to only one of the ions, or is there a similar tolerance to all the ions? Discuss the patterns you find.***

The two varieties differ in their patterns of tolerance. Tolerance to one metal ion does not equip the plant with tolerance to another. The overall patterns may disguise the fact that within one species some individuals may be more tolerant than others, and that individual plants in the same population may be tolerant to different metals.

d ***Explain how you could demonstrate the usefulness of either of these varieties for the reclamation of spoil heaps.***

Select individuals from the experiments which grow well in high concentrations of metal ions. Transfer these resistant seedlings to good soil. When they set seed, sow their seeds on modified mine waste soil.

e ***How would you produce a strain of plants tolerant to the presence of these metal ions? You will need to read about the propagation of grasses.***

Many grasses are self-incompatible, and in any case tolerance may be reduced if a tolerant grass is pollinated predominantly by non-tolerant pollen from grasses of the same species growing on unpolluted soil nearby.

Thus to produce large numbers of rooted plants of the same strain (some would say the same genetic individual!), remove plants growing on spoil heaps and grow them to maturity in the laboratory. Allow each separate plant to produce numerous tillers (stems with leaves). Then plant out these plants on spoil heaps. Select the fastest-growing individuals and repeat the process.

f ***It has been suggested that the addition of humus to polluted waste reduces the toxicity of metallic ions to plants. How would you test the hypothesis that the humus detoxifies the waste rather than simply diluting the ions?***

In an experiment similar to this investigation (investigation 29D), compare the effect on the growth of grasses of culture solution on the one hand, and culture solution which has been passed through humus before use, on the other.

## PART III Bibliography


ANDERSON, J. M. 'Life in the soil is a ferment of little rotters'. *New Scientist*, **100**(1378), 1983. (A lengthy and beautifully-illustrated summary of decomposition in the soil.)

ANDERSON, J. W. Studies in Biology No. 126, *Bioenergetics of autotrophs and heterotrophs*. Edward Arnold, 1980. (Concentrates on chemical energetics of ecosystems. A complex but rewarding read for those whose biochemistry is sound.)

BAYLISS-SMITH, T. *The ecology of agricultural systems*. Cambridge University Press, 1982. (Fascinating account of the energetics of seven types of crop-producing systems from different parts of the world and periods in history.)

BEEDHAM, G. E. *Identification of the British Mollusca*. Hulton Educational Publications, 1973.

BELCHER, H. and SWALE, E. *A beginner's guide to freshwater algae*. H.M.S.O., 1977.

BELCHER, H. and SWALE, E. *An illustrated guide to river phytoplankton*. H.M.S.O., 1980.

BLACKSELL, M. 'Spare that redwood tree'. *Geographical magazine*, **54**(4), 1982. (Describes attempts to conserve the giant redwoods in California.)

BRADSHAW, A. and MCNEILLY, T. Studies in Biology No. 130. *Evolution and pollution*. Edward Arnold, 1981.

CHAMAIDES, W. L. 'Increasing atmospheric methane'. *Nature* **301**(5901), 568, 1983.

CLARK, J. T. *Stick and leaf insects*. Barry Shurlock, 1974.

CLOUDSLEY-THOMPSON, J. L. and SANKEY, J. *Land invertebrates*. Methuen, 1961.

DOWDESWELL, W. H. *Ecology: principles and practice*. Heinemann Educational, 1984.

EVANS, L. T. 'The natural history of crop yield'. *American Scientist*, **68**(4), 1980. (Describes the environmental constraints on world crop yield and the prospects for overcoming them.)

HALL, D. O. 'Solar energy conversion through biology – could it be a practical energy source?' *Biologist*, **26**(1), 1979. (Concise description of factors affecting primary production in crop plants.)

JEFFERIES, P. *Harris micro-ecology tubes*. Philip Harris Biological, 1981.

LEADLEY BROWN, A. *Ecology of soil organisms*. Heinemann, 1978.

MACAN, T. T. *A guide to freshwater invertebrate animals*. Longman, 1959.

MACKEAN, D. G. Experimental work in Biology 7. *Respiration and gaseous exchange*. Murray, 1975.

MELLANBY, H. *Animal life in fresh water*. 6th edn. Chapman and Hall, 1963.

MELLANBY, K. Studies in Biology No. 38, *The biology of pollution*. Edward Arnold, 1972. (A well-written introductory account of the effects of pollutants on humans or wildlife.)

NUFFIELD ADVANCED BIOLOGICAL SCIENCE. *Key to pond organisms.* Longman, 1970.
NUFFIELD REVISED ADVANCED BIOLOGICAL SCIENCE, *Mathematics for biologists.* Longman, 1987.
ODUM, E. P. *Fundamentals of Ecology.* 3rd edn. Saunders, 1971.
OTTOWAY, J. H. Studies in Biology No. 123, *The biochemistry of pollution.* Edward Arnold, 1980. (A valuable up-to-date complement to Mellanby's book. Particularly suitable for sound chemists and biochemists.)
PORTER, K. G. 'The plant-animal interface in freshwater ecosystems'. *American Scientist* **65**, 1977.
QUIGLEY, M. *Invertebrates of streams and rivers: a key to identification.* Edward Arnold, 1977.
REVELLE, R. 'Carbon dioxide and world climate'. *Scientific American* **247**(2), 1982. (Excellent analysis of the possible climatic impact of increased carbon dioxide levels in the atmosphere.)
SCHOOL NATURAL SCIENCE SOCIETY. *Key to pond organisms.* 1964.
SLATTER, R. J. 'The energy budget of the stick insect'. *School Science Review* **62**(219), 1980.
SPELLERBERG, I. H. Studies in Biology No. 133, *Ecological evaluation for conservation.* Edward Arnold, 1981. (Describes various techniques for evaluating land for its conservation value in competition with other possible uses.)
WHITTAKER, R. H. *Communities and Ecosystems.* 2nd edn. Macmillan, 1975. (A standard introductory text, with a considerable emphasis on ecosystems.)

# CHAPTER 30 EVOLUTION

*A review of the chapter's aims and content*

1 An important aim of this chapter is to show students that there is no universally agreed neo-Darwinist theory of evolution.
2 The historical background to evolution theory, up to the development of neo-Darwinism, is given (**30.1**).
3 Other interesting developments in, or related to, evolution theory are considered, such as punctuated equilibria (**30.2**) and cladistics (**30.3**).
4 It is emphasized that similarities of organisms are as fundamental to an understanding of classification and evolution as differences are (**30.4**).
5 Another aim is to stimulate students to ask questions about the 'laws' that cause such a degree of order in the biological realm.

# PART I The *Study guide*

## 30.1 Two centuries of evolution theory

*Assumptions*

1 An appreciation of variation in living organisms.
2 Some knowledge of the geographical distribution of organisms.
3 A basic understanding of the fundamental principles of Mendelian genetics.

*Principles*

1 The concept having most influence on early evolution theory was the *scala naturae*.
2 There was widespread interest in evolution in the thirty years before the *Origin of species* appeared.
3 Darwin's work promoted a general acceptance of evolution theory, but not of his theory of natural selection.
4 The concept of natural selection was only accepted by most biologists after about 1930, when neo-Darwinism had been developed.
5 Influences outside the scientific community may sometimes play a role in scientific development – for example, there are certain analogies between natural selection and free enterprise economy.

It is hoped the passage will promote discussion and the questions have been designed with this end in view.

*Questions and answers*

a ***How does the sequence of major groups in* figure [S] 382 *differ from a modern evolutionary series?***

Higher plants are not on the same evolutionary line as animals; the order of fish and reptiles is reversed; mammals are not descended from birds, nor birds from fish, nor reptiles from molluscs. By looking closely at figure (S)382, students should see that the concept of the *scala* is very different from that of an evolutionary series. It was a substantial theory in its own right, and we should not think of it merely as a faltering step along our road to modern evolution theory.

**b** ***What other intermediate kinds could have been regarded as evidence for the existence of the* scala naturae*?***

Whales and seals link fish with mammals (quadrupeds). *Archaeopteryx* links reptiles with birds (though it was discovered only in 1861). Around 1800 considerable interest focused on the recent discovery of the duck-billed platypus or *Ornithorhynchus*. Similarly the ostrich family, as Chambers wrote in his 1844 *Vestiges of the natural history of Creation*, 'form a link between birds and mammalia, and in them we find the wings imperfectly or not at all developed, a diaphragm and urinary sac (organs wanting in other birds), and feathers approaching the nature of hair. Again, *Ornithorhynchus* belongs to a class at the bottom of the mammalia, and approximating to birds, and in it behold the bill and web feet of that order!'
A problem with discovering too many intermediates between major groups is that the linear *scala* comes to look more like a network. Finding convincing intermediates proved to be a mixed blessing to those wishing to sustain the *scala* concept.

**c** ***Do you think there could have been any connection between the theory of natural selection and the opinions prevailing in politics at the time of its formulation?***

There is undoubtedly a close analogy between the concept of natural selection and the nineteenth-century liberal philosophy of human progress through free enterprise economics. In both cases competition allows the more successful to emerge and determine the future. But it is difficult for historians to say how close was the connection between biology and politics. Students will notice that both Wallace and Darwin were inspired by Malthus' book on human society. In addition Karl Marx remarked, after reading the *Origin of species*, how closely Darwin had portrayed a free enterprise economy in his description of natural selection. On the other hand, advocates of 'social Darwinism' in the late nineteenth century drew support for their encouragement of human competition by pointing out that Darwin had shown that this was how nature made progress. (For the 'social Darwinists', the accumulation of money and economic power was equivalent to the population increase of a naturally selected variant.) Although many similar connections between evolution theory and politics can be cited, their influence on people's opinions is not easy to assess. Unlike Chambers, Lyell, and Darwin, who were liberal in their outlook, Wallace was a socialist yet advocated the theory of natural selection perhaps more strongly than Darwin.

**d** ***Why is it misleading to think of invertebrates → fish → amphibia → reptiles → mammals as an evolutionary series?***

Evolutionary series such as this are frequently used, not only in biology teaching but also in research. They bear a close resemblance to the *scala naturae* and are often used in the same way. Another manifestation of this is the frequency with which adjectives such as 'higher', 'lower', 'primitive', and 'advanced' are found in evolutionist literature. Such a linear arrangement does not fit very easily with the theory of natural selection, which is much better at explaining how adaptive radiation occurs. Students are unlikely to raise this point spontaneously, although it can be introduced into the discussion.

A more immediate reason for objecting to the evolutionary sequence in the question is that it suggests, for instance, that *all* fish are ancestral to, or more 'primitive' than *all* amphibia. It ignores the great diversity and variation in degree of complexity among fish, and the fact that some fish may be much more complex or 'advanced' than some amphibia are.

Further discussion could tackle words like 'higher', 'lower', 'primitive', and 'advanced'. Do they really mean anything, or are they just a way of pretending that organisms can be arranged in a straight line from *Amoeba* to *Homo sapiens*?

Finally, are we still clinging to a surprisingly Lamarckian concept of life ascending the ladder of nature? Perhaps we are, and for this reason have never really understood what Darwin was trying to say in his theory of natural selection.

**e** ***How many of these objections to Darwin's theory might still hold water today?***

**1** Christian objections: Christians must presumably still object to natural selection as the origin of Man, understanding by 'Man' his body, mind, and soul.

**2** Age of the earth: Kelvin's calculations were based on erroneous assumptions (about the energy source in the Sun, for instance). Physicists are not always right.

**3** Gradualism and jumps: this is still a matter of debate, and is dealt with further in section 30.2. However, some if not most punctuated equilibrists believe that the speciation events are simply the result of rapid natural selection.

**4** Advantages of intermediate stages: with a bit of ingenuity it is usually possible to think up possible functions of half-formed organs. A low-voltage electric organ might, for example, be used for direction-finding in muddy water. In any case, demonstrating that a given organ had no function or could have had no function is next to impossible.

**5** Orthogenesis: although claims about unopenable oysters are generally dismissed as fanciful, some palaeontologists still think persistent trends in fossil lineages cannot be explained by natural selection.

**6** Heredity: the resolution of this problem is described in the next section of the *Study guide* text.

**f** ***Would the neo-Darwinist theory of evolution have been established sooner if Darwin had never been born?***

To be provocative someone might argue that the twenty years before 1858 were wasted, because Darwin kept his theory secret; that by insisting on slow gradual change, and by advocating the inheritance of acquired characters, he diverted attention from the discontinuous variation which was the clue to Mendelian genetics; that Wallace anyway had discovered natural selection in 1858, and advocated it in a purer form through the rest of the century than Darwin did; and that Wallace was a better popularizer than Darwin, and would have had more influence if he had not generously given Darwin the major credit. So one might argue that Darwin was more of a hindrance than a help. This of course is a one-sided selection of the facts, but it might make for a lively discussion.

## 30.2 Punctuated equilibria

*Assumption*

1 Some understanding of the geological time table.

*Principle*

1 The punctuated equilibrium model has promoted palaeontologists to think seriously about biological processes responsible for both changes and the stasis seen in fossils.

The punctuated equilibrium model is one of the most exciting developments in evolutionary theory since the modern synthesis around 1940. For decades, evolution as conceived by palaeontologists had borne little resemblance to that of population geneticists; moreover, there was hardly any effective dialogue on this disparity between the two groups of biologists.

Students should also understand how confusion has arisen from misunderstanding the two time scales involved. Six hundred million years are required for the fossil record since the beginning of the Cambrian, and now it has been extended much further back. Only decades are involved in population geneticists' studies of micro-evolution. So although ten thousand years may appear to the traditional palaeontologists as no time at all, it allows adequate time for speciation to occur by the neo-Darwinist mechanism.

### STUDY ITEM

### 30.21 Species selection

*Questions and answers*

Students are referred to figure (S)395 with the aim of deepening their understanding of the species selection concept.

**a** ***How many speciation events occur in the 60 million years?***

Thirty-eight.

**b** ***How many of these are to the left (producing smaller antler size) and how many to the right (increasing antler size)?***

Nineteen to the left, nineteen to the right.

**c** ***How could you test whether the trend was caused by the average increase in antler size (in speciations increasing antler size) being greater than the average decrease in antler size in the other speciations?***

By measuring the horizontal lines and comparing the total of leftward with the total of rightward speciations. One way to get an approximate comparison is to mark off leftward horizontal lines along the straight edge of a piece of paper, joining them end to end, and then compare the total length with the result obtained along another edge using the rightward lines.

**d** ***How many speciations resulting in smaller antlers produce a species which becomes extinct before it in its turn gives rise to a new species? Make the same count for speciations resulting in increased antler size.***

Fourteen; three. The exercise merely involves counting first the L shapes in which no branches come off the vertical line, and then the similar shapes based on the mirror image of L. □

## 30.3 Cladistics and evolution theory

*Assumption*

1 Some appreciation of the purpose of classification.

*Principle*

1 Classification can serve several purposes: **a** to group similar things and separate dissimilar ones;
**b** to provide a filing system for handling biological data;
**c** to provide sets about which generalizations can be made;
**d** to allow predictions to be made about individual members of sets;
**e** to reconstruct phylogeny;
**f** to reveal God's plan of creation;
and so on.

Should more than one purpose be pursued at once, or should there be different classifications for different purposes? In addition to the above list, special-purpose classifications have been constructed of fossil fragments, to help palaeontologists to date strata in oil exploration, and of populations of plants to depict the degree of gene flow between the populations.

The phylogenies in figure 397 of this section of the *Study guide*, and in the figure in *Systematics and classification*, are not identical. If students notice this it can be explained that the different diagrams are alternative hypotheses. If they do not notice it, they can be ticked off for being unobservant!

## STUDY ITEM

## 30.31 An example of cladistic analysis

*Questions and answers*

This exercise should be easy, at least up to question **c**; it simply requires the application of the procedure given in the panel. If students attempt both this and the exercise in *Systematics and classification*, Study item 1 of 1.3, 'Classifying by a numerical method', they should note the differences between the two methods. For instance, cladistics does not involve a similarity matrix or phenogram; it does not group organisms on the basis of negative (absence) characters.

**a** ***Construct a table to show the state of each character in each plant.***

See *figure 60.*

| Character: | Taxa: daisy | fern | tulip | moss |
|---|---|---|---|---|
| 1 photosynthesis by gametophyte | | ■ | | ■ |
| 2 seed production | ■ | | ■ | |
| 3 chlorophyll | ■ | ■ | ■ | ■ |
| 4 spore dispersion | | ■ | | ■ |
| 5 vascular tissue | ■ | ■ | ■ | |
| 6 flowers | ■ | | ■ | |
| 7 sporophyte is dominant generation | ■ | ■ | ■ | |
| 8 true roots | ■ | ■ | ■ | |

**Figure 60**
Study item 30.31: the character matrix.

**b** ***Find the cladogram using most characters.***

See *figure 61.*

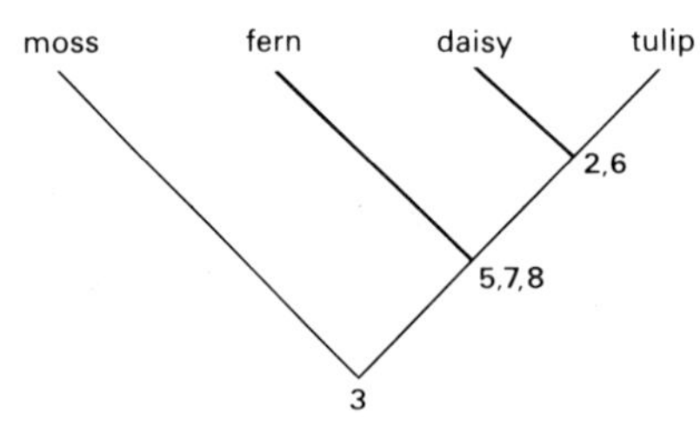


**Figure 61**

**c** ***How many characters are not used?***

Characters 1 (photosynthesis by gametophyte) and 4 (spore dispersion) are not used. Students could run through this exercise a second time, omitting characters 7 and 8. This could be used to start a discussion on how much information should be used.

**d** ***Consider the various evolutionary trees that fit the cladogram, and give reasons for the one you think most likely.***

This is the most difficult part of the Study item, and students may need help with understanding the differences between the trees in figure (S)398. They could be asked, for instance, to say whether they think any of the four plants is an ancestor of any of the others (like B in tree **b** of figure (S)398) and if not why not. It is important that they understand that for every cladogram in this four-taxon problem twelve evolutionary trees can be drawn; they could be asked to work out for themselves the possible cladograms and trees for a hypothetical three-taxon problem (*figure 62*).

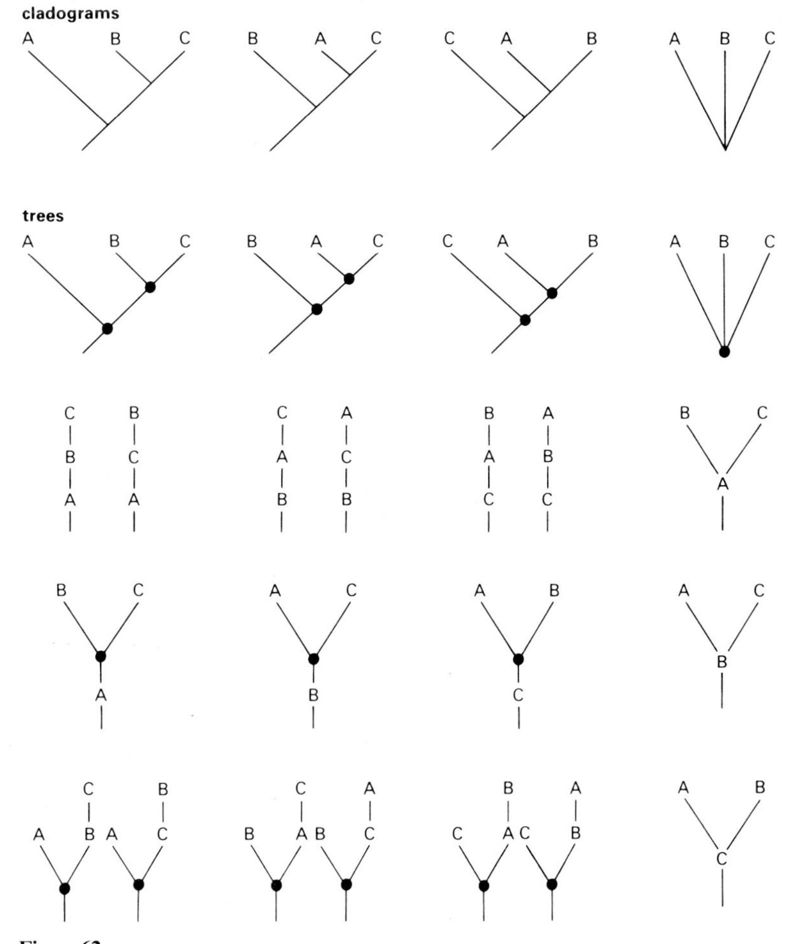


**Figure 62**
Cladograms and tree permutations for a three-taxon problem. The black circles represent hypothetical or unknown ancestors.
*From Platnick, N.,* Systematic zoology, **26**(*4*), *1977, 441.*

One aim of this section is to give students enough understanding of the cladistic method to appreciate why it has become so controversial. Just two of these controversial points are discussed at the end in a fairly even-handed way. □

## 30.4 Looking for laws in biology

*Principles*

1 Different kinds of theory are needed to arrive at a comprehensive understanding of certain sets of objects (in this case organisms).

2 One kind of theory accounts for the differences between the members of the set; the other kind explains why only certain common structures are possible.

3 Because of the astounding success of genetics since 1900, the method of which forces us to concentrate on phenotypic differences, we have neglected the laws determining the limited range of structures which organisms can assume.

## STUDY ITEM
### 30.41 Differences and similarities

*Questions and answers*

If students attempt this on their own they are unlikely to see the point of it. It is intended to be a framework for a teacher-guided class discussion, and so full instructions have not been given to the students. Although some examples show a closer analogy with organisms than others, they all share the following features:

**1** A legitimate field of study could or does exist to account for the differences between the members of the set.

**2** Another approach could concentrate instead on the structure common to all the members, in order to discover what structure characterizes their essential nature.

**3** It is useful to think in one's imagination of hypothetical members of the set which do not in fact exist; their very non-existence may help to distinguish essential structures from inessential or impossible ones.

In this section we have contrasted the 'similarities' and 'differences' aspects of organisms. For instance, the genetic differences between organisms are contrasted with the features they share, such as polarity, cleavage patterns in embryos, and bone patterns in limbs. For each of the following 'differences' aspects of a set of objects, suggest what the corresponding 'similarities' aspects might be:

**a** ***differences in colour, shape, density, hardness, and chemical composition of the mineral crystals found in rocks;***

It might interest students to be told that minerals are classified into species and higher-rank taxa in much the same way as plants and animals are. As in biology, this was developed by Linnaeus and has been followed ever since. It means that a hierarchical classification in no way demonstrates that evolution has occurred. There is a close analogy with organisms, however, in so far as there are definite laws explaining the crystalline structures common to minerals. These specify what balance of electric charges must be maintained in a crystal, what spatial arrangements are possible, given that the atoms composing the crystal have certain sizes, and so on. If these laws were not already well known, we might have been helped to arrive at them by considering hypothetical non-existent crystals, such as one composed entirely of sodium ions ($Na^+$).

**b** ***differences between languages in vocabulary and sentence structure;***

In vocabulary, certain rules are presumably obeyed, such as pronounceability. Thus the word 'pktloo' is unlikely to occur in any natural language. So suitability for oral communication is an essential condition for natural words. On the other hand, 'pktloo' could of course be a word in a non-spoken language.
Similarly, 'cat hit me eat dog climb worm' is unlikely to mean much in any language. All languages have some kind of grammatical structure in which words must be arranged if any kind of meaning is to be conveyed; this example does not agree with any known grammar.

**c** ***differences in the structure of human societies in various parts of the World, ranging from tribal societies lacking any kind of visible leadership to modern bureaucratic dictatorships.***

Various branches of sociology and anthropology are devoted to discovering the laws of political organization. For example, any society that is to survive must presumably involve the acquiescence of the majority of the people in that society, and also a fairly stable system for sharing out resources such as property and wealth. Students can be asked to suggest a structure of society which could not possibly work.
Unlike **a**, the examples in **b** and **c** offer an opportunity to consider history. If language and societies evolve by gradual modification of pre-existing structures, part of the explanation of their similarity may lie in their common historical origin. This is analogous to phylogeny in organisms. But similarities due to a common origin, seen, for example, in the pentadactyl limb, may fail to indicate essential structures: the tridactyl and hexadactyl limbs might be equally viable.

In his anniversary address in 1981 the President of the Royal Society, Sir Andrew Huxley, commented on the controversy and discussion which had recently arisen concerning the implications made by some people that many distinguished scientists, particularly palaeontologists, had given up believing in evolution. The arguments concerned both the punctuated equilibria model and cladistics. The extracts from this address given below could be used in discussing this chapter with students.

**Punctuated equilibria**
'Most of the arguments have depended, directly or indirectly, on contrasting the "sudden" evolutionary change postulated in Eldredge and Gould's "punctuated equilibria" with the gradual change postulated by Darwin. On this issue, my first point is that this is a debate within the Darwinian framework. Darwin's achievement in the *Origin of species* was twofold: first, he presented massive evidence that evolution by descent with modification had actually occurred, and second, he not only proposed natural selection as the principal agent of this evolution but made it appear highly probable. "Punctuated equilibria" contradicts neither of these propositions; it is a proposal on the time course of the evolutionary process, in a degree of detail that was not accessible in the fossil evidence in Darwin's time.

'Second, how sudden are Eldredge and Gould's "punctuations", and how gradual is Darwinian gradualism?'

**Cladistics and evolution theory**

'Coming to the connections between "cladistics" and evolutionary theory, each of the branchings in a cladistic classification – a "cladogram" – is of course naturally interpretable as the evolutionary splitting of a parent species into two daughter species, and this interpretation was made explicitly in the original proposal of cladistics by Hennig, whose book is entitled *Phlyogenetic systematics.* However, some of its practitioners, known as "transformed cladists", go further and insist that classifications should be made without reference to ideas – necessarily conjectural – about the evolutionary processes and the selective pressures which may have brought about the state of affairs that we are now able to study in living animals and in fossils. This is in no sense a denial of evolution: it is a method for avoiding the danger of arguing in a circle if evolutionary hypotheses are used in forming a classification which is itself subsequently used as evidence on the course that evolution has taken.'

Sir Andrew Huxley closed his remarks on the subject as follows:

'Apart from this progressive reduction of the difficulties in evolution by natural selection, it is easy to forget the positive indications from the great developments of biology during this century. In relation to the question of common ancestry, biochemical investigation has shown the ubiquity of the main metabolic pathways and especially of the mechanism of replication of the genetic material, and of the genetic code. As regards natural selection, the Mendelian mechanism of heredity, which preserves variability from each generation to the next, is precisely what Darwin needed, and the power of natural selection – far beyond what one would guess intuitively – was shown in 1930 by Sir Ronald Fisher in his famous book *The genetical basis of natural selection* and in the innumerable theoretical studies which have followed.

'Of course, plenty of difficulties still exist. There is almost no fossil evidence of the origin of the major divisions of the animal kingdom which appear early in the Palaeozoic. Sterility between closely related species seems in some cases to be due to chromosomal accidents – translocations, inversions, and changes in chromosome number – but it is not easy to see how an event of this kind can become incorporated in a population. The importance of genetic drift is still far from being clear, though neither it nor any other known process can substitute for natural selection as a mechanism for producing adaptive change. The question of the origin of life on Earth – barely touched on by Darwin – necessarily lies in the realms of speculation and analogy. And the biggest of all problems for biology – too often swept under the carpet – is the existence of consciousness, for which present-day physics and chemistry do not contain so much as the necessary dimension.

'The present position in relation to evolution seems to me analogous to the position which existed in physiology and biochemistry say fifty years ago; steady progress was being made toward explaining biological processes but there were enormous gaps, and a pessimist might have asserted that these would never be filled. But most of those gaps have now been largely filled – notable exceptions being the control of embryological development and the nervous events underlying complex behaviour – and progress is still accelerating.

'The work needed for progress on the mechanisms of evolution is exeptionally laborious. In palaeontology, not only is fine time-resolution needed, but also sampling of coeval sediments at many points on the Earth's surface, in order to have a good chance of catching speciation in progress. Studies of evolution in action not only need to be pursued continuously for many years but may need extreme precision in order to be significant – as Fisher showed in 1930, even a 1 per cent selective advantage is extremely powerful, and this is likely to be too small to be detected directly.'

Finally, there will be some who will consider that some discussion of creationism should have been included in this chapter. The reason for this omission is that creationism is not scientific because it rests on a metaphysical belief (untestable against observation) that supernatural causes operate. Science, however, rests on the equally metaphysical and untestable belief that supernatural causes do not operate. Students should at least realize that creationists and scientists each live in their own metaphysical world.

Please note that there are no Practical investigations accompanying this chapter.

## PART II Bibliography


BERRY, R. J. *Inheritance and natural history.* Collins, 1977.

BODER, W. F., and CAVALLI-SFORZA, L. L. *Genetics, evolution and man.* W. H. Freeman, 1976.

DOBZHANSKY, T., AYALA, F. J., STEBBINS, G. L., and VALENTINE, J. W. *Evolution.* W. H. Freeman, 1977.

MAYNARD SMITH, J. *The theory of evolution.* 3rd edn. Penguin Books, 1975.

STANSFIELD, W. D. *The science of evolution.* Collier-Macmillan, 1977.

YOUNG, J. Z. *Introduction to the study of man.* Oxford, Clarendon Press, 1972.

# APPENDIX A SYSTEMATICS AND CLASSIFICATION

The following notes refer to the book *Systematics and classification*, which is part of this scheme.

## Aims

The aims of this section of the book are:

1 To introduce the principles on which biological classifications are based and to give the historical background to their development.

2 To familiarize students with the terminology and methods used in producing classifications.

3 To draw attention to the usefulness and relevance of systematics in contemporary biology.

## The use of Systematics in the scheme

An appreciation of the diversity of organisms and how they are classified is an essential feature of biological education and training.

This section of the book provides an introduction to systematics which can be used largely for self-instruction. Some study items are included for students to work through if they wish to, or the topic could be the subject for a group discussion.

However, systematics is not a subject which can just be read about. Students need to be given the experience of classifying organisms in order to appreciate and understand the principles involved. There are many possibilities for providing this experience, two of which are as follows.

**1** As a variety of organisms is encountered during the course, students could be asked, when it is appropriate, to note the features and to discover where a particular organism fits into a given classification and assign it to its major group.

**2** Special collections of a wide variety of organisms could be used. These would be given to the students to sort into groups, which they decide upon by using the principles explained in the booklet, and noting the features they use to do it. The students should then compare their groupings with those in a given classification.

Classification, which is essentially about using features to group organisms together, should not be confused with identifying organisms by using keys. Although you may wish to introduce this at some time during the course, it should be seen as a quite different exercise.

## STUDY ITEM

### 1 Classifying by a numerical method

As mentioned in the students' booklet, this is an optional exercise.

*Questions and answers*

a ***Looking at A–J, find ten characters that can be scored in two states on a table such as the one below.***

| Character / Creature | 1 *Body circular* | 2 *Hairs* | 3 *Stripes* | 4 *Spots* | 5 *Tail* | 6 *Single antennae* | 7 *Inner body wall* | 8 *Eye* | 9 *Abdomen* | 10 *Nose* |
|---|---|---|---|---|---|---|---|---|---|---|
| A | + | + | – | – | + | + | + | + | – | – |
| B | + | + | + | – | + | + | – | + | + | + |
| C | + | + | – | – | + | + | – | + | + | + |
| D | + | + | + | – | + | + | + | + | – | – |
| E | + | + | + | – | – | + | – | + | + | + |
| F | + | + | – | + | – | + | – | – | + | + |
| G | + | + | + | – | + | + | + | + | – | – |
| H | + | + | – | + | + | + | – | + | + | – |
| I | + | + | – | + | + | – | – | + | + | – |
| J | – | – | – | + | – | + | – | + | + | – |

**Table 29**
Completion of the data matrix in *Systematics and classification.*

b ***When the data matrix is complete, draw a table like the one below and compare each pair of 'creatures' (A with B, A with C, A with D and so on), counting up the number of characters in which they match. Both positive and negative count. Before putting the score into the table, it needs conversion to a percentage (in this case you multiply by 10).***

| *Creature* | *A* | *B* | *C* | *D* | *E* | *F* | *G* | *H* | *I* | *J* |
|---|---|---|---|---|---|---|---|---|---|---|
| A | 100 | | | | | | | | | |
| B | 60 | 100 | | | | | | | | |
| C | 70 | 90 | 100 | | | | | | | |
| D | 90 | 70 | 60 | 100 | | | | | | |
| E | 50 | 90 | 80 | 60 | 100 | | | | | |
| F | 40 | 60 | 70 | 30 | 70 | 100 | | | | |
| G | 90 | 70 | 60 | 100 | 60 | 30 | 100 | | | |
| H | 70 | 70 | 80 | 60 | 60 | 70 | 60 | 100 | | |
| I | 60 | 60 | 70 | 50 | 50 | 60 | 50 | 90 | 100 | |
| J | 40 | 40 | 50 | 30 | 50 | 60 | 30 | 70 | 60 | 100 |

**Percentage similarity**

**Table 30**
Completed similarity matrix in *Systematics and classification.*

A regrouping of the similarity matrix would bring together G, D, A and B, E, C and H, I.
The phenogram (figure (S)14) is produced by a process of clustering based on average linkage.

c ***Which 'creature' is least like any other?***

'Creature' J.

It is suggested that students may regard the phenogram as representing a conventional taxonomic hierarchy.

**d** ***If you do this, how many genera and how many orders have you produced?***

Five genera:
one containing creatures A, D, G;
two containing creatures B, E, C;
three containing creatures H, I;
four containing creature F only;
five containing creature J only.
Two orders:
one containing creatures A, D, G;
two containing creatures B, E, C, H, I, F, J.

**e** ***Construct a box-in-box classification to illustrate this hierarchy. (Assume that each 'creature' represents a different species except D and G which are identical.)***

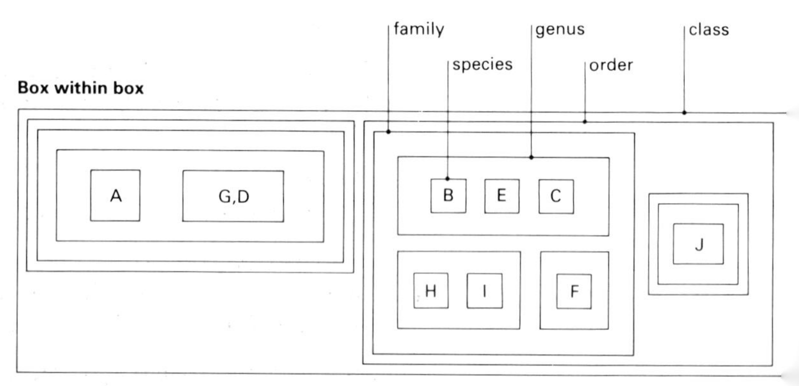


**Figure 63**
Box-within-box classification.

## STUDY ITEM
## 2 An application of chromatography in taxonomy

*Questions and answers*

**a** ***Which compound suggests 'China' ancestry?***

Compound C.

**b** ***Suggest a possible ancestry for tea type e.***

A cross between Assam and a tea containing compound C, perhaps China, could have given tea type e.

**c** ***Suggest how tea type f could have arisen.***

Hybridization of an Assam type tea with *Camellia taliensis* could have given rise to tea type f.

d ***Suggest a possible ancestry for tea type h.***

Tea type h combines Assam, China, and Southern form characters. It could be a hybrid of any types which combine these characters: for example, f × b or g × e (unlikely since e is from Ceylon).

## STUDY ITEM
## 3 Electrophoresis in taxonomy

*Questions and answers*

a ***Could you use this method to determine whether an unknown species of* Corynebacterium *was a human pathogen?***

Yes, possession of one catalase with no esterases or peroxidases is unique to human pathogens. *C. haemolyticum* is different with three esterases and one catalase, but can be accurately identified by this method.

b ***Can species of* Corynebacterium *which are plant pathogens be distinguished from those species which are saprophytes by this method?***

No, not in all cases. *C. tritici* has the same pattern as *C. manihot*, and *C. betae* is the same as *C. aquaticum*.

## STUDY ITEM
## 4 DNA hybridization of some cereal crops

*Questions and answers*

a ***Do these results support the placing of oats in a separate tribe from other cereals in the traditional classification?***

Yes, oats seems widely separated from the other three genera.

b ***Do these results support the placing of wheat and rye in a different sub-tribe from barley in the traditional classification?***

Wheat and rye appear more similar to each other than each is to barley, so they do not contradict the placing of wheat and rye in a different sub-tribe from barley. (Also see the graph of thermal stability of barley DNA hybrids in *Systematics and classification*.)

c ***What are the $T_m$ values for each of the DNA hybrids shown on the graph?***

| | |
|---|---|
| Barley–barley | 72.5 |
| Barley–rye | 69.0 |
| Barley–wheat | 68.5 |
| Barley–oats | 67.0 |

d ***If a 1 °C reduction in $T_m$ represents a 1.5 % difference in bases, what would be the percentage base differences of the other cereals from barley?***

| | |
|---|---|
| Barley–rye | 5.25 % |
| Barley–wheat | 6.0 % |
| Barley–oats | 7.0 % |

It should be noted that DNA hybridizations do not always produce the same percentage binding values. The reason for this is probably because repetitious DNA in some species is more heterogeneous than it is in others. This fact has to be allowed for when taxonomists interpret the data.

**STUDY ITEM**

5 Evidence from cytochrome c

*Question and answer*

a ***Can you see any relationships which do not agree with traditional classifications?***

Turtle seems closer to birds than snake. Humans and monkey split off from mammals before marsupials do. Chicken is rather close to penguin!

# A classification of living organisms

## Aims

The aims of this section of the book are to provide an inexperienced student with:

1 A comprehensible overall view of the major groups of living organisms.
2 A simple account of the features which are shared by the members of these groups.

We have used a Five-Kingdom classification based on Barnes, and Margulis and Schwartz (see the bibliography). In it we have attempted to give equal treatment to each Kingdom and a balanced view of the whole living World.

As far as possible photographs and diagrams are used to illustrate the important features of the groups.

We accept that no classification will be universally accepted. One such as this which has many omissions may not satisfy experienced biologists. We have become aware that it is impossible to achieve total agreement among biologists on the classification of all groups of living organisms. This classification is the best we could do!

# Bibliography

## PART I Systematics

*Evolution: a Scientific American* book, Freeman, 1978. Originally published as articles in *Scientific American*, September, 1978.)
HEYWOOD, V. H. *Plant taxonomy*. 2nd edn. Edward Arnold, 1976.
HEYWOOD, V. H. (ed.) *Modern methods in plant taxonomy*. Botanical Society of the British Isles, 1967.
JEFFREY, C. *Introduction to plant taxonomy*. 2nd edn. Cambridge University Press, 1982.
MAYR, E. *Principles of systematic zoology*. McGraw Hill, 1969.
ROSS, H. H. *Biological systematics*. Addison Wesley, 1974.
SIMPSON, G. G. *Principles of animal taxonomy*. Oxford University Press, 1961.
SMITH, P. M. *Chemotaxonomy of plants*. Edward Arnold, 1976.
SNEATH, P. H. A. and SOKAL, R. R. *Numerical taxonomy: the principles and practice of numerical classification*. Freeman, 1973.
SOKAL, R. R. and SNEATH, P. H. A. *Numerical taxonomy*. Freeman, 1963.
STACE, C. A. *Plant taxonomy and biosystematics*. Edward Arnold, 1980.

## PART II Classification

ALEXOPOULOS, C. J. (revd by Mims, C. W.) *Introductory mycology*. 3rd edn. Wiley, 1979.
BARNES, R. D. *Invertebrate zoology*. Saunders, 1980.
BARNES, R. S. K. (ed.) *A Synoptic classification of living organisms*, Blackwell Scientific Publications, 1984.
BUCHSBAUM, R. *Animals without backbones*. 2nd revd edn. Chicago University Press, 1972.
FROBISHER, M., HINSDILL, R. D., CRABTREE, K. T., and GOODHEART, C. R. *Fundamentals of microbiology*. Saunders, 1974.
GROUNDS, R. *The ferns*. Pelham, 1974.
KIRK, D. L. *Biology today*. Random House, 1980.
MARGULIS, L. and SCHWARTZ, K. V. *Five kingdoms: An illustrated guide to the phyla of life on Earth*. Freeman, 1982.
MASH, K. *How invertebrates live*. Phaidon, 1975.
RAMSBOTTOM, J. *Mushrooms and toadstools*, Collins, 1979.
RAVEN, P. H., EVERT, R. F., and CURTIS, H. *Biology of plants*. Worth, 1976.
ROSS, I. K. *Biology of the fungi*. McGraw Hill, 1979.
SCAGEL, R. F. *et al. Plant diversity: an evolutionary approach*. Wadsworth, 1969.
YOUNG, J. Z. *The life of vertebrates*. 3rd revd edn. Oxford University Press, 1981.

# APPENDIX B INCLUDING MICROCOMPUTERS IN THE COURSE MATERIAL

Although little mention has been made of it in the texts, there are various ways in which use of computing material or techniques could enhance and enrich the study of some topics within this course. The extent to which this is possible depends on the particular circumstances; the interests and experience of the teachers, the availability of appropriate equipment, and access to suitable software.

It is not suggested that the use of microcomputers would replace the Practical investigations or Study items described, but they may in some circumstances provide an alternative method or approach or additional support material. In such ways an awareness and appreciation of the role of microcomputers as a 'scientists' tool' can be introduced.

Research biologists increasingly make use of the new technologies in revising and upgrading their methods; more precise experimentation, easier management of large quantities of data, better analysis tools, quicker calculation, and so forth. Such changes can be reflected in the classroom by familiarizing students with some of these techniques and skills during their normal course of study.

There are a number of ways in which this can be done:

## 1 Experimentation

Biological experiments are frequently limited in the school laboratory or in field work because of the time-scales involved or the nature of the measurements to be recorded. The use of microelectronic equipment can sometimes improve such investigations, enabling a more satisfactory control of conditions, and recording results at times and levels and to degrees of accuracy not normally possible. An understanding of the way that a microcomputer can be connected to external devices such as switches, timers, and sensors is valid and relevant practical knowledge. A wide range of interface devices and programmable instruments are now available, and they have particular value in physiological, ecological, and behavioural investigations.

A popular example is the use of sensory devices to monitor the breathing rate of human subjects after different levels of activity (*Study guide I*, Chapter 2 'Breathing and gas exchange in Man'), and to compare this with another recording measured simultaneously, such as a pulse rate. Readings can be continuously displayed in a graphical form on a computer monitor, so allowing interesting biofeedback to occur. Similarly, light sensors can be used to monitor changes in colour of hydrogen carbonate indicator solutions in gas exchange experiments (*Study guide I*, Chapter 1 'Gas exchange' and Chapter 7 'Photosynthesis'). Electronic manometers can record water loss over a period of time (*Study guide I*, Chapter 8 'The plant and water'). Many such ideas are regularly described in the science notes in the Association for Science Education's journal, *School science review*.

**Figure 64**
Using VELA to take measurements during field work.
*Photograph, Andrew Lambert.*

Another common example is the collection of field data (*Study guide II*, Chapters 26 to 29 and *Practical guide* 7 'Organisms and the environment') with portable equipment designed for the purpose (*figure 64*). Various environmental sensors such as temperature probes, light meters, and humidity and oxygen level measurers can be used to record automatically information about the physical conditions at various points where the different species are being sampled. Such data is stored and can later be fed into a laboratory computer for further analysis.

## 2 Data management and analysis

The ability to store and manipulate large quantities of information is of significance to biologists in searching for trends and relationships through extensive quantitative data. A computer database where such stored data are categorized enables fast retrieval of specified combinations for comparison and analysis (*figure 65*). Understanding of the type of structures of such databases, and the ability to create and use them, are an increasingly important scientific skill. As communication systems develop to enable easy transfer between different data stores, students may want to access data recorded by other people in different places and times and to compare these with locally recorded results, apart from just maintaining detailed long-term results of their own.

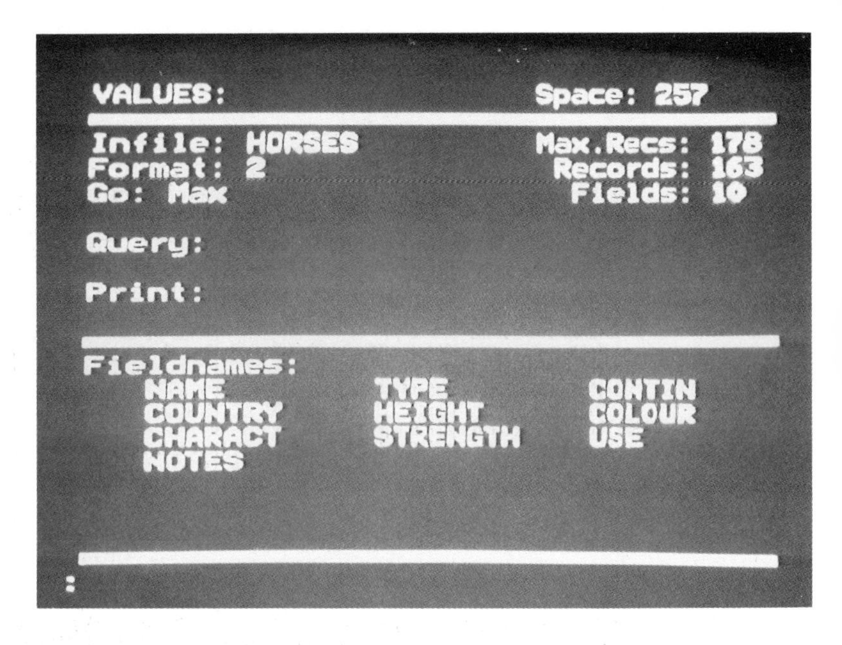


**Figure 65**
QUEST, an example of a database system designed for school use which can be used for many different purposes.

If collected field data are accumulated within a computer database in school, then students' results can be combined and/or compared from one year to the next. Eventually, a body of detailed data can be collected over a number of years showing changes and trends in the environment, and can be compared with data collected by other schools in ecologically similar but geographically different places.

Many commercial software packages are available, but a number have been designed for educational use. Some are quite open-ended and can be used for any purpose, for example a record set up by the teacher of relevant articles or books for the course. Others are more specific and related directly to certain areas, a database of dietary requirements and nutritional value of foods for example (*Study guide I*, Chapter 6 'Heterotrophic nutrition').

Similarly, spreadsheets are well known commercially as a means of handling and analysing large amounts of information in a dynamic visual way. The spreadsheet format in educational software provides an easy and convenient way of comparing data and seeking trends and patterns. For example, analysing collected data along a transect can be done in this way because numbers of species at different sampling points can be entered, as well as environmental measurements (*figure 66*).

Frequently, such packages are linked with statistical analysis functions, using the computer in one of its oldest research roles—as a 'calculator'. In this way it can combine standard statistical testing facilities with effective visual displays, producing graphs, bar charts, scattergrams, pie charts, and kite diagrams (*figure 67*).

Software systems such as databases and spreadsheets are now tools of modern scientists, but students need to learn how to use them.

## 3 Exploration

Combining microcomputing facilities such as data storage, calculation

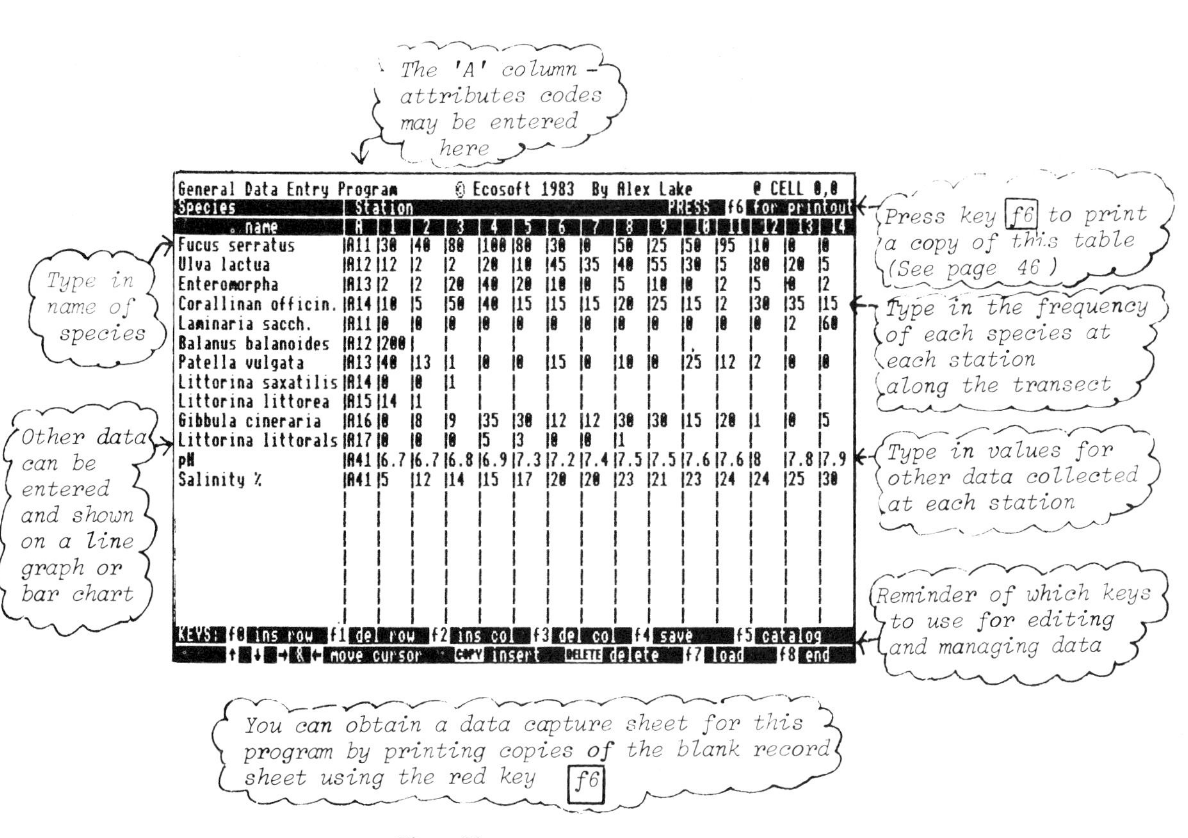

General Data Entry Program © Ecosoft 1983 By Alex Lake @ CELL 0,0

Species Station PRESS f6 for printout

| name | A | 1 | 2 | 3 | 4 | 5 | 6 | 7 | 8 | 9 | 10 | 11 | 12 | 13 | 14 |
|---|---|---|---|---|---|---|---|---|---|---|---|---|---|---|---|
| Fucus serratus | A11 | 30 | 40 | 80 | 100 | 80 | 30 | 0 | 50 | 25 | 50 | 95 | 10 | 0 | 0 |
| Ulva lactua | A12 | 12 | 2 | 2 | 20 | 10 | 45 | 35 | 40 | 55 | 30 | 5 | 80 | 20 | 5 |
| Enteromorpha | A13 | 2 | 2 | 20 | 40 | 20 | 10 | 0 | 5 | 10 | 0 | 2 | 5 | 0 | 2 |
| Corallinan officin. | A14 | 10 | 5 | 50 | 40 | 15 | 15 | 15 | 20 | 25 | 15 | 2 | 30 | 35 | 15 |
| Laminaria sacch. | A11 | 0 | 0 | 0 | 0 | 0 | 0 | 0 | 0 | 0 | 0 | 0 | 0 | 2 | 60 |
| Balanus balanoides | A12 | 200 | | | | | | | | | | | | | |
| Patella vulgata | A13 | 40 | 13 | 1 | 0 | 0 | 15 | 0 | 10 | 0 | 25 | 12 | 2 | 0 | 0 |
| Littorina saxatilis | A14 | 0 | 0 | 1 | | | | | | | | | | | |
| Littorina littorea | A15 | 14 | 1 | | | | | | | | | | | | |
| Gibbula cineraria | A16 | 0 | 8 | 9 | 35 | 30 | 12 | 12 | 30 | 30 | 15 | 20 | 1 | 0 | 5 |
| Littorina littorals | A17 | 0 | 0 | 0 | 5 | 3 | 0 | 0 | 1 | | | | | | |
| pH | A41 | 6.7 | 6.7 | 6.8 | 6.9 | 7.3 | 7.2 | 7.4 | 7.5 | 7.5 | 7.6 | 7.6 | 8 | 7.8 | 7.9 |
| Salinity % | A41 | 5 | 12 | 14 | 15 | 17 | 20 | 20 | 23 | 21 | 23 | 24 | 24 | 25 | 30 |

KEYS: f0 ins row f1 del row f2 ins col f3 del col f4 save f5 catalog

↑ ↓ → & ← move cursor COPY insert DELETE delete f7 load f8 end



**Figure 66**
The 'General data entry program' of ECOSOFT uses a spreadsheet format. Data can be transferred between ECOSOFT and QUEST.

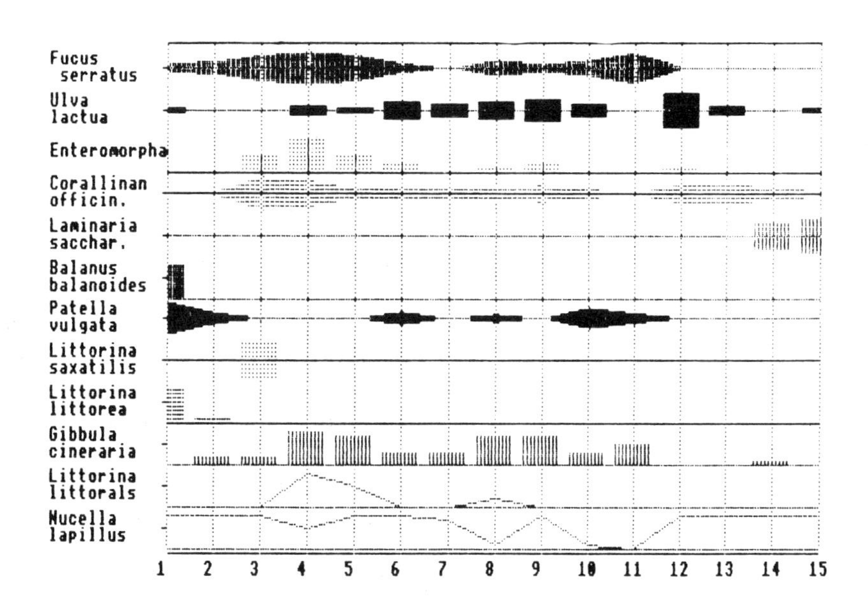


**Figure 67**
Kite diagrams produced by the program ECOSOFT.

and visual display offers the opportunity of creating learning environments for exploration and investigation which can take students beyond the range of experience afforded within the laboratory. This has been effectively demonstrated by the use of computer simulations where a 'situation' is modelled, by defining the known variables and the relationships between them. Various examples of biological simulations are available, covering many different topics in this course. They can have a valuable role to play in encouraging students to extend their practical investigations by giving them the freedom to manipulate conditions and factors outside the normal practical limits, and perhaps in ways not usually possible or over impossible time scales. Such simulations can be used to illustrate the complexity of inter-relationships, to highlight significant controlling factors and their effects, or just simply to provide a better 'feel' for the contributing aspects. Examples are the concept of limiting factors in photosynthesis (*Study guide I*, Chapter 7 'Photosynthesis'), flowering in plants and the consideration of different theories related to day length and the significance of internal rhythms (*Study guide II* Chapter 25, 'Development and the external environment'), and the homeostatic control of the blood sugar level (*Study guide I*, Chapter 6 'Heterotrophic nutrition', Study item 6.81) (*figure 68*).

Similarly, computer simulations can handle manipulative models such as the bead model (*Practical guide 5*, investigation 20A 'Models of a gene pool') or provide practice in genetic mapping (*Practical guide 5*, investigation 17D 'Linkage and linkage mapping in tomato') (*figure 69*). Frequently, computer genetic simulations provide a valuable source of large quantities of results from a wide range of different examples which could not be obtained practically, and can normally only be provided as recorded secondhand data. Various simulations exist to illustrate different types of inheritance, providing results in many different forms (*figure 70*).

Some simulations are, however, set within the context of a goal-oriented activity. The student is encouraged to play a role, design strategies, and make decisions to achieve a particular end result. Examples of this range from managing a reserve, running a fish farm, or

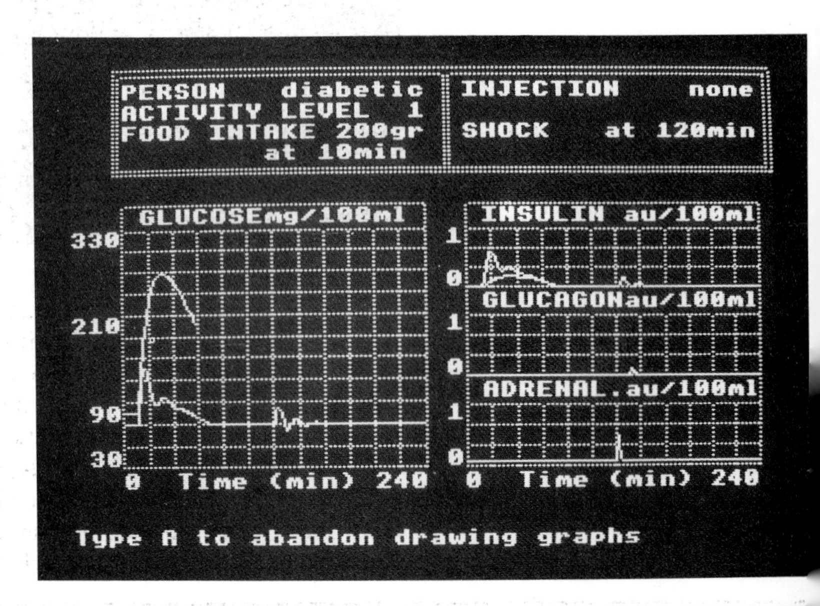


**Figure 68**
A simulation of the control of the blood sugar level from the program BLOOD SUGAR.

Figure 69
Setting up test crosses to work out the position of the genes along the 'computer chromosomes' in the program LINKOVER.

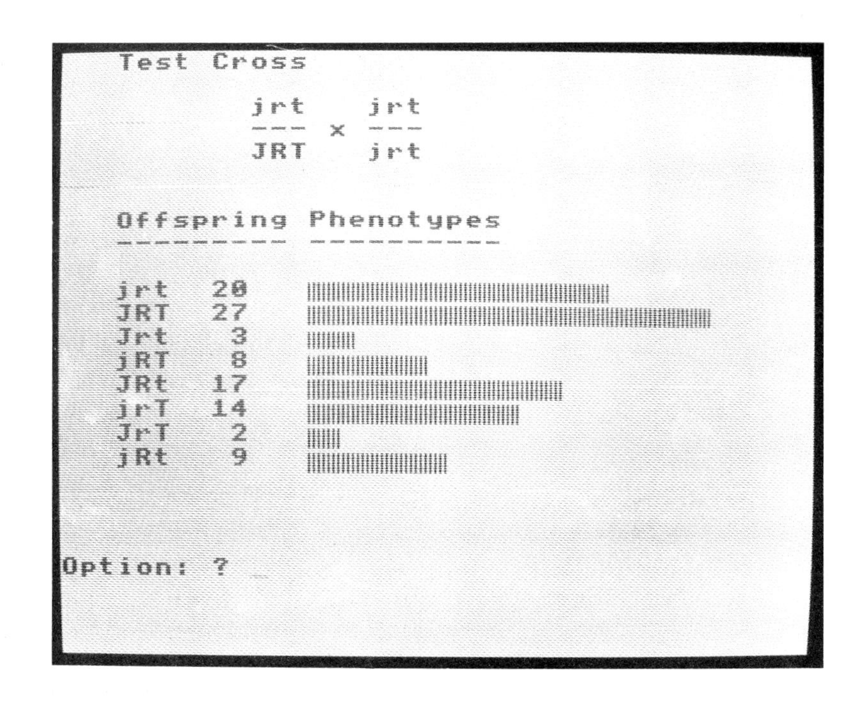


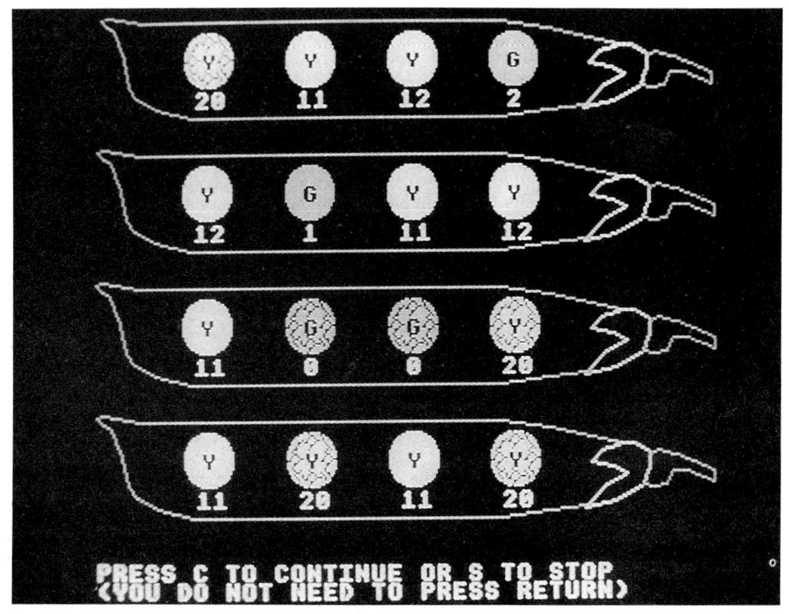


Figure 70
A pictorial display of the results of a test cross with peas and pods – WARWICK SCIENCE SIMULATIONS, BIOLOGY PACK 1, DIHYBRID.

controlling a disease, to developing new breeds of plants and animals through selective breeding. Students not only have to apply their understanding of the biology of the situation, but may also be faced with real life issues related to attitudes and values: 'What compromises must be made in conservation management?', 'How can long-term measures be implemented within a tight budgetary control?'

Another approach to simulation is where students go beyond using predefined models hidden within carefully designed packages and actually become involved in developing and modifying the model for themselves, to try to replicate situations as realistically as possible. The bead model can be seen as going partly along this pathway. Using the simple probability model, students can select conditions of dominance and type of selection to set up situations representing, say, natural selection operating in a particular way, heterozygote advantage, genetic drift, and so forth. However, most modelling requires the initial identification of all the significant contributory factors and not just manipulating them and establishing, usually by trial and error, the relationship between them. Such an activity may make use of a number of different modelling methods or systems (*figures 71–3*). It may be

R = rabbit population
F = fox population
A = growth rate of rabbit population
B = chance of a fox catching a rabbit if they meet
G = generation time
C and D are constants

```
VALUES                          MODEL
G=0                             R=(A*R-B*R*F)+R
A=0.6                           F=(C*R*F-D*F)+F
B=0.02                          G=G+1
C=0.004
D=0.8
F=15
R=300

^↑↓ uses     ←→↑↓ cursor     f2 open line
^←→ swop                     f3 line up
TAB size                     f4 line down
^H  help     f0 rub char
ESC new/     f1 rub line     f5 read file
    stop
```

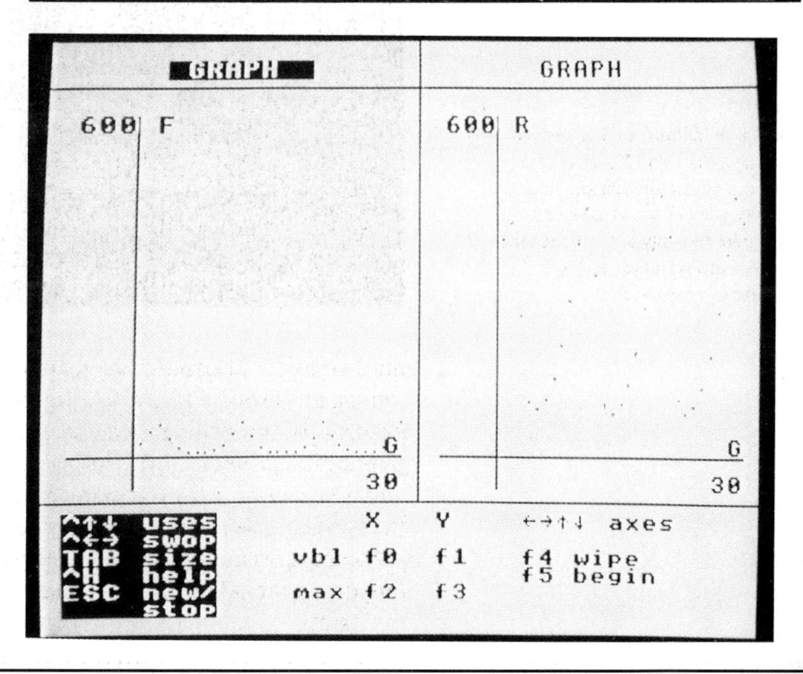


**Figure 71**
Attempting to model a predator-prey relationship using the dynamic modelling system.

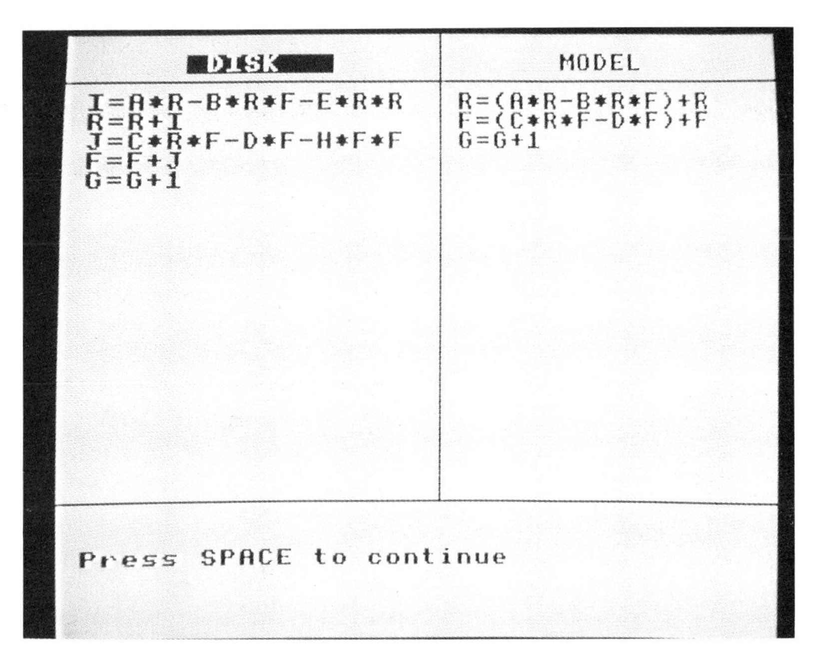


**Figure 72**
Redefining the model to include a finite food supply and living space.

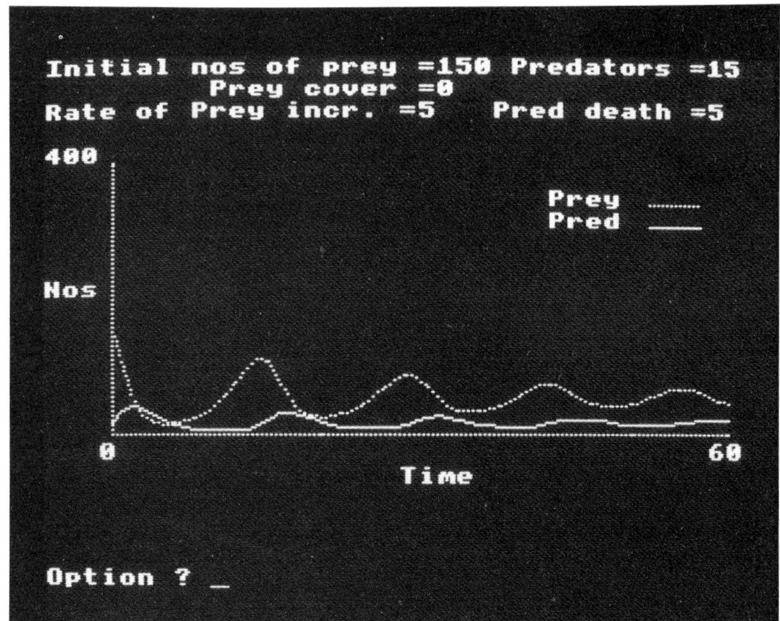


**Figure 73**
Use of a predefined model derived from the Lotka and Voltera equations in the simulation PREDATOR–PREY RELATIONSHIPS. The student is only able to alter the values of the variables and is unaware of the model.

mathematical or descriptive in approach, but always involves a gradual refinement process to produce the best fit to practical observation. This process of modelling, whether related to ecological relationships, population growth, or physiological systems, involves a detailed analysis and breakdown of the component elements and hence an indepth look at the system itself. This type of exercise is familiar to physicists, but software tools are now being developed based on various programming languages including LOGO which are creating new opportunities for biologists.

The microcomputer may therefore have many different roles to play by extending traditional practice, providing new thinking tools, and creating different types of learning environments. A range of good software and useful devices is available to schools and designed for use within courses such as this. A list of the main suppliers has been provided. Computing is, however, a rapidly changing area, and new material is continually being published, so a detailed list of titles would soon be out of date. The publishers listed are regularly updating catalogues and most now send out inspection copies. Software can vary enormously in quality and approach. It needs care in selection.

Most of the areas and ideas described do not need an extensive range or amount of equipment. More important are the planning and organization to incorporate the use of what is available within the normal range of activities, perhaps as an occasional reference point for information, a handy tool for analysis, a versatile device for experimentation, or sometimes as the focus of an activity either for demonstration or group work. Most important are the ideas and discussions which are stimulated and the different activities which are generated *away* from the microcomputer itself.

### Some suppliers of microelectronics laboratory equipment

Educational Electronics
Griffin & George
Philip Harris

### Some suppliers of software for Biology

Acornsoft, Cambridge Technopark, 645 Newmarket Road, Cambridge, CB5 8PD.
For general purpose databases, spreadsheets, wordprocessing (BBC only).

AUCBE, Advisory Unit for Computer-Based Education, Endymion Road, Hatfield, Herts, AL10 8AU.
For QUEST utility database package and ECOSOFT data handling for fieldwork.

BBC Publications, 35 Marylebone High Street, London W1M 4AA.
For simulations and some utility packages related to television series, *e.g.* Science Topics Software (BBC only).

Edward Arnold, 41 Bedford Square, London WC1 3DQ.
For Chelsea Science Simulations.

Garland Computing, 35 Dean Hill, Plymouth PL9 9AF.
For a range of simulations and electronic display packages.

Longman Microsoftware, 62 Hallfield Road, Layerthorpe, York YO3 7XQ.
For Computers in the Curriculum and Warwick Science Simulations, and for biological simulations, statistical analysis, and modelling packages.

Nelson Computer Assisted Learning, Nelson House, Mayfield Road, Walton-on-Thames, Surrey, KT12 5PL.
For NELCAL – a range of simulations and statistical analysis packages (STATPACK).

### Further reading on microcomputers in Biology teaching

*Exploring biology with microcomputers.* MEP Reader 4, Council for Educational Technology, 1983.

MARGHAM, J. P. and HALE W. G., *Basic biology.* Collins Educational, 1983.
LEWIS, A. AND H., *Microcomputers and school biology.* Pitman, 1983.
McCORMICK, S., *Microcomputers and teaching biology.* Longman Micro Guide, 1984.
BELL, F., *Printout biology.* Edward Arnold, 1983.

# INDEX